T0202698

Numerical Methods for Engineers

Second Edition

Numerical Methods for Engineers

Second Edition

D.V. Griffiths
Colorado School of Mines
Golden, Colorado

I.M. Smith
University of Manchester
Manchester, UK

CRC Press
Taylor & Francis Group
Boca Raton London New York

CRC Press is an imprint of the
Taylor & Francis Group, an **informa** business

A CHAPMAN & HALL BOOK

CRC Press
Taylor & Francis Group
6000 Broken Sound Parkway NW, Suite 300
Boca Raton, FL 33487-2742

First issued in paperback 2019

© 2006 by Taylor & Francis Group, LLC
CRC Press is an imprint of Taylor & Francis Group, an Informa business

No claim to original U.S. Government works

ISBN-13: 978-1-58488-401-9 (hbk)
ISBN-13: 978-0-367-39066-2 (pbk)

This book contains information obtained from authentic and highly regarded sources. Reasonable efforts have been made to publish reliable data and information, but the author and publisher cannot assume responsibility for the validity of all materials or the consequences of their use. The authors and publishers have attempted to trace the copyright holders of all material reproduced in this publication and apologize to copyright holders if permission to publish in this form has not been obtained. If any copyright material has not been acknowledged please write and let us know so we may rectify in any future reprint.

Except as permitted under U.S. Copyright Law, no part of this book may be reprinted, reproduced, transmitted, or utilized in any form by any electronic, mechanical, or other means, now known or hereafter invented, including photocopying, microfilming, and recording, or in any information storage or retrieval system, without written permission from the publishers.

For permission to photocopy or use material electronically from this work, please access www.copyright.com (http://www.copyright.com/) or contact the Copyright Clearance Center, Inc. (CCC), 222 Rosewood Drive, Danvers, MA 01923, 978-750-8400. CCC is a not-for-profit organization that provides licenses and registration for a variety of users. For organizations that have been granted a photocopy license by the CCC, a separate system of payment has been arranged.

Trademark Notice: Product or corporate names may be trademarks or registered trademarks, and are used only for identification and explanation without intent to infringe.

Visit the Taylor & Francis Web site at
http://www.taylorandfrancis.com

and the CRC Press Web site at
http://www.crcpress.com

To Valerie, Will and James

List of Programs

2.1	Gaussian elimination for linear simultaneous equations	17
2.2	Gaussian elimination using $[\mathbf{L}][\mathbf{U}]$ factorization	23
2.3	Gaussian elimination using $[\mathbf{L}][\mathbf{D}][\mathbf{L}]^T$ factorization	28
2.4	Cholesky $[\mathbf{L}][\mathbf{L}]^T$ factorization using banded storage	34
2.5	Cholesky $[\mathbf{L}][\mathbf{L}]^T$ factorization using skyline storage	36
2.6	$[\mathbf{L}][\mathbf{U}]$ factorization with pivoting	40
2.7	Cholesky $[\mathbf{L}][\mathbf{L}]^T$ factorization using skyline storage prescribed solutions by penalty method	43
2.8	Jacobi iteration for linear simultaneous equations	49
2.9	Gauss-Seidel iteration for linear simultaneous equations	54
2.10	Successive overrelaxation for linear simultaneous equations	58
2.11	Steepest descent for linear simultaneous equations	62
2.12	Conjugate gradients for linear simultaneous equations	65
2.13	Stabilized bi-conjugate gradients for linear simultaneous equations	70
2.14	Preconditioned conjugate gradients for linear simultaneous equations	74
2.15	BiCGSTAB scheme (left preconditioned)	77
2.16	BiCGSTAB scheme (right preconditioned)	79
3.1	Iterative substitution for a single root	92
3.2	Bisection method for a single root	98
3.3	False position method for a single root	102
3.4	Newton-Raphson method for a single root	106
3.5	Modified Newton-Raphson method for a single root	109
3.6	Iterative substitution for systems of equations	114
3.7	Newton-Raphson for systems of equations	119
3.8	Modified Newton-Raphson for systems of equations	123

List of Programs (continued)

4.1	Vector iteration for "largest" eigenvalue and its eigenvector	137
4.2	Shifted vector iteration for eigenvalue and its eigenvector	140
4.3	Shifted inverse iteration for nearest eigenvalue and its eigenvector	145
4.4	Vector iteration for $[\mathbf{K}]\{\mathbf{x}\} = \lambda[\mathbf{M}]\{\mathbf{x}\}$	151
4.5	Conversion of $[\mathbf{K}]\{\mathbf{x}\} = \lambda[\mathbf{M}]\{\mathbf{x}\}$ to symmetrical standard form	155
4.6	Jacobi diagonalization for eigenvalues of symmetrical matrices	163
4.7	Householder reduction of symmetrical matrix to tridiagonal form	169
4.8	Lanczos reduction of symmetrical matrix to tridiagonal form	173
4.9	$[\mathbf{L}][\mathbf{R}]$ transformation for eigenvalues	177
4.10	Characteristic polynomial method for eigenvalues of symmetrical tridiagonal matrix	184
5.1	Interpolation by Lagrangian polynomials	196
5.2	Interpolation by forward differences	204
5.3	Interpolation by cubic spline functions	211
5.4	Curve fitting by least squares	232
6.1	Repeated Newton-Cotes rules	262
6.2	Repeated Gauss-Legendre rules	275
6.3	Adaptive Gauss-Legendre rules	280
6.4	Gauss-Laguerre rules	287
6.5	Multiple integrals by Gauss-Legendre rules	302
7.1	One-step methods for systems of ODEs	337
7.2	Theta-method for linear ODEs	345
7.3	Fourth order predictor-corrector methods	356
7.4	Shooting method for second order ODEs	371
8.1	Explicit finite differences in 1D	424
8.2	Simple FE analysis of Example 8.3	432

Contents

1 **Introduction and Programming Preliminaries**   1
   1.1   Introduction 1
   1.2   Running programs 1
   1.3   Hardware 2
   1.4   External Fortran subprogram libraries 2
   1.5   A simple Fortran program 4
   1.6   Some simple Fortran constructs 7
   1.7   Intrinsic functions 7
   1.8   User-supplied functions and subroutines 8
   1.9   Errors and accuracy 9
      1.9.1   Roundoff 9
      1.9.2   Truncation 10
      1.9.3   Cancellation 11
      1.9.4   Intrinsic and library-supplied precision routines 11
   1.10   Graphical output 12
   1.11   Conclusions 13

2 **Linear Algebraic Equations**   15
   2.1   Introduction 15
   2.2   Gaussian elimination 15
      2.2.1   Observations on the elimination process 20
   2.3   Equation solution using factorization 20
      2.3.1   Observations on the solution process by factorization . 25
   2.4   Equations with a symmetrical coefficient matrix 25
      2.4.1   Quadratic form and positive definiteness 30
      2.4.2   Cholesky's method 31
   2.5   Banded equations 33
   2.6   Compact storage for variable bandwidths 35
   2.7   Pivoting 38
      2.7.1   Ill-conditioning 41
   2.8   Equations with prescribed solutions 42
   2.9   Iterative methods 46
      2.9.1   The iterative process 46
      2.9.2   Very sparse systems 52
      2.9.3   The Gauss-Seidel method 52
      2.9.4   Successive overrelaxation 57
   2.10   Gradient methods 61

		2.10.1	The method of 'steepest descent'	61
		2.10.2	The method of 'conjugate gradients'	64
		2.10.3	Convergence of iterative methods	68
	2.11	Unsymmetrical systems		68
	2.12	Preconditioning		72
	2.13	Comparison of direct and iterative methods		81
	2.14	Exercises		82

3 Nonlinear Equations **89**

	3.1	Introduction		89
	3.2	Iterative substitution		91
	3.3	Multiple roots and other difficulties		94
	3.4	Interpolation methods		97
		3.4.1	Bisection method	97
		3.4.2	False position method	100
	3.5	Extrapolation methods		103
		3.5.1	Newton-Raphson method	104
		3.5.2	A modified Newton-Raphson method	107
	3.6	Acceleration of convergence		112
	3.7	Systems of nonlinear equations		112
		3.7.1	Iterative substitution for systems	113
		3.7.2	Newton-Raphson for systems	116
		3.7.3	Modified Newton-Raphson method for systems	121
	3.8	Exercises		125

4 Eigenvalue Equations **131**

	4.1	Introduction		131
		4.1.1	Orthogonality and normalization of eigenvectors	132
		4.1.2	Properties of eigenvalues and eigenvectors	134
		4.1.3	Solution methods for eigenvalue equations	136
	4.2	Vector iteration		136
		4.2.1	Shifted vector iteration	140
		4.2.2	Shifted inverse iteration	143
	4.3	Intermediate eigenvalues by deflation		148
	4.4	The generalized eigenvalue problem $[\mathbf{K}]\{\mathbf{x}\} = \lambda[\mathbf{M}]\{\mathbf{x}\}$		150
		4.4.1	Conversion of generalized problem to symmetrical standard form	154
	4.5	Transformation methods		158
		4.5.1	Comments on Jacobi diagonalization	167
		4.5.2	Householder's transformation to tridiagonal form	167
		4.5.3	Lanczos transformation to tridiagonal form	171
		4.5.4	**LR** transformation for eigenvalues of tridiagonal matrices	176
	4.6	Characteristic polynomial methods		180
		4.6.1	Evaluating determinants of tridiagonal matrices	180

      4.6.2   The Sturm sequence property 181

      4.6.3   General symmetrical matrices, e.g., band matrices . . 187

  4.7   Exercises 188

5 Interpolation and Curve Fitting   **193**

  5.1   Introduction 193

  5.2   Interpolating polynomials 193

      5.2.1   Lagrangian polynomials 194

      5.2.2   Difference methods 198

      5.2.3   Difference methods with equal intervals 199

  5.3   Interpolation using cubic spline functions 207

  5.4   Numerical differentiation 214

      5.4.1   Interpolating polynomial method 215

      5.4.2   Taylor series method 219

  5.5   Curve fitting 226

      5.5.1   Least squares 226

      5.5.2   Linearization of data 229

  5.6   Exercises 237

6 Numerical Integration   **245**

  6.1   Introduction 245

  6.2   Newton-Cotes rules 247

      6.2.1   Introduction 247

      6.2.2   Rectangle rule, $(n = 1)$ 247

      6.2.3   Trapezoid rule, $(n = 2)$ 248

      6.2.4   Simpson's rule, $(n = 3)$ 250

      6.2.5   Higher order Newton-Cotes rules $(n > 3)$ 252

      6.2.6   Accuracy of Newton-Cotes rules 253

      6.2.7   Summary of Newton-Cotes rules 254

      6.2.8   Repeated Newton-Cotes rules 255

      6.2.9   Remarks on Newton-Cotes rules 264

  6.3   Gauss-Legendre rules 265

      6.3.1   Introduction 265

      6.3.2   Midpoint rule, $(n = 1)$ 265

      6.3.3   Two-point Gauss-Legendre rule, $(n = 2)$ 267

      6.3.4   Three-point Gauss-Legendre rule, $(n = 3)$ 270

      6.3.5   Changing the limits of integration 271

      6.3.6   Accuracy of Gauss-Legendre rules 277

  6.4   Adaptive integration rules 278

  6.5   Special integration rules 284

      6.5.1   Gauss-Chebyshev rules 288

      6.5.2   Fixed weighting coefficients 289

      6.5.3   Hybrid rules 290

      6.5.4   Sampling points outside the range of integration . . . 290

  6.6   Multiple integrals 292

	6.6.1	Introduction	292
	6.6.2	Integration over a general quadrilateral area	299
6.7	Exercises		307

7 Numerical Solution of Ordinary Differential Equations   **317**

7.1	Introduction	317
7.2	Definitions and types of ODE	317
7.3	Initial value problems	319
	7.3.1 One-step methods	321
	7.3.2 Reduction of high order equations	330
	7.3.3 Solution of simultaneous first order equations	332
	7.3.4 θ-methods for linear equations	343
	7.3.5 Predictor-corrector methods	349
	7.3.6 Stiff equations	359
	7.3.7 Error propagation and numerical stability	360
	7.3.8 Concluding remarks on initial value problems	361
7.4	Boundary value problems	362
	7.4.1 Finite difference methods	362
	7.4.2 Shooting methods	368
	7.4.3 Weighted residual methods	376
7.5	Exercises	386

8 Introduction to Partial Differential Equations   **393**

8.1	Introduction	393
8.2	Definitions and types of PDE	393
8.3	First order equations	394
8.4	Second order equations	399
8.5	Finite difference method	401
	8.5.1 Elliptic systems	404
	8.5.2 Parabolic systems	417
	8.5.3 Hyperbolic systems	427
8.6	Finite element method	430
8.7	Exercises	434

A Descriptions of Library Subprograms   **445**

B Fortran 95 Listings of Library Subprograms   **447**

C References and Additional Reading   **469**

Index   **475**

Chapter 1

Introduction and Programming Preliminaries

1.1 Introduction

There are many existing texts aimed at introducing engineers to the use of the numerical methods which underpin much of modern engineering practice. Some contain "pseudocode" to illustrate how algorithms work, while others rely on commercial "packaged" software such as "Mathematica" or "MATLAB". But the vast majority of the large computer programs which lie behind the design of engineering systems are written in the Fortran language, hence the use of the latest dialect, Fortran 95, in this book. Nearly fifty entire programs are listed, many being made more concise through the use of "libraries" of approximately twenty five subprograms which are also described and listed in full in Appendices A and B. Free Fortran 95 compilers are also widely available and users are therefore encouraged to modify the programs and develop new ones to suit their particular needs.

1.2 Running programs

Chapters 2-8 of this textbook describe 49 Fortran 95 programs covering a wide range of numerical methods applications. Many of the programs make use of a subprogram library called `nm_lib` which holds 23 subroutines and functions. In addition, there is a module called `precision` which controls the precision of the calculations.

Detailed instructions for running the programs described in this textbook are to be found at the web site:

`www.mines.edu/~vgriffit/NM`

After linking to this site, consult the `readme.txt` file for information on:

1) how to download all the main programs, subprograms and sample data files,

2) how to download a free Fortran 95 compiler, e.g., from the Free Software Foundation at `www.g95.org`,

3) how to compile and execute programs using a simple batch file

In the interests of generality and portability, all programs in the book assume that the data and results files have the generic names `nm95.dat` and `nm95.res` respectively. A batch file that can be downloaded from the web with the main programs and libraries called `run2.bat` copies the actual data file (say `fred.dat`) to `nm95.dat` before execution. Finally, after the program has run, the generic results file `fe95.res` is copied to the actual results file name (say `fred.res`). If users wish to use the `run2.bat` batch file, data files and results files must have the extensions `*.dat` and `*.res` respectively. See the `readme.txt` file as described above for more details.

1.3 Hardware

The use of Fortran means that the numerical methods described can be "ported" or transferred, with minimal alteration, from one computer to another. For "small" problems and teaching purposes a PC will suffice whereas Smith and Griffiths (2004) show how the same philosophy can be used to solve "large" problems, involving many millions of unknowns, on parallel "clusters" of PCs or on "supercomputers".

1.4 External Fortran subprogram libraries

Another advantage of the use of Fortran is the existence of extensive subprogram libraries already written in that language. These can therefore easily be inserted into programs such as those described herein, avoiding unnecessary duplication of effort. Some libraries are provided by manufacturers of computers, while others are available commercially, although academic use is often free.

A good example is the NAG subroutine library (`www.nag.co.uk`) which contains over 600 subroutines. These cover a wide range of numerical applications, and are organized into "Chapters" which are listed in Table 1.1. The NAG "Chapters" can very broadly be classified into "deterministic" numerical analyses, statistical analyses and "utility" routines. The present book deals only with the first of these classes and even then with a subset. In the first column of Table 1.1 the number in parentheses indicates the chapter of the current text which forms an introduction to the same topic.

Other Fortran libraries with which readers may be familiar include HSL, IMSL, LINPACK and EISPACK, the last two being sublibraries dealing with linear algebra (Chapter 2 in this book) and eigenvalue problems (Chapter 4) respectively. LINPACK and EISPACK have been largely superseded by the more modern LAPACK[1]. It can be seen that the majority of deterministic analysis methods will be dealt with in the following chapters. The selection is governed by limitations of space and of teaching time in typical courses. Attention is directed specifically towards coverage of probably the most important areas of numerical analysis concerning engineers, namely the solution of ordinary and partial differential equations. In the chapters that follow, it will be illustrated how subprograms are constructed and how these are assembled to form small computer programs to address various numerical tasks. This will serve as an introduction for students and engineers to the use of more comprehensive software such as the NAG mathematical subroutine library.

TABLE 1.1: Contents of NAG mathematical subroutine library

Chapter in present book	NAG "Chapter"	Subject area
	1	Utilities
	3	Special Functions
	4	Matrix and Vector Operations
(2)	5	Linear Equations
(4)	6	Eigenvalues and Least-Squares Problems
	7	Transforms
(5)	8	Curve and Surface Fitting
	9	Optimization
(3)	10	Nonlinear Equations
(6)	11	Quadrature
(7)	12	Ordinary Differential Equations
(8)	13	Partial Differential Equations
	19	Operations Research
	20	Statistical Distribution Functions
	21	Random Number Generation
	22	Basic Descriptive Statistics
	25	Correlation and Regression Analysis
	28	Multivariate Analysis
	29	Time Series Analysis

---

[1]See Appendix C for Web references to all these libraries.

1.5 A simple Fortran program

For the details of Fortran 90/95 see for example Smith(1995). Shown in the following pages are two variants of Program 4.1 from Chapter 4, involving the calculation of the numerically largest eigenvalue of a matrix a(n,n).

```fortran
PROGRAM nmex
!---Vector Iteration for "Largest" Eigenvalue and Eigenvector---
USE nm_lib; USE precision; IMPLICIT NONE
INTEGER::i,iters,limit,n; REAL(iwp)::big,l2,tol,zero=0.0_iwp
LOGICAL::converged; REAL(iwp),ALLOCATABLE::a(:,:),x(:),x1(:)
READ*,n; ALLOCATE(a(n,n),x(n),x1(n))
READ*,a; READ*,x; READ*,tol,limit
a=TRANSPOSE(a); iters=0
DO; iters=iters+1; x1=MATMUL(a,x); big=zero
  DO i=1,n; IF(ABS(x1(i))>ABS(big))big=x1(i); END DO; x1=x1/big
  converged=checkit(x1,x,tol)
  IF(converged.OR.iters==limit)EXIT; x=x1
END DO
l2=norm(x1); x1=x1/l2
PRINT*,"Iterations to Convergence:",iters
PRINT*,"Largest Eigenvalue:",big
PRINT*,"Corresponding Eigenvector:"; PRINT*,x1
END PROGRAM nmex
```

Every Fortran program has a name, in this case nmex. There are rules for the construction of names - basically they must consist of letters and digits and begin with a letter. The first line of every program is PROGRAM followed by the program's name, and the last line is END PROGRAM followed again by the program's name. PROGRAM could be written program or Program etc. but we hold to the convention that all words specific to the Fortran language are written in capitals. Spaces are necessary where confusion could occur, for example after PROGRAM and after END. The symbol ! denotes a "comment" and text preceded by it can be used to describe what is going on, or to render code inoperable, since symbols on the line after ! are ignored by the compiler.

The statements USE precision and USE nm_lib (note that the underscore can be used in names) mean that the program is making use of "libraries" called precision and nm_lib. Such libraries (see also Section 1.4) contain subprograms, which thus become available to many main programs which can call on them. The statement IMPLICIT NONE means that all variables used in the program - i,n,big etc. must be declared by statements such as INTEGER::i,n,iters,limit. The IMPLICIT NONE statement is optional but is helpful in debugging programs.

There are four TYPEs of variable declared in the example program, namely INTEGER, REAL, LOGICAL and REAL,ALLOCATABLE. Integers are mainly used for counting while logical variables are used for discriminating and can only take the values .TRUE. or .FALSE.

Arithmetic is usually carried out using real numbers, which cannot be represented exactly in any actual computer. These numbers are therefore held to a given precision (if attainable by that computer), represented here by the appendage _iwp . We return to the questions of arithmetic, precision and associated errors in Section 1.9.

Many of our computations will involve arrays of real numbers, which are declared as REAL,ALLOCATABLE and their names followed by (:), for one-dimensional arrays, (:,:) for two-dimensional and so on. The alternative versions of the program illustrate two methods for getting information into and out of the calculation. The first uses the simpler READ* and PRINT* input and output statements. These read information typed in on the keyboard, and output information direct to the screen, in what is called "free" format.

In the present example n is the size of the $n \times n$ matrix a whose largest eigenvalue we seek, tol is the numerical tolerance to which the result is sought and limit is the maximum number of iterations we will allow in the iterative process. If these are 3, 10^{-5} and 100 respectively we can type in : 3   1.E-5 100 or 03 0.00001   100 or any other correct representation of the input numbers separated by one or more spaces. "Free" format means that what appears as a result on the screen, although correct, cannot be controlled by the user. On a particular PC, the following appeared on the screen:

```
Iterations to Convergence: 19
Largest Eigenvalue: 33.70929144206253
Corresponding Eigenvector:
0.30015495757009153 0.36644583305469774 0.8806954370739892
```

In order to avoid re-typing input numbers on the keyboard, they can be saved in a FILE. Similarly results can be saved and printed later. So in the second version of the example program called nmey we OPEN two FILEs called nm95.dat and nm95.res for input data and output results respectively. The numbers 10 and 11 are essentially arbitrary UNIT numbers (but avoid low numbers like 1 or 2). Input can then be read from UNIT 10 using READ(10,*) to replace the previous READ*. The * means we are retaining the convenience of free format for the input numbers.

```
PROGRAM nmey
!---Vector Iteration for "Largest" Eigenvalue and Eigenvector---
USE nm_lib; USE precision; IMPLICIT NONE
INTEGER::i,iters,limit,n; REAL(iwp)::big,l2,tol,zero=0.0_iwp
LOGICAL::converged; REAL(iwp),ALLOCATABLE::a(:,:),x(:),x1(:)
OPEN(10,FILE='nm95.dat'); OPEN(11,FILE='nm95.res')
```

```
READ(10,*)n; ALLOCATE(a(n,n),x(n),x1(n))
READ(10,*)a; READ(10,*)x; READ(10,*)tol,limit
a=TRANSPOSE(a); iters=0
DO; iters=iters+1; x1=MATMUL(a,x); big=zero
  DO i=1,n; IF(ABS(x1(i))>ABS(big))big=x1(i); END DO; x1=x1/big
  converged=checkit(x1,x,tol)
  IF(converged.OR.iters==limit)EXIT; x=x1
END DO
12=norm(x1); x1=x1/12
WRITE(11,'(A,/,I5)')"Iterations to Convergence",iters
WRITE(11,'(/,A,/,E12.4)')"Largest Eigenvalue",big
WRITE(11,'(/,A,/,6E12.4)')"Corresponding Eigenvector",x1
END PROGRAM nmey
```

When it comes to output, using in this case UNIT 11, it is convenient to be able to control the appearance of the information, and (/,A,/,6E12.4) and (A,/,I5) are examples of "format specifications". A format is used for text, / signifies a new line, E format is for REAL numbers in scientific notation (e.g., 0.3371E+02) and I format is for INTEGERs. The 6 and 4 in the first example signify the maximum number of REAL numbers to be displayed on a single line, and the number of significant figures respectively. The 12 and 5 in these examples specify the "field" width which the number has to occupy. I5 format can therefore not display an integer like 222222 for example. Another useful "format specification" not used in this example is F format. A format specification such as (F16.4) would display a REAL number with a field width of 16 and 4 numbers after the decimal point (e.g., 33.7093).

The second version of the program yields the results:

```
Iterations to Convergence
  19

Largest Eigenvalue
  0.3371E+02

Corresponding Eigenvector
  0.3002E+00   0.3664E+00   0.8807E+00
```

Returning to the example program, after n is known, arrays a, x and x1 can be ALLOCATEd, that is, space can be reserved in the computer for them.

Within the program, variables can be assigned values by means of statements like iters=0 or x1=x1/big using the normal symbols of arithmetic with the exception of * for multiplication.

1.6 Some simple Fortran constructs

The above programs employ two of the most fundamental constructs of any programming language, namely loops and conditions. A loop is enclosed by the Fortran words DO and END DO. The version DO i=1,n will be completed n times but if the i=1,n part is missing the loop is left after EXIT if some condition is satisfied. (Be careful, in case this is never true.) Conditional statements can in general take the form:

```
IF(condition) THEN
   Statements
ELSE IF(condition) THEN
   Statements
END IF
```

But often the simpler form:

```
IF(condition) Statement
```

suffices, as in the present program.

In IF statements, "condition" will often involve the "relational operators" <, >, <=, /= and so on, following the usual mathematical convention.

A useful alternative conditional statement, used for discriminating on the value of simple quantities, for example an integer nsp, reads:

```
SELECT CASE(nsp)
CASE(1)
   Statements
CASE(2)
   Statements
CASE(etc.)
   Statements
END SELECT
```

1.7 Intrinsic functions

Both programs make use of the "intrinsic" Fortran FUNCTION ABS, which delivers the absolute value of its argument. There are about 40 such FUNCTIONs in Fortran 95. These undertake commonly used mathematical tasks like finding a square root (SQRT), or a logarithm to the base e (LOG), or, for an angle expressed in radians, its sine, cosine (SIN, COS) and so on. "Intrinsic" means that these are part of the language and are automatically available to all

programs. A further set of intrinsic FUNCTIONs operates on arrays and there
are about a dozen of these in Fortran 95. Both programs also make use of
MATMUL for multiplying two arrays or matrices together. Also widely used are
DOT_PRODUCT, TRANSPOSE, MAXVAL (which finds the maximum array element)
and SUM (which returns the sum of all the elements of an array).

1.8   User-supplied functions and subroutines

Fortran has always encouraged the use of subprograms so that users can
encapsulate code which may be used many times in a single program or in
other programs. The form of these subprograms has been copied in other
languages - compare the following Fortran and MATLAB versions of **LU**
factorization (see Chapter 2):

```
Fortran 95                      MATLAB

FUNCTION lu(a)                  function A=lu(A)
! LU factorization              % LU factorization
 REAL::a(:,:)                    [n,m]=size(A)
 REAL::lu(SIZE(a,1),SIZE(a,2))   for k=1,n-1
 INTEGER::i,j,k,n; n=SIZE(a,1)    for i=k+1,n
 DO k=1,n-1                        A(i,k)=A(i,k)/A(k,k)
 DO i=k+1,n                        for j=k+1,n
   a(i,k)=a(i,k)/a(k,k)             A(i,j)=A(i,j)-A(i,k)*A(k,j)
   DO j=k+1,n                       end
     a(i,j)=a(i,j)-a(i,k)*a(k,j)  end
   END DO                        end
 END DO                         end
 END DO
 lu=a
END FUNCTION lu
```

Whereas a FUNCTION delivers a single quantity (REAL, LOGICAL etc.) as its
result, a SUBROUTINE can deliver several quantities to the program which
CALLs it. Subprograms are collected together in a "library" such as precision
and nm_lib in our example and are made available to the main program by
statements like USE nm_lib.

Programs nmex and nmey employ two user-supplied subprograms from library nm_lib, namely checkit which checks whether arrays $\{x\}$ and $\{x_1\}$
are nearly the same, and norm which calculates the "L2 norm" of vector $\{x_1\}$.

Library precision contains instructions about the precision with which
REAL numbers (employing the suffix _iwp as in zero=0.0_iwp) are held in the

computer, so it is appropriate now to consider the whole question of errors and accuracy in numerical computations (see Higham (2002) for a full discussion).

1.9 Errors and accuracy

In the chapters which follow, it is assumed that calculations in numerical methods will be made using a digital computer. It will be very rare for computations to be made in exact arithmetic and, in general, "real" numbers represented in "floating point" form will be used. This means that the number 3 will tend to be represented by $2.99999\ldots9$ or $3.0000\ldots1$. Therefore, all calculations will, in a strict sense, be erroneous and what will concern us is that these errors are within the range that can be tolerated for engineering purposes.

Employing the Fortran 95 intrinsic function `SELECTED_REAL_KIND` in library `precision` we attempt in the calculations in this book to hold `REAL` numbers to a precision of about 15 decimal places. The most significant measure of error is not the "absolute" error, but the "relative" error. For example, if x_0 is exact and x an approximation to it, then what is significant is not $x - x_0$ (absolute) but $(x - x_0)/x_0$ (relative). In engineering calculations it must always be remembered that excessive computational accuracy may be unjustified if the data (for example some physical parameters) are not known to within a relative error far larger than that achievable in the calculations.

Modern computers (even "hand" calculators) can represent numbers to a high precision. Nevertheless, in extreme cases, errors can arise in calculations due to three main sources, called "roundoff", "truncation" and "cancellation" respectively.

1.9.1 Roundoff

The following examples illustrate how accuracy can be affected by rounding to two decimal places. First consider

$$0.56 \times 0.65 = 0.36 \quad \text{hence} \quad (0.56 \times 0.65) \times 0.54 = 0.19$$

whereas

$$0.65 \times 0.54 = 0.35 \quad \text{hence} \quad 0.56 \times (0.65 \times 0.54) = 0.20$$

In this case the order in which the rounding was performed influences the result.

Now consider the sum,

$$51.4 \times 23.25 - 50.25 \times 22.75 + 1.25 \times 10.75 - 1.2 \times 9.8 = 53.54 \text{ (exactly)}$$

If this calculation is worked to zero decimal places, two decimal places, three significant figures and four significant figures respectively, adding forwards (F) or backwards (B), rounding *after* multiplication (R) or *before* multiplicatione (C), the answers given in Table 1.2 are found. These represent extremes

TABLE 1.2:   Influence on accuracy of precision level and order of calculations

	FR	FC	BR	BC
0 decimal places	53	24	53	24
2 decimal places	53.54	53.54	53.54	53.54
3 significant figures	61.6	51.7	60	50
4 significant figures	53.68	53.68	54	54

by modern standards, since calculations tend to be done on computers, but illustrate the potential pitfalls in numerical work.

In doing floating point arithmetic, computers hold a fixed number of digits in the "mantissa", that is the digits following the decimal point and preceding the "exponent" $\times 10^6$ in the number 0.243875×10^6. The effect of roundoff in floating point calculations can be seen in these alternative calculations to six digits of accuracy,

$$(0.243875 \times 10^6 + 0.412648 \times 10^1) - 0.243826 \times 10^6 = 0.530000 \times 10^2$$

and

$$(0.243875 \times 10^6 - 0.243826 \times 10^6) + 0.412648 \times 10^1 = 0.531265 \times 10^2$$

It should however be emphasized that many modern computers represent floating point numbers by 64 binary digits, which retain about 15 significant figures, and so roundoff errors are less of a problem than they used to be, at least on scalar computers. For engineering purposes, sensitive calculations on machines with only 32-bit numbers should always be performed in "double precision".

1.9.2   Truncation

Errors due to this source occur when an infinite process is replaced by a finite one. For example, some of the intrinsic FUNCTIONs in Fortran 95 mentioned earlier compute their result by summing a series where, for example,

$$S = \sum_{i=0}^{\infty} a_i x^i \quad \text{would be replaced by the finite sum} \quad \sum_{i=0}^{N} a_i x^i$$

Consider the errors that might be introduced in the computation of e^x with $N = 5$, thus

$$e^x \approx 1 + \frac{x}{1!} + \frac{x^2}{2!} + \frac{x^3}{3!} + \frac{x^4}{4!}$$

Suppose we wish to calculate $e^{1/3}$ using the intrinsic function EXP(0.3333). Even before considering truncation, we have introduced a rounding error by replacing $1/3$ by 0.3333, thus

$$\epsilon_1 = e^{0.3333} - e^{1/3}$$
$$= -.0000465196$$

Then we might truncate the series after five terms, leading to a further error given by

$$\epsilon_2 = -\left(\frac{0.3333^5}{5!} + \frac{0.3333^6}{6!} + \frac{0.3333^7}{7!} + \frac{0.3333^8}{8!} + \frac{0.3333^9}{9!} + \ldots\right)$$
$$= -.0000362750$$

Finally we might sum with values rounded to 4 decimal places, thus

$$1 + 0.3333 + 0.0555 + 0.0062 + 0.0005 = 1.3955$$

where the propagated error from rounding is -0.0000296304, leading to a final total error of -0.0001124250 to 10 decimal places.

1.9.3 Cancellation

The quadratic equation $x^2 - 2ax + \epsilon = 0$ has the roots $x_1 = a + \sqrt{a^2 - \epsilon}$ and $x_2 = a - \sqrt{a^2 - \epsilon}$. If a is 100 and ϵ is 1, $x_2 = 100 - \sqrt{10000 - 1}$. In extreme cases, this could lead to dangerous cancellation and it would be safer to write
$$x_2 = \frac{\epsilon}{\left(a + \sqrt{a^2 - \epsilon}\right)}.$$

1.9.4 Intrinsic and library-supplied precision routines

To assist users in questions of error and accuracy Fortran 95 provides several intrinsic FUNCTIONs, for example EPSILON, CEILING, FLOOR, HUGE, RADIX, RANGE and TINY which return properties of REAL numbers attainable on a particular processor.

Mathematical subroutine libraries often contain useful routines which help users to appreciate the possibilities of numerical errors in digital computations. For example, the NAG library (see Table 1.1) contains a utility MODULE nag_error_handling which controls how errors are to be handled by the Library and the information communicated to a user's program.

1.10 Graphical output

We expect that users of the programs and libraries described in this book will wish to create graphical interpretations of their results in the form of "$x-y$ plots" and contour maps. While we do not provide any plotting software, by listing the Fortran 95 source code users are invited to modify the output to produce the results they require in a convenient format. Plotting can then be performed using any of the numerous graphics packages available.

As an example, consider the widely used Excel plotting system, which is part of Microsoft®Office Suite. Figure 7.4 described in Chapter 7 shows the numerical solution of a second order differential equation describing the motion of a simple damped oscillator from Example 7.7. The results produced by Program 7.1 give three columns of numbers representing respectively, from left to right, time, displacement and velocity of the oscillator. A typical set of results (truncated here for the sake of brevity) is given in Results 1.1.

```
---One-Step Methods for Systems of ODEs---


***** 4TH ORDER RUNGE-KUTTA METHOD ****


      x             y(i) , i = 1, 2
  0.00000E+00 -0.25000E-01 -0.10000E+01
  0.50000E-02 -0.29713E-01 -0.88435E+00
  0.10000E-01 -0.33835E-01 -0.76358E+00
  0.15000E-01 -0.37343E-01 -0.63939E+00
  0.20000E-01 -0.40226E-01 -0.51347E+00
  0.25000E-01 -0.42478E-01 -0.38744E+00
  0.30000E-01 -0.44103E-01 -0.26284E+00
  0.35000E-01 -0.45111E-01 -0.14111E+00
        .
        .
        .
  88 lines removed
        .
        .
        .
  0.48000E+00  0.51851E-02 -0.16502E-01
  0.48500E+00  0.50729E-02 -0.28215E-01
  0.49000E+00  0.49045E-02 -0.39013E-01
  0.49500E+00  0.46845E-02 -0.48810E-01
  0.50000E+00  0.44182E-02 -0.57534E-01
```

Results 1.1: Typical results from Program 7.1 (Example 7.7 in Chapter 7)

In order to plot time vs. displacement using Excel we could follow the following steps:

1) "Copy" (ctrl-c) lines 6-106 of the output file (excluding the first five lines of text and blank lines) and "Paste" (ctrl-v) them into cell A1 of the spreadsheet.

2) Select cell A1 and then on the Excel Tool Bar go to Data|Text to Columns...

3) Check on "Delimited" and click on "Next>"

4) Check "Space", click on "Next>" and then "Finished".

5) This procedure should result in each column of data having its own lettered column in the spreadsheet. Since there was white space before the first column, it is likely that the first column of numbers appears in column B.

6) If any cell appears as "########" the width needs adjusting, in which case highlight the cell by clicking on it and on the Excel Tool Bar go to Format|Column|Autofit Selection.

7) In order to plot Column B against Column C, highlight both columns by clicking on "B" and then while holding down "ctrl" click on "C".

8) On the Excel Tool Bar click on the "Chart Wizard", select "XY(Scatter)", select the "Chart sub-type" and click on "Next>", "Next>", "Next>" and "Finish".

9) The Graph should now be visible on the screen. There is great flexibility available for customizing the appearance of the graph by right-clicking on the white space and selecting "Chart Options..."

Figure 7.4 and many other graphs presented in this text were obtained using a similar approach to that described above.

1.11   Conclusions

A style of programming using portable subprograms (SUBROUTINEs and FUNCTIONs in Fortran 95) has been outlined. Using this strategy, in subsequent chapters the same subprograms will find use in many different contexts of numerical analysis. Attention has to be devoted to the possibility of calculation errors, for example due to roundoff, which will vary from one machine to another. Due to the relatively large "wordlengths" commonly used in modern computers, this may seem to be a forgotten issue, but see Smith and Margetts (2006) in the context of parallel computing. Chapters 2 to 8 go on to describe concise programs built up using subprograms wherever possible. These chapters cover a subset of, and form an introduction to, more comprehensive subroutine libraries such as NAG.

Chapter 2 deals with the numerical solution of sets of linear algebraic equations, while Chapter 3 considers roots of single nonlinear equations and sets of nonlinear equations. In Chapter 4, eigenvalue equations are considered, while Chapter 5 deals with interpolation and curve fitting. Chapter 6 is devoted to numerical "quadrature", that is to say to numerical evaluation of integrals, while Chapter 7 introduces the solution of ordinary differential equations by numerical means. Chapter 8 is an introduction to the solution of partial differential equations, using finite difference and finite element approaches. In all chapters, mathematical ideas and definitions are introduced as they occur, and most numerical aspects are illustrated by compact computer programs. The "library" routines are described and listed in Appendices A and B respectively.

All software described in the text can be downloaded from the web site `www.mines.edu/~vgriffit/NM`

Chapter 2

Linear Algebraic Equations

2.1 Introduction

One of the commonest numerical tasks facing engineers is the solution of sets of linear algebraic equations of the form,

$$a_{11}x_1 + a_{12}x_2 + a_{13}x_3 = b_1$$
$$a_{21}x_1 + a_{22}x_2 + a_{23}x_3 = b_2 \qquad (2.1)$$
$$a_{31}x_1 + a_{32}x_2 + a_{33}x_3 = b_3$$

commonly written

$$[\mathbf{A}]\{\mathbf{x}\} = \{\mathbf{b}\} \qquad (2.2)$$

where $[\mathbf{A}]$ is a "matrix" and $\{\mathbf{x}\}$ and $\{\mathbf{b}\}$ are "vectors".

In these equations the a_{ij} are constant known quantities, as are the b_i. The problem is to determine the unknown x_i. In this chapter we shall consider two different solution techniques, usually termed "direct" and "iterative" methods. The direct methods are considered first and are based on row by row "elimination" of terms, a process usually called "Gaussian elimination".

2.2 Gaussian elimination

We begin with a specific set of equations

$$10x_1 + x_2 - 5x_3 = 1 \qquad \text{(a)}$$
$$-20x_1 + 3x_2 + 20x_3 = 2 \qquad \text{(b)} \qquad (2.3)$$
$$5x_1 + 3x_2 + 5x_3 = 6 \qquad \text{(c)}$$

To "eliminate" terms, we could, for example, multiply equation (a) by two and add it to equation (b). This would produce an equation from which the term in x_1 had been eliminated. Similarly, we could multiply equation (a) by 0.5 and subtract it from equation (c). This would also eliminate the term in x_1 leaving an equation in (at most) x_2 and x_3.

We could formally write this process as

$$(b) - \left(\frac{-20}{10}\right) \times (a) \longrightarrow 5x_2 + 10x_3 = 4 \qquad (d)$$

$$(2.4)$$

$$(c) - \left(\frac{5}{10}\right) \times (a) \longrightarrow 2.5x_2 + 7.5x_3 = 5.5 \qquad (e)$$

One more step of the same procedure would be

$$(e) - \left(\frac{2.5}{5}\right) \times (d) \longrightarrow 2.5x_3 = 3.5 \qquad (2.5)$$

Thus, for sets of n simultaneous equations, however big n might be, after n steps of this process a single equation involving only the unknown x_n would remain. Working backwards from equation (2.5), a procedure usually called "back-substitution", x_3 can first be found as $3.5/2.5$ or 1.4. Knowing x_3, substitution in equation 2.4(d) gives x_2 as 2.0 and finally substitution in equation 2.3(a) gives x_1 as 1.0. Writing the back-substitution process in terms of matrices and vectors, we have

$$\begin{bmatrix} 10 & 1 & -5 \\ 0 & 5 & 10 \\ 0 & 0 & 2.5 \end{bmatrix} \begin{Bmatrix} x_1 \\ x_2 \\ x_3 \end{Bmatrix} = \begin{Bmatrix} 1 \\ 4 \\ 3.5 \end{Bmatrix} \qquad (2.6)$$

or

$$[\mathbf{U}]\{\mathbf{x}\} = \{\mathbf{y}\} \qquad (2.7)$$

The matrix $[\mathbf{U}]$ is called an "upper triangular matrix" and it is clear that such matrices will be very convenient in linear equation work.

In a similar way, if we had the system of equations

$$\begin{bmatrix} l_{11} & 0 & 0 \\ l_{21} & l_{22} & 0 \\ l_{31} & l_{32} & l_{33} \end{bmatrix} \begin{Bmatrix} x_1 \\ x_2 \\ x_3 \end{Bmatrix} = \begin{Bmatrix} y_1 \\ y_2 \\ y_3 \end{Bmatrix} \qquad (2.8)$$

or

$$[\mathbf{L}]\{\mathbf{x}\} = \{\mathbf{y}\} \qquad (2.9)$$

it would be relatively easy to calculate $\{\mathbf{x}\}$ given $[\mathbf{L}]$ and $\{\mathbf{y}\}$. The matrix $[\mathbf{L}]$ is called a "lower triangular matrix", and the process of finding $\{\mathbf{x}\}$ in equations (2.9) is called "forward-substitution". The direct methods we shall discuss all involve, in some way or another, matrices like $[\mathbf{L}]$ and $[\mathbf{U}]$.

Example 2.1

Use Gaussian elimination to solve the following set of equations.

$$2x_1 - 3x_2 + x_3 = 7$$
$$x_1 - x_2 - 2x_3 = -2$$
$$3x_1 + x_2 - x_3 = 0$$

Solution 2.1

Eliminate the first column:

$$2x_1 - 3x_2 + x_3 = 7$$
$$0.5x_2 - 2.5x_3 = -5.5$$
$$5.5x_2 - 2.5x_3 = -10.5$$

Eliminate the second column:

$$2x_1 - 3x_2 + x_3 = 7$$
$$0.5x_2 - 2.5x_3 = -5.5$$
$$25x_3 = 50$$

Back-substitute:

$$x_3 = 2$$
$$x_2 = (-5.5 + 2.5(2))/0.5 = -1$$
$$x_1 = (7 - 2 + 3(-1))/2 = 1$$

Program 2.1: Gaussian elimination for linear simultaneous equations

```
PROGRAM nm21
!---Gaussian Elimination for Linear Simultaneous Equations---
 IMPLICIT NONE; INTEGER,PARAMETER::iwp=SELECTED_REAL_KIND(15,300)
 INTEGER::i,j,k,n; REAL(iwp)::x,zero=0.0_iwp
 REAL(iwp),ALLOCATABLE::a(:,:),b(:)
 OPEN(10,FILE='nm95.dat'); OPEN(11,FILE='nm95.res')
 READ(10,*)n; ALLOCATE(a(n,n),b(n)); READ(10,*)a; READ(10,*)b
 WRITE(11,'(A)')                                                &
```

```
"---Gaussian Elimination for Linear Simultaneous Equations---"
WRITE(11,'(/,A)')"Coefficient Matrix"
a=TRANSPOSE(a); DO i=1,n; WRITE(11,'(6E12.4)')a(i,:); END DO
WRITE(11,'(/,A,/,6E12.4)')"Right Hand Side Vector",b
!---Convert to Upper Triangular Form---
DO k=1,n-1
  IF(ABS(a(k,k))>1.E-6_iwp)THEN
    DO i=k+1,n
      x=a(i,k)/a(k,k); a(i,k)=zero
      DO j=k+1,n; a(i,j)=a(i,j)-a(k,j)*x; END DO
      b(i)=b(i)-b(k)*x
    END DO
  ELSE; WRITE(11,*)"Zero pivot found in row",k; STOP
  END IF
END DO
WRITE(11,'(/,A)')"Modified Matrix"
DO i=1,n; WRITE(11,'(6E12.4)')a(i,:); END DO
WRITE(11,'(/,A,/,6E12.4)')"Modified Right Hand Side Vector",b
!---Back-substitution---
DO i=n,1,-1
  x=b(i)
  IF(i<n)THEN
    DO j=i+1,n; x=x-a(i,j)*b(j); END DO
  END IF
  b(i)=x/a(i,i)
END DO
WRITE(11,'(/,A,/,6E12.4)')"Solution Vector",b
END PROGRAM nm21
```

Number of equations	n		
	3		
Coefficient matrix	(a(i,:),i=1,n)		
	10.	1.	-5.
	-20.	3.	-20.
	5.	3.	5.
Right hand side	b		
	1.	2.	6.

Data 2.1: Gaussian Elimination

List 2.1:

Scalar integers:
i   simple counter
j   simple counter
k   simple counter
n   number of equations to be solved

Scalar reals:
x   elimination factor
zero   set to zero

Dynamic real arrays:
a $n \times n$ matrix of coefficients
b   "right hand side" vector of length n, overwritten by solution

---

```
---Gaussian Elimination for Linear Simultaneous Equations---

Coefficient Matrix
   0.1000E+02   0.1000E+01  -0.5000E+01
  -0.2000E+02   0.3000E+01   0.2000E+02
   0.5000E+01   0.3000E+01   0.5000E+01

Right Hand Side Vector
   0.1000E+01   0.2000E+01   0.6000E+01

Modified Matrix
   0.1000E+02   0.1000E+01  -0.5000E+01
   0.0000E+00   0.5000E+01   0.1000E+02
   0.0000E+00   0.0000E+00   0.2500E+01

Modified Right Hand Side Vector
   0.1000E+01   0.4000E+01   0.3500E+01

Solution Vector
   0.1000E+01  -0.2000E+01   0.1400E+01
```

Results 2.1: Gaussian Elimination

In equations (2.4) and (2.5) it can be seen that the elements of the original arrays [**A**] and {**b**} are progressively altered during the calculation. In Program 2.1, once the terms in {**x**} have been calculated, they are stored in {**b**} since the original {**b**} has been lost anyway.

In passing it should be noted that equations (2.4) involve division by the coefficient a_{11} (equal to 10 in this case) while equations (2.5) involve division by the modified coefficient a_{22} (equal to 5 in this case).

These coefficients a_{kk} where $1 \leq k \leq n$ are called the "pivots" and it will be clear that they might be zero, either at the beginning of the elimination process, or during it. We shall return to this problem later, but for the moment shall merely check whether a_{kk} is or has become zero and stop the calculation if this is so. Input details are shown in Data 2.1 with output in Results 2.1.

The program begins by reading in the number of equations n, the array $[\mathbf{A}]$ (stored in a) and the right-hand side vector $\{\mathbf{b}\}$ (stored in b). Because of the way Fortran READ works, a must first be transposed. A check is first made to see if diagonal a_{kk} is greater than "zero" (a small number in this case). Rows 2 to n are then processed according to equations (2.4) and (2.5) and the modified $[\mathbf{A}]$ and $\{\mathbf{b}\}$ printed out for comparison with equations (2.6). The back-substitution calculation is then performed, leaving the original unknowns $\{\mathbf{x}\}$ stored in $\{\mathbf{b}\}$ which is printed out.

2.2.1 Observations on the elimination process

Since only the upper triangle of the modified matrix is needed for back-substitution, it is not necessary to compute the zero terms below the pivot at each stage of the elimination process. In order to emphasize the upper triangular nature of the modified matrix however, Program 2.1 explicitly sets the terms below the pivot to zero. Further, during the conversion of $[\mathbf{A}]$ to upper triangular form, it was necessary to operate also on $\{\mathbf{b}\}$. Therefore if equations with the same coefficients $[\mathbf{A}]$ have to be solved for different $\{\mathbf{b}\}$, which are not known in advance as is often the case, the conversion of $[\mathbf{A}]$ to triangular form would be necessary for every $\{\mathbf{b}\}$ using this method.

2.3 Equation solution using factorization

We therefore seek a way of implementing Gaussian elimination so that multiple right-hand side $\{\mathbf{b}\}$ vectors can be processed after only a single "decomposition" of $[\mathbf{A}]$ to triangular form. Such methods involve "factorization" of $[\mathbf{A}]$ into triangular matrix components. For example, it can be shown that matrix $[\mathbf{A}]$ can always be written as the product

$$[\mathbf{A}] = [\mathbf{L}][\mathbf{U}] \tag{2.10}$$

where $[\mathbf{L}]$ is a lower triangular matrix and $[\mathbf{U}]$ an upper triangular matrix, in the forms

$$[\mathbf{L}] = \begin{bmatrix} l_{11} & 0 & 0 \\ l_{21} & l_{22} & 0 \\ l_{31} & l_{32} & l_{33} \end{bmatrix} \tag{2.11}$$

and

$$[\mathbf{U}] = \begin{bmatrix} u_{11} & u_{12} & u_{13} \\ 0 & u_{22} & u_{23} \\ 0 & 0 & u_{33} \end{bmatrix} \tag{2.12}$$

The diagonal terms l_{kk} and u_{kk} are arbitrary except that their product is known. For example,

$$l_{kk}u_{kk} = a_{kk} \qquad (2.13)$$

so it is conventional to assume that either l_{kk} or u_{kk} is unity, hence

$$\begin{bmatrix} a_{11} & a_{12} & a_{13} \\ a_{21} & a_{22} & a_{23} \\ a_{31} & a_{32} & a_{33} \end{bmatrix} = \begin{bmatrix} 1 & 0 & 0 \\ l_{21} & 1 & 0 \\ l_{31} & l_{32} & 1 \end{bmatrix} \begin{bmatrix} u_{11} & u_{12} & u_{13} \\ 0 & u_{22} & u_{23} \\ 0 & 0 & u_{33} \end{bmatrix} \qquad (2.14)$$

is a typical statement of $[\mathbf{L}][\mathbf{U}]$ factorization.

When the triangular factors $[\mathbf{L}]$ and $[\mathbf{U}]$ in equation (2.10) have been computed, equation solution proceeds as follows:

$$[\mathbf{A}]\{\mathbf{x}\} = \{\mathbf{b}\}$$

or

$$[\mathbf{L}][\mathbf{U}]\{\mathbf{x}\} = \{\mathbf{b}\} \qquad (2.15)$$

We now let

$$[\mathbf{U}]\{\mathbf{x}\} = \{\mathbf{y}\} \qquad (2.16)$$

hence

$$[\mathbf{L}]\{\mathbf{y}\} = \{\mathbf{b}\} \qquad (2.17)$$

Since $[\mathbf{L}]$ and $\{\mathbf{b}\}$ are known, and $[\mathbf{L}]$ does not depend on $\{\mathbf{b}\}$, this process is simply the "forward-substitution" we saw in equation (2.9). Once equation (2.17) has been solved for $\{\mathbf{y}\}$, equation (2.16) is then the "back-substitution" described previously by equation (2.7). A solution algorithm will therefore consist of three phases, namely a factorization (equation 2.15) followed by a forward-substitution (equation 2.17) and a back-substitution (equation 2.16). The procedures of factorization and forward- and back-substitution will be used in other contexts in this book and elsewhere, so that it makes sense to code them as library subroutines. They are called lufac, subfor and subbac respectively and their actions and parameters are described in Appendix A with full listings in Appendix B.

Equations (2.14) are evaluated as follows:

Row 1: $u_{11} = a_{11}$, $u_{12} = a_{12}$, $u_{13} = a_{13}$

This shows that with unity on the diagonal of $[\mathbf{L}]$, the first row of $[\mathbf{U}]$ is simply a copy of the first row of $[\mathbf{A}]$. Subroutine lufac therefore begins by nulling $[\mathbf{L}]$ (called lower) and $[\mathbf{U}]$ (called upper) and by copying the first row of $[\mathbf{A}]$ into upper.

Row 2: $l_{21}u_{11} = a_{21}$, hence $l_{21} = \dfrac{a_{21}}{u_{11}}$

Having found l_{21}, u_{22} and u_{23} can be computed from

$$l_{21}u_{12} + u_{22} = a_{22}, \quad \text{hence} \quad u_{22} = a_{22} - l_{21}u_{12} \quad \text{and}$$

$$l_{21}u_{13} + u_{23} = a_{23}, \quad \text{hence} \quad u_{23} = a_{23} - l_{21}u_{13}$$

Row 3: $\quad l_{31}u_{11} = a_{31}, \quad \text{hence} \quad l_{31} = \dfrac{a_{31}}{u_{11}}$

$$l_{31}u_{12} + l_{32}u_{22} = a_{32}, \quad \text{hence} \quad l_{32} = \dfrac{a_{32} - l_{31}u_{12}}{u_{22}}$$

Having found l_{31} and l_{32}, u_{33} can be computed from

$$l_{31}u_{13} + l_{32}u_{23} + u_{33} = a_{33}, \quad \text{hence} \quad u_{33} = a_{33} - l_{31}u_{13} - l_{32}u_{23}$$

Subroutine `lufac` carries out these operations in two parts, commented "Lower Triangular Factors" and "Upper Triangular Factors" respectively. A "zero pivot" is tested for in the same way as was done in Program 2.1.

Example 2.2

*Use [**L**][**U**] factorization to solve the following set of equations*

$$2x_1 - 3x_2 + x_3 = 7$$
$$x_1 - x_2 - 2x_3 = -2$$
$$3x_1 + x_2 - x_3 = 0$$

Solution 2.2

Factorize the coefficient matrix into upper and lower triangular matrices, hence [**A**] = [**L**][**U**],

$$\begin{bmatrix} 2 & -3 & 1 \\ 1 & -1 & -2 \\ 3 & 1 & -1 \end{bmatrix} = \begin{bmatrix} 1 & 0 & 0 \\ l_{21} & 1 & 0 \\ l_{31} & l_{32} & 1 \end{bmatrix} \begin{bmatrix} u_{11} & u_{12} & u_{13} \\ 0 & u_{22} & u_{23} \\ 0 & 0 & u_{33} \end{bmatrix}$$

Solving for l_{ij} and u_{ij} gives

$$[\mathbf{L}] = \begin{bmatrix} 1. & 0. & 0. \\ 0.5 & 1. & 0. \\ 1.5 & 11. & 1. \end{bmatrix}$$

$$[\mathbf{U}] = \begin{bmatrix} 2. & -3. & 1. \\ 0. & 0.5 & -2.5 \\ 0. & 0. & 25. \end{bmatrix}$$

Forward-substitution gives $\quad [\mathbf{L}]\{\mathbf{y}\} = \{\mathbf{b}\}$

$$\begin{bmatrix} 1. & 0. & 0. \\ 0.5 & 1. & 0. \\ 1.5 & 11. & 1. \end{bmatrix} \begin{Bmatrix} y_1 \\ y_2 \\ y_3 \end{Bmatrix} = \begin{Bmatrix} 7 \\ -2 \\ 0 \end{Bmatrix}$$

hence

$$y_1 = 7, \quad y_2 = -2 - 0.5(7) = -5.5$$

$$y_3 = -1.5(7) + 11(5.5) = 50$$

Back-substitution gives $\quad [\mathbf{U}]\{\mathbf{x}\} = \{\mathbf{y}\}$

$$\begin{bmatrix} 2. & -3. & 1. \\ 0. & 0.5 & -2.5 \\ 0. & 0. & 25. \end{bmatrix} \begin{Bmatrix} x_1 \\ x_2 \\ x_3 \end{Bmatrix} = \begin{Bmatrix} 7. \\ -5.5 \\ 50. \end{Bmatrix}$$

hence

$$x_3 = 50/25 = 2, \quad x_2 = (-5.5 + 2.5(2))/0.5 = -1$$

$$x_1 = (7 - 2 + 3(-1))/2 = 1$$

Program 2.2: Gaussian elimination using [L][U] factorization

```
PROGRAM nm22
!---Gaussian Elimination using LU Factorization---
 USE nm_lib; USE precision; IMPLICIT NONE
 INTEGER::i,n; REAL(iwp),ALLOCATABLE::a(:,:),b(:),lower(:,:),    &
    upper(:,:)
 OPEN(10,FILE='nm95.dat'); OPEN(11,FILE='nm95.res')
 READ(10,*)n; ALLOCATE(a(n,n),lower(n,n),upper(n,n),b(n))
 READ(10,*)a; READ(10,*)b
 WRITE(11,'(A)')                                                &
    "---Gaussian Elimination using LU Factorization---"
 WRITE(11,'(/,A)')"Coefficient Matrix"
 a=TRANSPOSE(a); DO i=1,n; WRITE(11,'(6E12.4)')a(i,:); END DO
 WRITE(11,'(/,A,/,6E12.4)')"Right Hand Side Vector",b
 CALL lufac(a,lower,upper)
 WRITE(11,'(/,A)')"Lower Triangular Factors"
 DO i=1,n; WRITE(11,'(6E12.4)')lower(i,:); END DO
 WRITE(11,'(/,A)')"Upper Triangular Factors"
 DO i=1,n; WRITE(11,'(6E12.4)')upper(i,:); END DO
 CALL subfor(lower,b); CALL subbac(upper,b)
 WRITE(11,'(/,A,/,6E12.4)')"Solution Vector",b
END PROGRAM nm22
```

List 2.2:

Scalar integers:
i   simple counter
n   number of equations to be solved

Dynamic real arrays:
a $n \times n$ matrix of coefficients
b   "right hand side" vector of length n, overwritten by solution
lower   lower triangular factor of a
upper   upper triangular factor of a

```
Number of equations          n
                             3

Coefficient matrix          (a(i,:),i=1,n)
                             10.   1.  -5.
                            -20.   3. -20.
                              5.   3.   5.


Right hand side               b
                              1.   2.   6.
```

Data 2.2: [L][U] Factorization

```
---Gaussian Elimination using LU factorization---

Coefficient Matrix
   0.1000E+02  0.1000E+01 -0.5000E+01
  -0.2000E+02  0.3000E+01  0.2000E+02
   0.5000E+01  0.3000E+01  0.5000E+01

Right Hand Side Vector
   0.1000E+01  0.2000E+01  0.6000E+01

Lower Triangular Factors
   0.1000E+01  0.0000E+00  0.0000E+00
  -0.2000E+01  0.1000E+01  0.0000E+00
   0.5000E+00  0.5000E+00  0.1000E+01

Upper Triangular Factors
   0.1000E+02  0.1000E+01 -0.5000E+01
   0.0000E+00  0.5000E+01  0.1000E+02
   0.0000E+00  0.0000E+00  0.2500E+01
```

```
Solution Vector
   0.1000E+01 -0.2000E+01  0.1400E+01
```

Results 2.2: [L][U] Factorization

The program simply consists of reading in n, [A] and {b} followed by three subroutine calls to lufac, subfor and subbac. The lower and upper triangular factors are printed out, followed by the solution which has overwritten the original right-hand side in {b}. Input data are as in Data 2.2, with output in Results 2.2.

2.3.1 Observations on the solution process by factorization

Comparison of outputs in Results 2.1 and 2.2 will show that the upper triangular factor of [A] in Results 2.2 is precisely the same as the modified upper triangular part of [A] in Results 2.1. Thus Programs 2.1 and 2.2 have much in common. However, if many {b} vectors were to be processed, it would merely be necessary to call lufac once in Program 2.2 and to create a small loop reading in each new {b} and calling subfor and subbac to produce the solutions. Since the time taken in lufac is substantially more than that taken in subfor and subbac, this yields great economies as n increases.

Inspection of the arithmetic in lufac will show that storage could be saved by overwriting [A] by lower and upper, and this will be done in subsequent programs. It has not been implemented in this first factorization program in an attempt to make the computational process as clear as possible.

2.4 Equations with a symmetrical coefficient matrix

If the coefficients of the [A] matrix satisfy the condition

$$a_{ij} = a_{ji} \qquad (2.18)$$

that matrix is said to be symmetrical. For example the matrix

$$[\mathbf{A}] = \begin{bmatrix} 16 & 4 & 8 \\ 4 & 5 & -4 \\ 8 & -4 & 22 \end{bmatrix} \qquad (2.19)$$

has symmetrical coefficients. If subroutine lufac is used to factorize this matrix, the result will be found to be

$$[\mathbf{L}] = \begin{bmatrix} 1. & 0. & 0. \\ 0.25 & 1. & 0. \\ 0.5 & -1.5 & 1. \end{bmatrix} \qquad (2.20)$$

$$[\mathbf{U}] = \begin{bmatrix} 16 & 4 & 8 \\ 0 & 4 & -6 \\ 0 & 0 & 9 \end{bmatrix} \qquad (2.21)$$

If the rows of $[\mathbf{U}]$ are then divided by u_{kk} we get

$$[\mathbf{U}^1] = \begin{bmatrix} 1 & 0.25 & 0.5 \\ 0 & 1 & -1.5 \\ 0 & 0 & 1 \end{bmatrix} \qquad (2.22)$$

and it can be seen that $[\mathbf{L}] = [\mathbf{U}^1]^T$. The scaling of $[\mathbf{U}]$ to $[\mathbf{U}^1]$ is accomplished in matrix terms by

$$[\mathbf{U}] = [\mathbf{D}][\mathbf{U}^1] \qquad (2.23)$$

where $[\mathbf{D}]$ is the diagonal matrix

$$[\mathbf{D}] = \begin{bmatrix} 16 & 0 & 0 \\ 0 & 4 & 0 \\ 0 & 0 & 9 \end{bmatrix} \qquad (2.24)$$

Thus, if $[\mathbf{A}]$ is a symmetrical matrix, we can write

$$[\mathbf{A}] = [\mathbf{L}][\mathbf{U}] = [\mathbf{L}][\mathbf{D}][\mathbf{U}^1] = [\mathbf{L}][\mathbf{D}][\mathbf{L}]^T \qquad (2.25)$$

Since the terms in $[\mathbf{L}]^T$ (the "transpose") can be inferred from the terms in $[\mathbf{L}]$, it will be sufficient to compute only $[\mathbf{L}]$ (or $[\mathbf{L}]^T$), involving approximately half the work in the $[\mathbf{L}][\mathbf{U}]$ factorization of unsymmetrical matrices.

Example 2.3

Use $[\mathbf{L}][\mathbf{D}][\mathbf{L}]^T$ *factorization to solve the symmetrical equations*

$$3x_1 - 2x_2 + x_3 = 3$$
$$-2x_1 + 3x_2 + 2x_3 = -3$$
$$x_1 + 2x_2 + 2x_3 = 2$$

Solution 2.3

The $[\mathbf{L}][\mathbf{D}][\mathbf{L}]^T$ *factors can be written in the form*

$$
\begin{aligned}
[\mathbf{L}][\mathbf{D}][\mathbf{L}]^T &= \begin{bmatrix} 1 & 0 & 0 \\ l_{21} & 1 & 0 \\ l_{31} & l_{32} & 1 \end{bmatrix} \begin{bmatrix} d_{11} & 0 & 0 \\ 0 & d_{22} & 0 \\ 0 & 0 & d_{33} \end{bmatrix} \begin{bmatrix} 1 & l_{21} & l_{31} \\ 0 & 1 & l_{32} \\ 0 & 0 & 1 \end{bmatrix} \\
&= \begin{bmatrix} 1 & 0 & 0 \\ l_{21} & 1 & 0 \\ l_{31} & l_{32} & 1 \end{bmatrix} \begin{bmatrix} d_{11} & d_{11}l_{21} & d_{11}l_{31} \\ 0 & d_{22} & d_{22}l_{32} \\ 0 & 0 & d_{33} \end{bmatrix}
\end{aligned}
$$

hence

$$[\mathbf{A}] = \begin{bmatrix} 3 & -2 & 1 \\ -2 & 3 & 2 \\ 1 & 2 & 2 \end{bmatrix} = \begin{bmatrix} 1 & 0 & 0 \\ l_{21} & 1 & 0 \\ l_{31} & l_{32} & 1 \end{bmatrix} \begin{bmatrix} d_{11} & d_{11}l_{21} & d_{11}l_{31} \\ 0 & d_{22} & d_{22}l_{32} \\ 0 & 0 & d_{33} \end{bmatrix}$$

Solving for the factors gives

$$d_{11} = 3, \;\; l_{21} = -2/3 = -0.6667, \;\; l_{31} = 1/3 = 0.3333$$

$$d_{22} = 3 - 3(-0.6667)^2 = 1.6667, \;\; l_{32} = (2 - 3(-0.6667)(0.3333))/1.6667 = 1.6$$

$$d_{33} = 2 - 3(0.3333)^2 - 1.6667(1.6)^2 = -2.6$$

thus,

$$[\mathbf{L}] = \begin{bmatrix} 1. & 0. & 0. \\ -0.6667 & 1. & 0. \\ 0.3333 & 1.6 & 1. \end{bmatrix}$$

and

$$[\mathbf{D}][\mathbf{L}]^T = \begin{bmatrix} 3. & -2. & 1. \\ 0. & 1.6667 & 2.6667 \\ 0. & 0. & -2.6 \end{bmatrix}$$

Forward-substitution gives

$$[\mathbf{L}]\{\mathbf{y}\} = \{\mathbf{b}\}$$

$$\begin{bmatrix} 1. & 0. & 0. \\ -0.6667 & 1. & 0. \\ 0.3333 & 1.6 & 1. \end{bmatrix} \begin{Bmatrix} y_1 \\ y_2 \\ y_3 \end{Bmatrix} = \begin{Bmatrix} 3 \\ -3 \\ 2 \end{Bmatrix}$$

hence,

$$y_1 = 3, \;\; y_2 = -3 + 3(0.6667) = -1$$

$$y_3 = 2 - 0.3333(3) + 1.6 = 2.6$$

Back-substitution gives

$$[\mathbf{D}][\mathbf{L}]^T\{\mathbf{x}\} = \{\mathbf{y}\}$$

$$\begin{bmatrix} 3. & -2. & 1. \\ 0. & 1.6667 & 2.6667 \\ 0. & 0. & -2.6 \end{bmatrix} \begin{Bmatrix} x_1 \\ x_2 \\ x_3 \end{Bmatrix} = \begin{Bmatrix} 3 \\ -1 \\ 2.6 \end{Bmatrix}$$

hence

$$x_3 = -1, \;\; x_2 = (-1 + 2.6667)/1.6667 = 1$$

$$x_1 = (3 + 1 + 2)/3 = 2$$

Program 2.3: Gaussian elimination using $[L][D][L]^T$ factorization

```
PROGRAM nm23
!---Gaussian Elimination Using LDLT Factorization---
 USE nm_lib; USE precision; IMPLICIT NONE
 INTEGER::i,j,n; REAL(iwp),ALLOCATABLE::a(:,:),b(:),d(:)
 OPEN(10,FILE='nm95.dat'); OPEN(11,FILE='nm95.res')
 READ(10,*)n; ALLOCATE(a(n,n),b(n),d(n))
 DO i=1,n; READ(10,*)a(i,i:n); a(i:n,i)=a(i,i:n); END DO
 READ(10,*)b; WRITE(11,'(A)')                                        &
    "---Gaussian Elimination using LDLT Factorization---"
 WRITE(11,'(/,A)')"Coefficient Matrix"
 DO i=1,n; WRITE(11,'(6E12.4)')a(i,:); END DO
 WRITE(11,'(/,A,/,6E12.4,/)')"Right Hand Side Vector",b
 CALL ldlt(a,d)
 WRITE(11,'(/A)')"Lower Triangular Factors"
 DO i=1,n; WRITE(11,'(6E12.4)')(a(i,j)/d(j),j=1,i); END DO
 WRITE(11,'(/,A)')"Diagonal Terms"
 WRITE(11,'(6E12.4)')d
 CALL ldlfor(a,b)
 DO i=1,n; a(i,:)=a(i,:)/d(i); END DO
 CALL subbac(a,b)
 WRITE(11,'(/,A,/,6E12.4)')"Solution Vector",b
END PROGRAM nm23
```

List 2.3:

Scalar integers:
i simple counter
j simple counter
n number of equations to be solved

Dynamic real arrays:
a $n \times n$ matrix of coefficients
b "right hand side" vector of length n, overwritten by solution
d diagonal matrix (stored as a vector)

```
Number of equations  ·        n
                               3

Coefficient matrix            (a(i,:),i=1,i)
                              16.    4.    8.
                                     5.   -4.
                                          22.


Right hand side               b
                              4.    2.    5.
```

Data 2.3: $[L][D][L]^T$ **Factorization**


```
---Gaussian Elimination using LDLT Factorization---

Coefficient Matrix
  0.1600E+02  0.4000E+01  0.8000E+01
  0.4000E+01  0.5000E+01 -0.4000E+01
  0.8000E+01 -0.4000E+01  0.2200E+02

Right Hand Side Vector
  0.4000E+01  0.2000E+01  0.5000E+01

Lower Triangular Factors
  0.1000E+01
  0.2500E+00  0.1000E+01
  0.5000E+00 -0.1500E+01  0.1000E+01

Diagonal Terms
  0.1600E+02  0.4000E+01  0.9000E+01

Solution Vector
 -0.2500E+00  0.1000E+01  0.5000E+00
```

Results 2.3: $[L][D][L]^T$ Factorization

The program reads in the number of equations to be solved, the upper-triangle of the symmetrical coefficient matrix and the right-hand side vector.

Subroutine ldlt forms the matrix $[U]$ from equation (2.21), which over-writes $[A]$, and the diagonal matrix $[D]$ in equation (2.24), which is stored as a vector. The program then infers $[L]$ from $[U]$ and prints out $[L]$ together with the diagonals from $[D]$.

In the program, forward-substitution operating on $[L]$ is accomplished us-ing the special subroutine for symmetrical matrices ldlfor. Conventional back-substitution operating on $[\mathbf{D}][\mathbf{L}]^T$ using subroutine subbac completes the process and the results are printed. Input data are as in Data 2.3, with output in Results 2.3.

A useful by-product of this factorization is that the determinant of the coefficient matrix $[\mathbf{A}]$ can be found as the product of the diagonal elements of $[\mathbf{D}]$, that is $16 \times 4 \times 9$ in this case, or 576.

2.4.1 Quadratic form and positive definiteness

A "quadratic form" can be defined as a second degree expression in n variables of the form

$$Q(x) = \sum_{i=1}^{n}\sum_{j=1}^{n} a_{ij}x_i x_j \quad \text{where} \quad a_{ij} = a_{ji} \tag{2.26}$$

This quadratic form is "positive" if it is equal to or greater than zero for all values of its variables x_i, $i = 1, 2, \cdots, n$. A positive form which is zero only for the values $x_1 = x_2 = \cdots = x_n = 0$ is said to be "positive definite". A positive quadratic form which is not positive definite is said to be "positive semi-definite".

With our usual definition of vectors and matrices, namely

$$\{\mathbf{x}\} = \begin{Bmatrix} x_1 \\ x_2 \\ \vdots \\ x_n \end{Bmatrix}, \quad [\mathbf{A}] = \begin{bmatrix} a_{11} & a_{12} & \cdots & a_{1n} \\ a_{21} & a_{22} & \cdots & a_{2n} \\ \vdots & \vdots & \vdots & \vdots \\ a_{n1} & a_{n2} & \cdots & a_{nn} \end{bmatrix}$$

the quadratic form can be written compactly as

$$Q(x) = \{\mathbf{x}\}^T[\mathbf{A}]\{\mathbf{x}\} \tag{2.27}$$

where $[\mathbf{A}]$ is the "matrix of the quadratic form $Q(x)$". $Q(x)$ is "singular" or "nonsingular" if $|\mathbf{A}|$ is zero or nonzero respectively.

For the quadratic form $\{\mathbf{x}\}^T[\mathbf{A}]\{\mathbf{x}\}$ to be positive definite, the determinants

$$a_{11}, \quad \begin{vmatrix} a_{11} & a_{12} \\ a_{21} & a_{22} \end{vmatrix}, \quad \begin{vmatrix} a_{11} & a_{12} & a_{13} \\ a_{21} & a_{22} & a_{23} \\ a_{31} & a_{32} & a_{33} \end{vmatrix}, \cdots, \quad \begin{vmatrix} a_{11} & a_{12} & \cdots & a_{1n} \\ a_{21} & a_{22} & \cdots & a_{2n} \\ \vdots & \vdots & \vdots & \vdots \\ a_{n1} & a_{n2} & \cdots & a_{nn} \end{vmatrix}$$

must all be positive.

Example 2.4

Show that the quadratic form

$$
\begin{bmatrix} x_1 & x_2 & x_3 \end{bmatrix}
\begin{bmatrix} 1 & 2 & -2 \\ 2 & 5 & -4 \\ -2 & -4 & 5 \end{bmatrix}
\begin{Bmatrix} x_1 \\ x_2 \\ x_3 \end{Bmatrix}
=
\begin{matrix}
x_1^2 + 2x_1x_2 - 2x_1x_3 \\
+2x_2x_1 + 5x_2^2 - 4x_2x_3 \\
-2x_3x_1 - 4x_3x_2 + 5x_3^2
\end{matrix}
$$

is positive definite.

Solution 2.4

The three determinants,

$$
1, \quad
\begin{vmatrix} 1 & 2 \\ 2 & 5 \end{vmatrix} = 1, \quad
\begin{vmatrix} 1 & 2 & -2 \\ 2 & 5 & -4 \\ -2 & -4 & 5 \end{vmatrix} = 1
$$

are all positive, hence the quadratic form

$$
Q(x) = (x_1 + 2x_2 - 2x_3)^2 + x_2^2 + x_3^2
$$

is positive definite, since it can only be zero if $x_1 = x_2 = x_3 = 0$.

2.4.2 Cholesky's method

When the coefficient matrix $[\mathbf{A}]$ is symmetrical and positive definite, a slightly different factorization can be obtained by forcing $[\mathbf{U}]$ to be the transpose of $[\mathbf{L}]$. This factorization can be written

$$
[\mathbf{A}] = [\mathbf{L}][\mathbf{L}]^T \tag{2.28}
$$

or

$$
\begin{bmatrix} a_{11} & a_{12} & a_{13} \\ a_{21} & a_{22} & a_{23} \\ a_{31} & a_{32} & a_{33} \end{bmatrix}
=
\begin{bmatrix} l_{11} & 0 & 0 \\ l_{21} & l_{22} & 0 \\ l_{31} & l_{32} & l_{33} \end{bmatrix}
\begin{bmatrix} l_{11} & l_{21} & l_{31} \\ 0 & l_{22} & l_{32} \\ 0 & 0 & l_{33} \end{bmatrix}
\tag{2.29}
$$

In this case $l_{11}^2 = a_{11}$ and so $l_{11} = \sqrt{a_{11}}$ and in each row a square root evaluation is necessary. For the $[\mathbf{A}]$ given by equation (2.19) the Cholesky factors are

$$
[\mathbf{A}] =
\begin{bmatrix} 16 & 4 & 8 \\ 4 & 5 & -4 \\ 8 & -4 & 22 \end{bmatrix}
=
\begin{bmatrix} 4 & 0 & 0 \\ 1 & 2 & 0 \\ 2 & -3 & 3 \end{bmatrix}
\begin{bmatrix} 4 & 1 & 2 \\ 0 & 2 & -3 \\ 0 & 0 & 3 \end{bmatrix}
= [\mathbf{L}][\mathbf{L}]^T
\tag{2.30}
$$

If the symmetrical matrix to be factorized is not positive definite, then the Cholesky factorization will fail as indicated by the need to take the square root of a negative number.

It may be noted that the Cholesky factor [L] *can also be retrieved from the* $[L][D][L]^T$ *factors (equation 2.25) by multiplying each of the columns of* [L] *by the square root of the corresponding diagonal term from* [D].

Example 2.5

Use Cholesky factorization to solve the system of equations

$$16x_1 + 4x_2 + 8x_3 = 16$$
$$4x_1 + 5x_2 - 4x_3 = 18$$
$$8x_1 - 4x_2 + 22x_3 = -22$$

Solution 2.5

Assuming the coefficient matrix is positive definite, factorize as

$$[\mathbf{A}] = [\mathbf{L}][\mathbf{L}]^T$$

$$\begin{bmatrix} 16 & 4 & 8 \\ 4 & 5 & -4 \\ 8 & -4 & 22 \end{bmatrix} = \begin{bmatrix} l_{11} & 0 & 0 \\ l_{21} & l_{22} & 0 \\ l_{31} & l_{32} & l_{33} \end{bmatrix} \begin{bmatrix} l_{11} & l_{21} & l_{31} \\ 0 & u_{22} & l_{32} \\ 0 & 0 & u_{33} \end{bmatrix}$$

Solving for l_{ij} gives

$$[\mathbf{L}] = \begin{bmatrix} 4 & 0 & 0 \\ 1 & 2 & 0 \\ 2 & -3 & 3 \end{bmatrix}$$

Forward-substitution gives

$$[\mathbf{L}]\{\mathbf{y}\} = \{\mathbf{b}\}$$

$$\begin{bmatrix} 4 & 0 & 0 \\ 1 & 2 & 0 \\ 2 & -3 & 3 \end{bmatrix} \begin{Bmatrix} y_1 \\ y_2 \\ y_3 \end{Bmatrix} = \begin{Bmatrix} 16 \\ 18 \\ -22 \end{Bmatrix}$$

hence

$$y_1 = 16/4 = 4, \quad y_2 = (18 - 1(4))/2 = 7$$
$$y_3 = (-22 + 3(7) - 2(4))/3 = -3$$

Back-substitution gives

$$[\mathbf{L}]^T\{\mathbf{x}\} = \{\mathbf{y}\}$$

$$\begin{bmatrix} 4 & 1 & 2 \\ 0 & 2 & -3 \\ 0 & 0 & 3 \end{bmatrix} \begin{Bmatrix} x_1 \\ x_2 \\ x_3 \end{Bmatrix} = \begin{Bmatrix} 4 \\ 7 \\ -3 \end{Bmatrix}$$

hence

$$x_3 = -3/3 = -1, \quad x_2 = (7 - 3(1))/2 = 2$$
$$x_1 = (4 - 1(2) - 2(-1))/4 = 1$$

2.5   Banded equations

In many engineering applications, equations have to be solved when the coefficients have a "banded" structure (see, for example, Chapter 8). This means that the nonzero coefficients are clustered around the diagonal stretching from the top left-hand corner of the matrix $[\mathbf{A}]$ to the bottom right-hand corner. A typical example is shown in Figure 2.1 where there are never more than two nonzero coefficients to either side of the "leading diagonal" in any row. The "bandwidth" of this system is said to be 5. If the coefficients are symmetrical, only the leading diagonal and two more coefficients per row need to be stored and operated on. In this case the "half bandwidth" is said to be 2 (not including the leading diagonal).

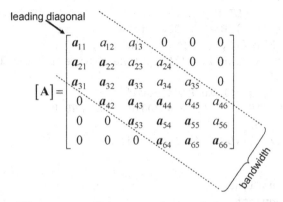

Figure 2.1: Structure of a banded matrix.

If $[\mathbf{A}]$ is symmetrical and we wish to store only the nonzero terms in the lower triangle as indicated by the bold font in Figure 2.1, only 15 terms are involved as opposed to 36 (if we stored the entire matrix). A reasonably economical method of storage is to keep the band in a rectangular array, by shifting the rows to obtain the structure

$$\begin{bmatrix} 0 & 0 & a_{11} \\ 0 & a_{21} & a_{22} \\ a_{31} & a_{32} & a_{33} \\ a_{42} & a_{43} & a_{44} \\ a_{53} & a_{54} & a_{55} \\ a_{64} & a_{65} & a_{66} \end{bmatrix} \tag{2.31}$$

This is still slightly inefficient since we have had to store 18 terms by including zeros which are not required in the first two rows. This storage method has the advantage however that there are three terms (the half bandwidth plus 1) in each row, which makes programming rather easy. The diagonal terms or "pivots" are also conveniently located in the third column.

Program 2.4: Cholesky $[L][L]^T$ factorization using banded storage

```
PROGRAM nm24
!---Cholesky LLT Factorization Using Banded Storage---
USE nm_lib; USE precision; IMPLICIT NONE
INTEGER::i,iw,iwp1,j,n; REAL(iwp),ALLOCATABLE::b(:),lb(:,:)
OPEN(10,FILE='nm95.dat'); OPEN(11,FILE='nm95.res')
READ(10,*)n,iw; iwp1=iw+1; ALLOCATE(b(n),lb(n,iwp1))
READ(10,*)(lb(i,:),i=1,n); READ(10,*)b
WRITE(11,'(A)')                                                    &
   "---Cholesky LLT Factorization Using Banded Storage---"
WRITE(11,'(/,A)')"Banded Coefficient Matrix"
DO i=1,n; WRITE(11,'(6E12.4)')lb(i,:); END DO
WRITE(11,'(/,A,/,6E12.4)')"Right Hand Side Vector",b
CALL cholin(lb)
WRITE(11,'(/,A)')"L in Band Form"
DO i=1,n; WRITE(11,'(6E12.4)')lb(i,:); END DO
CALL chobac(lb,b)
WRITE(11,'(/,A,/,6E12.4)')"Solution Vector",b
END PROGRAM nm24
```

Number of equations	n	iw
and half bandwidth	3	2

	(lb(i,:),i=1,n)		
Coefficient matrix	0.	0.	16.
(lower triangle in band form)	0.	4.	5.
	8.	-4.	22.

Right hand side	b		
	4.	2.	5.

Data 2.4: Cholesky $[L][L]^T$ Factorization Using Banded Storage

```
---Cholesky LLT Factorization Using Banded Storage---

Banded Coefficient Matrix
   0.0000E+00   0.0000E+00   0.1600E+02
   0.0000E+00   0.4000E+01   0.5000E+01
   0.8000E+01  -0.4000E+01   0.2200E+02
```

```
Right Hand Side Vector
  0.4000E+01  0.2000E+01  0.5000E+01

L in Band Form
  0.0000E+00  0.0000E+00  0.4000E+01
  0.0000E+00  0.1000E+01  0.2000E+01
  0.2000E+01 -0.3000E+01  0.3000E+01

Solution Vector
 -0.2500E+00  0.1000E+01  0.5000E+00
```

Results 2.4: Cholesky $[\mathbf{L}][\mathbf{L}]^T$ Factorization Using Banded Storage

---

List 2.4:

Scalar integers:

i	simple counter
iw	half-bandwidth
iwp1	half-bandwidth plus 1
j	simple counter
n	number of equations to be solved

Dynamic real arrays:

b	"right hand side" vector of length n, overwritten by solution
lb	lower triangle coefficients stored as a band (see equation 2.31)

The number of equations and half bandwidth are read in, followed by the lower "half" of $[\mathbf{A}]$ in the appropriate form of equation (2.31). Subroutine cholin computes the lower triangular factor which is printed out in band form. A call to chosub completes the substitution phase and the results, held in b, are printed. Input data are as in Data 2.4, with output in Results 2.4.

2.6 Compact storage for variable bandwidths

It is quite common to encounter symmetrical equation coefficient matrices which have the structure indicated in Figure 2.2.

The system is banded, with a half bandwidth of 4, but also has a significant number of zeros within the band. If the progress of factorization is monitored,

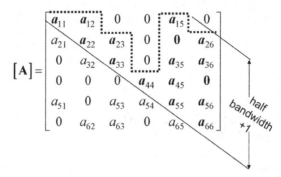

Figure 2.2: Structure of a skyline matrix.

it will be found that, in the lower triangle, zeros lying closer to the diagonal than a nonzero coefficient in that row will become nonzero during the calculation (called "fill in") whereas zeros lying further away from the diagonal than the outermost nonzero coefficient in that row will remain zero throughout the calculation and need not be stored or processed.

In the upper triangle, the same can be said of columns, and in Figure 2.2 a dashed line delineates the outermost extent of coefficients which need to be processed. Because of the appearance of this line in the upper triangle it is sometimes called the "skyline" of the coefficients.

In the case of the coefficients shown in Figure 2.2, only 16 actually need be stored (marked in bold), compared with 30 in a fixed bandwidth method such as that used in Program 2.4. The "skyline" coefficients would typically be stored in a one-dimensional array of the form

$$\begin{bmatrix} a_{11} & a_{12} & a_{22} & a_{23} & a_{33} & a_{44} & a_{15} & 0 & a_{35} & a_{45} & a_{55} & a_{26} & a_{36} & 0 & a_{56} & a_{66} \end{bmatrix}^T \quad (2.32)$$

The penalty incurred by using a variable bandwidth technique is that additional information must be kept of the number of coefficients to be processed in each column (row). This information is conveniently provided by a second one-dimensional (integer) array that holds the locations of the diagonal terms in the array of equation (2.32), thus

$$\begin{bmatrix} 1 & 3 & 5 & 6 & 11 & 16 \end{bmatrix}^T \quad (2.33)$$

Program 2.5: Cholesky $[L][L]^T$ factorization using skyline storage

```
PROGRAM nm25
!---Cholesky LLT Factorization Using Skyline Storage---
 USE nm_lib; USE precision; IMPLICIT NONE
 INTEGER::ir,n; INTEGER,ALLOCATABLE::kdiag(:)
```

```
REAL(iwp),ALLOCATABLE::a(:),b(:)
OPEN(10,FILE='nm95.dat'); OPEN(11,FILE='nm95.res')
READ(10,*)n; ALLOCATE(kdiag(n),b(n))
READ(10,*)kdiag; ir=kdiag(n); ALLOCATE(a(ir)); READ(10,*)a
READ(10,*)b; WRITE(11,'(A)')                                    &
   "---Cholesky LLT Factorization Using Skyline Storage---"
WRITE(11,'(/,A,/,6E12.4)')"Coefficient Vector",a
WRITE(11,'(/,A)')"Diagonal Locations"; WRITE(11,'(8I5)')kdiag
WRITE(11,'(/,A)')"Right Hand Side Vector"; WRITE(11,'(6E12.4)')b
CALL sparin(a,kdiag); CALL spabac(a,b,kdiag)
WRITE(11,'(/,A,/,6E12.4)')"Solution Vector",b
END PROGRAM nm25
```

In the program, n is first read in, followed by the diagonal location vector called kdiag as described in equation (2.33). Then the equation coefficients called a are then read in,

followed by the right-hand side vector b. Calls to the "skyline" factorization and substitution routines sparin and spabac complete the solution process and the results, held in b, can be printed. To test the program, we return to our familiar set of three simultaneous equations, leading to the input of Data 2.5 and the output of Results 2.5.

List 2.5:

Scalar integers:
ir   length of coefficient vector a
n   number of equations to be solved

Dynamic integer array:
kdiag   addresses of diagonal components of a

Dynamic real arrays:
a   skyline coefficients stored as a vector or length ir
b   "right hand side" vector of length n, overwritten by solution

---

```
Number of equations          n
                             3

Skyline diagonal locations   kdiag
                             1  3  6

Coefficient matrix           a
(in skyline form)            16.  4.  5.  8.  -4.  22.
```

Right hand side   b
                                    4.   2.   5.

Data 2.5: Cholesky $[L][L]^T$ Factorization Using Skyline Storage

---Cholesky LLT Factorization Using Skyline Storage---

Coefficient Vector
  0.1600E+02   0.4000E+01   0.5000E+01
  0.8000E+01 -0.4000E+01   0.2200E+02

Diagonal Locations
     1   3   6

Right Hand Side Vector
  0.4000E+01   0.2000E+01   0.5000E+01

Solution Vector
 -0.2500E+00   0.1000E+01   0.5000E+00

Results 2.5: Cholesky $[L][L]^T$ Factorization Using Skyline Storage

---

2.7 Pivoting

In the solution of unsymmetrical equations using conventional Gaussian elimination (Program 2.1) or $[L][U]$ factorization (Program 2.2) we sidestepped the problem of what to do should a leading diagonal component of the coefficient matrix be zero to start with, or become zero during the solution process. In the next program, we illustrate how to cope with this by using a row interchange technique usually called "pivoting". We saw that the crucial terms or pivots lay on the leading diagonal and so the technique involves searching for the "largest" (absolute) coefficient in the rows not so far processed, and moving it into the leading diagonal position by row and column interchange.

For example, returning to equation (2.3) the "largest" coefficient is -20 in row 2, so this is interchanged with row 1 to give

$$\begin{aligned}
-20x_1 + 3x_2 + 20x_3 &= 2 \\
10x_1 + x_2 - 5x_3 &= 1 \\
5x_1 + 3x_2 + 5x_3 &= 6
\end{aligned} \qquad (2.34)$$

After one step of elimination (or factorization) we would have

$$-20x_1 + \quad 3x_2 + 20x_3 = 2$$
$$2.5x_2 + \quad 5x_3 = 2 \qquad (2.35)$$
$$3.75x_2 + 10x_3 = 6.5$$

The largest coefficient in rows 2 to 3 is now 10 in row 3/column 3, so row 3 is interchanged with row 2 and column 3 interchanged with column 2 to give

$$-20x_1 + 20x_3 + \quad 3x_2 = 2$$
$$10x_3 + 3.75x_2 = 6.5 \qquad (2.36)$$
$$5x_3 + \quad 2.5x_2 = 2$$

leading to the final factorization,

$$-20x_1 + 20x_3 + \quad 3x_2 = 2$$
$$10x_3 + \quad 3.75x_2 = 6.5 \qquad (2.37)$$
$$0.625x_2 = -1.25$$

Back-substitution leads to

$$x_2 = -1.25/0.625 = -2, \quad x_3 = (6.5 - 3.75(-2))/10 = 1.4$$

$$x_1 = (2 - 3(-2) - 20(1.4))/(-20) = 1$$

which is the same solution as before, but the unknowns are obtained in a different order dictated by the preceding column interchanges.

In an effective but somewhat less robust approach involving only row interchanges, the search is called "partial pivoting". In this approach, the search for the largest absolute term is limited to the column below the diagonal terms.

For equations with unsymmetrical coefficients, pivoting is to be recommended. Even if a pivot does not become zero in a conventional elimination, a "small" pivot is undesirable because of potential "round-off" errors (see Chapter 1). Should the pivot become "zero", the set of equations is singular or very nearly so. Fortunately, for equations with symmetrical coefficients which are "positive definite" (see Section 2.4.1), pivoting is not necessary and Programs 2.3 to 2.5 can be used. It is also the case that systems of equations developed by engineering analysis techniques, notably the finite difference and finite element methods (see Chapter 8), are often "diagonally dominant".

The process in Program 2.6 is available in the nm_lib library as function eliminate, so that the statement eliminate(a,b) will return the desired result (see, e.g., Program 4.6).

Program 2.6: [L][U] factorization with pivoting

```
PROGRAM nm26
!---LU Factorization With Pivoting---
 USE nm_lib; USE precision; IMPLICIT NONE
 INTEGER::i,n; INTEGER,ALLOCATABLE::row(:)
 REAL(iwp),ALLOCATABLE::a(:,:),b(:),sol(:)
 OPEN(10,FILE='nm95.dat'); OPEN(11,FILE='nm95.res')
 READ(10,*)n; ALLOCATE(a(n,n),b(n),row(n),sol(n)); READ(10,*)a
 READ(10,*)b; WRITE(11,'(A)')                                            &
    "---LU Factorization With Pivoting---"
 WRITE(11,'(/,A)')"Coefficient Matrix"
 a=TRANSPOSE(a); DO i=1,n; WRITE(11,'(6E12.4)')a(i,:); END DO
 WRITE(11,'(/,A,/,6E12.4)')"Right Hand Side Vector",b
 CALL lupfac(a,row); CALL lupsol(a,b,sol,row)
 WRITE(11,'(/,A,/,6I5)')"Back-Substitution Order",row
 WRITE(11,'(/,A,/,6E12.4)')"Solution Vector",sol
END PROGRAM nm26
```

```
Number of equations            n
                               3

Coefficient matrix             (a(i,:),i=1,n)
                               10.   1.  -5.
                              -20.   3. -20.
                                5.   3.   5.


Right hand side                b
                               1.    2.   6.
```

Data 2.6: [L][U] Factorization with Pivoting

```
---LU Factorization With Pivoting---

Coefficient Matrix
  0.1000E+02  0.1000E+01 -0.5000E+01
 -0.2000E+02  0.3000E+01  0.2000E+02
  0.5000E+01  0.3000E+01  0.5000E+01

Right Hand Side Vector
  0.1000E+01  0.2000E+01  0.6000E+01
```

```
Back-Substitution Order
     2   3   1
```

```
Solution Vector
  0.1000E+01 -0.2000E+01  0.1400E+01
```

Results 2.6: [L][U] Factorization with Pivoting

List 2.6:

Scalar integers:
i   simple counter
n   number of equations to be solved

Dynamic integer array:
row   column interchange numbers

Dynamic real arrays:
a $n \times n$ matrix of coefficients
b   "right hand side" vector of length n
sol   solution vector of length n

In this program the interchanges are carried out in a special [L][U] factorization using subroutine lupfac. This stores the new row order in an integer array row for use in the substitution phase, carried out by subroutine lupsol. The factorization arithmetic remains the same. The data are exactly the same as for Program 2.2. The program reads in n, a and b, calls the factorization and substitution routines and outputs the order in which the unknowns are computed in the back-substitution phase. The solution is output in the conventional order. Input data are as in Data 2.6, with output in Results 2.6.

2.7.1   Ill-conditioning

Despite our best efforts to select optimum pivots, a set of linear equations may not be solvable accurately by any of the methods we have just described, especially when "hand" calculation techniques are employed. When this is so, the set of equations is said to be "ill-conditioned". Fortunately, modern electronic calculators are capable of working to many decimal places of accuracy and so the conditioning of sets of equations is not of such great importance as it was when "hand" calculations implied only a few decimal places of accuracy.

A very well-known example of ill-conditioned equations arises in the form of the "Hilbert matrix" obtained from polynomial curve fitting. The set of equations so derived take the form

$$
\begin{bmatrix}
\frac{1}{2} & \frac{1}{3} & \frac{1}{4} & \frac{1}{5} & \frac{1}{6} \\[4pt]
\frac{1}{3} & \frac{1}{4} & \frac{1}{5} & \frac{1}{6} & \frac{1}{7} \\[4pt]
\frac{1}{4} & \frac{1}{5} & \frac{1}{6} & \frac{1}{7} & \frac{1}{8} \\[4pt]
\frac{1}{5} & \frac{1}{6} & \frac{1}{7} & \frac{1}{8} & \frac{1}{9} \\[4pt]
\frac{1}{6} & \frac{1}{7} & \frac{1}{8} & \frac{1}{9} & \frac{1}{10}
\end{bmatrix}
\begin{Bmatrix}
x_1 \\ x_2 \\ x_3 \\ x_4 \\ x_5
\end{Bmatrix}
=
\begin{Bmatrix}
1 \\ 1 \\ 1 \\ 1 \\ 1
\end{Bmatrix}
\tag{2.38}
$$

In Table 2.1, numerical solutions rounded off to 4, 5, 6, 7 and 8 decimal places are compared with the true solution. It can be seen that something like 8 decimal places of accuracy are necessary for an answer adequate for engineering purposes. When using computers with 32-bit words, engineers are advised to use "double precision", i.e., 64-bit words, which should achieve something like 15 decimal places of accuracy and be sufficient for most practical purposes.

The precision of calculation performed by programs described in this text is typically in "double precision" and controlled by the module `precision` (see Chapter 1).

TABLE 2.1: Influence of precision on accuracy of numerical solutions to equation (2.38)

True Solution	$x_1 = 30$	$x_2 = -420$	$x_3 = 1680$	$x_4 = -2520$	$x_5 = 1260$
4DP	-8.8	90.5	-196.0	58.9	75.6
5DP	55.3	746.5	2854.9	4105.2	1976.9
6DP	13.1	229.9	1042.7	1696.4	898.1
7DP	25.9	373.7	1524.4	2318.6	1171.4
8DP	29.65	416.02	1666.62	2502.69	1252.39

Although mathematical measures of ill-conditioning, called "condition numbers," can be calculated, this process is expensive and not used in engineering practice. By far the best guides to solution accuracy are independent checks on the physical consistency of the results.

2.8 Equations with prescribed solutions

Consider a common engineering occurrence of the system of equations $[\mathbf{A}]\{\mathbf{x}\} = \{\mathbf{b}\}$ where $[\mathbf{A}]$ represents the stiffness of an elastic solid, $\{\mathbf{b}\}$ the forces applied to the solid and $\{\mathbf{x}\}$ the resulting displacements. In some cases we will know all the components of $\{\mathbf{b}\}$ and have to calculate all the components of $\{\mathbf{x}\}$, but in others we may be told that some of the displacements (or solutions) are known in advance. In this event we could eliminate the known components of $\{\mathbf{x}\}$ from the system of equations, but this involves some quite elaborate coding to reshuffle the rows and columns of $[\mathbf{A}]$. A simple scaling procedure can be used instead to produce the desired result without modifying the structure of $[\mathbf{A}]$.

Suppose that in the symmetrical system

$$\begin{bmatrix} 16 & 4 & 8 \\ 4 & 5 & -4 \\ 8 & -4 & 22 \end{bmatrix} \begin{Bmatrix} x_1 \\ x_2 \\ x_3 \end{Bmatrix} = \begin{Bmatrix} b_1 \\ b_2 \\ b_3 \end{Bmatrix} \tag{2.39}$$

we know that $x_2 = 5$, while $b_1 = 4$ and $b_3 = 5$. We can force x_2 to be 5 by adding a very large number, say 10^{20}, to the corresponding diagonal term a_{22}. The purpose of this is to make the diagonal term in row 2 several orders of magnitude larger than the off-diagonal terms in that row. Following this modification, row 2 will appear as

$$4x_1 + (5 + 10^{20})x_2 - 4x_3 = b_2 \tag{2.40}$$

and the term in x_2 swamps the other two terms. To get the desired result of $x_2 = 5$, it is clear that the right-hand side should be set to $b_2 = (5 + 10^{20})5$. We could also use this technique to fix x_2 to "zero" by setting $b_2 = 0$. The large number is a simple example of what is more generally called a "penalty method" applied to that equation.

Program 2.7: Cholesky $[\mathbf{L}][\mathbf{L}]^T$ factorization using skyline storage, prescribed solutions by penalty method

```
PROGRAM nm27
!---Cholesky LLT Factorization With Skyline Storage---
!---Prescribed Solutions by Penalty Method---
 USE nm_lib; USE precision; IMPLICIT NONE
 INTEGER::ir,n,nfix; REAL(iwp)::penalty=1.E20_iwp
 REAL(iwp),ALLOCATABLE::a(:),b(:),val(:)
 INTEGER,ALLOCATABLE::kdiag(:),no(:)
 OPEN(10,FILE='nm95.dat'); OPEN(11,FILE='nm95.res')
 READ(10,*)n; ALLOCATE(kdiag(n),b(n))
 READ(10,*)kdiag; ir=kdiag(n); ALLOCATE(a(ir))
 READ(10,*)a; READ(10,*)b
 READ(10,*)nfix; ALLOCATE(no(nfix),val(nfix))
```

```
READ(10,*)no; READ(10,*)val
WRITE(11,'(A)')"-Cholesky Factorization With Skyline Storage-"
WRITE(11,'(A)')"---Prescribed Solutions by Penalty Method---"
WRITE(11,'(/,A,/,6E12.4)')"Coefficient Vector",a
WRITE(11,'(/,A,/,8I5)')"Diagonal Locations",kdiag
WRITE(11,'(/,A,/,6E12.4)')"Right Hand Side Vector",b
WRITE(11,'(/,A,/,8I5)')"Prescribed Solution Number(s)",no
WRITE(11,'(/,A,/,6E12.4)')"Prescribed Solution Value(s)",val
a(kdiag(no(:)))=a(kdiag(no(:)))+penalty
b(no)=a(kdiag(no(:)))*val(:)
CALL sparin(a,kdiag); CALL spabac(a,b,kdiag)
WRITE(11,'(/,A,/,6E12.4)')"Solution Vector",b
END PROGRAM nm27
```

Program 2.7 is essentially the same as Program 2.5 with the option of including fixed solutions. The addresses of the terms in the solution vector to be fixed and their values are read into arrays no and val respectively. The leading diagonal terms in the skyline storage vector $\{a\}$ are located using kdiag, and the appropriate entries in $\{a\}$ augmented by the "penalty" number 10^{20}. The appropriate terms in $\{b\}$ are then set to the corresponding augmented diagonal terms multiplied by the required values of the solutions. The program is demonstrated using the same set of equations considered by Program 2.5 but with the solution x_2 set equal to 5. Input and output are shown in Data 2.7 and Results 2.7 respectively.

List 2.7:

Scalar integers:
ir	length of coefficient vector a
n	number of equations to be solved
nfix	number of fixed "solution"

Scalar reals:
penalty	set to 1×10^{20}

Dynamic integer arrays:
kdiag	addresses of diagonal components of a
no	addresses of fixed "solutions"

Dynamic real arrays:
a	skyline coefficients stored as a vector or length *ir*
b	"right hand side" vector of length *n*, overwritten by solution
val	values of fixed "solutions"

```
Number of equations              n
                                 3

Skyline diagonal locations       kdiag
                                 1   3   6

Coefficient matrix               a
(in skyline form)                16.  4.  5.  8.  -4.  22.

Right hand side                  b
                                 4.   2.  5.

Number of prescribed solutions   nfix
                                 1

Prescribed solution number(s)    no
                                 2

Prescribed solution value(s)     val
                                 5.
```

Data 2.7: Cholesky Factorization with Prescribed Solutions

```
-Cholesky Factorization With Skyline Storage-
---Prescribed Solutions by Penalty Method---

Coefficient Vector
   0.1600E+02  0.4000E+01  0.5000E+01
   0.8000E+01 -0.4000E+01  0.2200E+02

Diagonal Locations
      1     3     6

Right Hand Side Vector
   0.4000E+01  0.2000E+01  0.5000E+01

Prescribed Solution Number(s)
   2

Prescribed Solution Value(s)
   0.5000E+01
```

```
Solution Vector
 -0.1917E+01   0.5000E+01   0.1833E+01
```

Results 2.7: Cholesky Factorization with Prescribed Solutions

---

2.9   Iterative methods

In the previous sections of this Chapter, "direct" solution methods for systems of linear algebraic equations have been described. By "direct" we meant that the solution method proceeded to the answer in a fixed number of arithmetic operations. Subject to rounding errors, the solutions were as "exact" as the computer hardware permitted. There was no opportunity for intervention by the user in the solution process.

In contrast, this section deals with iterative or "indirect" methods of solution. These proceed by the user first guessing an answer, which is then successively corrected iteration by iteration. Several methods will be described, which differ in the technique by which the corrections are made, and will be found to lead to different rates of convergence to the true solution.

The question whether convergence will be achieved at all is outside the scope of this book. In general it can be stated that equations with a coefficient matrix which is "diagonally dominant" are likely to have convergent iterative solutions. Roughly defined, this means that the diagonal term in any row is greater than the sum of the off-diagonal terms. As we saw earlier, it is fortunate that in many engineering applications this diagonal dominance exists.

Since it is very unlikely that the user's starting guess will be accurate, the solution will normally become gradually closer and closer to the true solution, and the opportunity is afforded for user intervention in that a decision as to whether the solution is "close enough" to the desired one is possible. For example, if the solution has in some sense hardly changed from the previous iteration, the process can be terminated. Thus, in general terms, the number of arithmetic operations to reach a solution is not known in advance for these methods.

2.9.1   The iterative process

Suppose we are looking for an iterative method of solving the equations

$$2x_1 + x_2 = 4 \qquad\qquad (2.41)$$
$$x_1 + 2x_2 = 5$$

The central feature of iterative processes is to arrange to have the unknowns occurring on both sides of the equations. For example, we could write equations (2.41) as

$$x_1 = 2 - 0.5x_2 \qquad (2.42)$$
$$x_2 = 2.5 - 0.5x_1$$

In the simplest approach called "Jacobi iteration", we make a guess at (x_1, x_2) on the right-hand side, say $(1, 1)$. Equations (2.42) then yield

$$\left\{ \begin{matrix} x_1 \\ x_2 \end{matrix} \right\} = \left\{ \begin{matrix} 1.5 \\ 2.0 \end{matrix} \right\} \qquad (2.43)$$

so the guess was wrong. However, if this simple iteration process were to converge, we could speculate that equations (2.43) represent a better solution than the initial guess, and substitute it on the right-hand side of equations (2.42) giving

$$\left\{ \begin{matrix} x_1 \\ x_2 \end{matrix} \right\} = \left\{ \begin{matrix} 1.0 \\ 1.75 \end{matrix} \right\} \qquad (2.44)$$

and the process is continued, with the results given in Table 2.2.

TABLE 2.2: Convergence of equations (2.42) by Jacobi iteration

Iteration	0	1	2	3	4	5	6
x_1	1.0	1.5	1.0	1.125	1.0	1.03125	1.0
x_2	1.0	2.0	1.75	2.0	1.9375	2.0	1.984375

In this trivial example, it is clear that the iterative process is converging on the true solution of $x_1 = 1.0$ and $x_2 = 2.0$.

In order to get started, the initial equations in the form $[\mathbf{A}]\{\mathbf{x}\} = \{\mathbf{b}\}$ are modified by dividing each row by the corresponding diagonal of $[\mathbf{A}]$, such that the leading diagonal terms of the modified coefficient matrix are unity, thus

$$\begin{bmatrix} 1 & a_{12} & a_{13} \\ a_{21} & 1 & a_{23} \\ a_{31} & a_{32} & 1 \end{bmatrix} \left\{ \begin{matrix} x_1 \\ x_2 \\ x_3 \end{matrix} \right\} = \left\{ \begin{matrix} b_1 \\ b_2 \\ b_3 \end{matrix} \right\} \qquad (2.45)$$

In the following explanation, the matrix $[\mathbf{A}]$ and the right-hand side vector $\{\mathbf{b}\}$ are assumed to hold the modified coefficients obtained after dividing through by the diagonals as described above.

The coefficient matrix may now be split as follows

$$[\mathbf{A}] = [\mathbf{I}] - [\mathbf{L}] - [\mathbf{U}] \qquad (2.46)$$

where $[\mathbf{I}]$ is the unit matrix

$$[\mathbf{L}] = \begin{bmatrix} 0 & 0 & 0 \\ -a_{21} & 0 & 0 \\ -a_{31} & -a_{32} & 0 \end{bmatrix} \tag{2.47}$$

and

$$[\mathbf{U}] = \begin{bmatrix} 0 & -a_{12} & -a_{13} \\ 0 & 0 & -a_{23} \\ 0 & 0 & 0 \end{bmatrix} \tag{2.48}$$

Note that these triangular matrices $[\mathbf{L}]$ and $[\mathbf{U}]$ have nothing to do with the $[\mathbf{L}][\mathbf{U}]$ factors we met previously in direct methods since they are additive components of $[\mathbf{A}]$, not multiplicative.

With these definitions, the modified system $[\mathbf{A}]\{\mathbf{x}\} = \{\mathbf{b}\}$ may be written as

$$[[\mathbf{I}] - [\mathbf{L}] - [\mathbf{U}]]\{\mathbf{x}\} = \{\mathbf{b}\} \tag{2.49}$$

or

$$\{\mathbf{x}\} = \{\mathbf{b}\} + [[\mathbf{L}] + [\mathbf{U}]]\{\mathbf{x}\} \tag{2.50}$$

in which the unknowns now appear on both sides of the equations, leading to the iterative scheme

$$\{\mathbf{x}\}_{k+1} = \{\mathbf{b}\} + [[\mathbf{L}] + [\mathbf{U}]]\{\mathbf{x}\}_k \tag{2.51}$$

where k represents the iteration number.

Example 2.6

Solve the equations

$$16x_1 + 4x_2 + 8x_3 = 4$$
$$4x_1 + 5x_2 - 4x_3 = 2$$
$$8x_1 - 4x_2 + 22x_3 = 5$$

by simple iteration.

Solution 2.6

Divide each equation by the diagonal term in the coefficient matrix and rearrange as

$$\{\mathbf{x}\}_{k+1} = \{\mathbf{b}\} + [[\mathbf{L}] + [\mathbf{U}]]\{\mathbf{x}\}_k$$

hence

$$\begin{Bmatrix} x_1 \\ x_2 \\ x_3 \end{Bmatrix}_{k+1} = \begin{Bmatrix} 0.25 \\ 0.4 \\ 0.2273 \end{Bmatrix} + \begin{bmatrix} 0.0 & -0.25 & -0.5 \\ -0.8 & 0.0 & 0.8 \\ -0.3636 & 0.1818 & 0.0 \end{bmatrix} \begin{Bmatrix} x_1 \\ x_2 \\ x_3 \end{Bmatrix}_k$$

As a starting guess, let

$$\left\{ \begin{array}{c} x_1 \\ x_2 \\ x_3 \end{array} \right\}_0 = \left\{ \begin{array}{c} 1 \\ 1 \\ 1 \end{array} \right\}$$

hence

$$\left\{ \begin{array}{c} x_1 \\ x_2 \\ x_3 \end{array} \right\}_1 = \left\{ \begin{array}{c} -0.5 \\ 0.4 \\ 0.0455 \end{array} \right\}$$

and

$$\left\{ \begin{array}{c} x_1 \\ x_2 \\ x_3 \end{array} \right\}_2 = \left\{ \begin{array}{c} 0.1273 \\ 0.8364 \\ 0.4848 \end{array} \right\} \longrightarrow \left\{ \begin{array}{c} -0.25 \\ 1.0 \\ 0.5 \end{array} \right\} \quad \textit{(after many iterations)}$$

In this case, convergence is very slow (see Program 2.8). To operate this scheme on a computer, we can see that the library modules we shall need are merely a matrix-vector multiply routine to compute $[[\mathbf{L}] + [\mathbf{U}]]\{\mathbf{x}\}$, a vector addition to add $\{\mathbf{b}\}$ and some means of checking on convergence.

Program 2.8: Jacobi iteration for linear simultaneous e-quations

```
PROGRAM nm28
!---Jacobi Iteration For Linear Simultaneous Equations---
USE nm_lib; USE precision; IMPLICIT NONE
INTEGER::i,iters,limit,n; REAL(iwp)::diag,tol,zero=0.0_iwp
REAL(iwp),ALLOCATABLE::a(:,:),b(:),x(:),xnew(:)
OPEN(10,FILE='nm95.dat'); OPEN(11,FILE='nm95.res')
READ(10,*)n; ALLOCATE(a(n,n),b(n),x(n),xnew(n))
READ(10,*)a; READ(10,*)b; READ(10,*)x; READ(10,*)tol,limit
WRITE(11,'(A)')                                             &
   "---Jacobi Iteration For Linear Simultaneous Equations---"
WRITE(11,'(/,A)')"Coefficient Matrix"
a=TRANSPOSE(a); DO i=1,n; WRITE(11,'(6E12.4)')a(i,:); END DO
WRITE(11,'(/,A,/,6E12.4)')"Right Hand Side Vector",b
WRITE(11,'(/,A,/,6E12.4)')"Guessed Starting Vector",x
DO i=1,n
   diag=a(i,i); a(i,:)=a(i,:)/diag; b(i)=b(i)/diag
END DO; a=-a
DO i=1,n; a(i,i)=zero; END DO
WRITE(11,'(/,A)')"First Few Iterations"; iters=0
DO; iters=iters+1
   xnew=b+MATMUL(a,x); IF(iters<5)WRITE(11,'(6E12.4)')x
```

```
   IF(checkit(xnew,x,tol).OR.iters==limit)EXIT; x=xnew
END DO
WRITE(11,'(/,A,/,I5)')"Iterations to Convergence",iters
WRITE(11,'(/,A,/,6E12.4)')"Solution Vector",x
END PROGRAM nm28
```

List 2.8:

Scalar integers:
i	simple counter
iters	iteration counter
limit	iteration limit
n	number of equations to be solved

Scalar reals:
diag	temporary store for diagonal components
tol	convergence tolerance
zero	set to 0.0

Dynamic real arrays:
a	$n \times n$ matrix of coefficients
b	"right hand side" vector of length n
x	approximate solution vector
xnew	improved solution vector

Number of equations	n		
	3		
Coefficient matrix	(a(i,:),i=1,n)		
	16.	4.	8.
	4.	5.	-4.
	8.	-4.	22.
Right hand side	b		
	4.	2.	5.
Initial guess	x		
	1.	1.	1.
Tolerance and iteration limit	tol	limit	
	1.E-5	100	

Data 2.8: Jacobi Iteration

```
---Jacobi Iteration For Linear Simultaneous Equations---

Coefficient Matrix
   0.1600E+02   0.4000E+01   0.8000E+01
   0.4000E+01   0.5000E+01  -0.4000E+01
   0.8000E+01  -0.4000E+01   0.2200E+02

Right Hand Side Vector
   0.4000E+01   0.2000E+01   0.5000E+01

Guessed Starting Vector
   0.1000E+01   0.1000E+01   0.1000E+01

First Few Iterations
   0.1000E+01   0.1000E+01   0.1000E+01
  -0.5000E+00   0.4000E+00   0.4545E-01
   0.1273E+00   0.8364E+00   0.4818E+00
  -0.2000E+00   0.6836E+00   0.3331E+00

Iterations to Convergence
    51

Solution Vector
  -0.2500E+00   0.9999E+00   0.5000E+00
```

Results 2.8: Jacobi Iteration

Input shown in Data 2.8 consists of the number of equations to be solved n followed by the coefficient matrix a and the right hand side vector b. The initial guess x is then read followed by the convergence tolerance tol and the iteration limit limit.

The program scales the original coefficients and right-hand sides, by dividing by the diagonal coefficient in each row (called diag in the program).

The intrinsic routine MATMUL performs the matrix-vector multiply called for by the right-hand side of equation (2.51) and puts the temporary result in xnew. This can then be added to right-hand side b to give a new solution, still called xnew. This can be compared with x by using the library function checkit which checks for convergence. If xnew agrees with x to the specified tolerance the iterations are terminated.

The program prints out the number of iterations taken and the values of the solution vector for the first 4 iterations and at convergence. Output shown in Results 2.8 indicates that 51 iterations are required to give a result accurate to about 4 places of decimals using this simple technique.

2.9.2 Very sparse systems

An obvious advantage of iterative methods occurs when the set of equation coefficients is very "sparse", that is there are very few nonzero entries. In the case of direct methods, we saw that "fill-in" occurred in the coefficient matrix during solution, except outside the "skyline" (Section 2.6), whereas in iterative methods, the coefficient matrix retains its sparsity pattern throughout the calculation. In such cases, one would not wish to retain the matrix form for [**A**] in programs like Program 2.8 but rather use specialized coding involving pointers to the nonzero terms in the coefficient matrix. In the Finite Element Method (e.g., Smith and Griffiths, 2004) [**A**] may never be assembled at all leading to very effective parallelizable algorithms.

2.9.3 The Gauss-Seidel method

In Jacobi's method, all of the components of $\{\mathbf{x}\}_{k+1}$ (called xnew in Program 2.8) are evaluated using all the components of $\{\mathbf{x}\}_k$ (x in the program). Thus the new information is obtained entirely in terms of the old.

However, after the first row of equations (2.51) has been evaluated, there is a new $(x_1)_{k+1}$ available which is presumably a better approximation to the solution than $(x_1)_k$. In the Gauss-Seidel technique, the new value $(x_1)_{k+1}$ is immediately substituted for $(x_1)_k$ in the "old" solution x. After evaluation of row two, $(x_2)_{k+1}$ is substituted for $(x_2)_k$ and so on. The convergence of this process operating on equation (2.42) is shown in Table 2.3.

TABLE 2.3:   Convergence of equations (2.42) by Gauss-Seidel

Iteration	0	1	2	3
x_1	1.0	1.5	1.125	1.03125
x_2	1.0	1.75	1.9375	1.984375

which is clearly better than than in Table 2.2.

The Gauss-Seidel iterative process is based as before on the additive factors given by equation (2.46), but has a different rearrangement as follows

$$[[\mathbf{I}] - [\mathbf{L}]]\{\mathbf{x}\}_{k+1} = \{\mathbf{b}\} + [\mathbf{U}]\{\mathbf{x}\}_k \qquad (2.52)$$

which is possible because in the operation $[\mathbf{U}]\{\mathbf{x}\}_k$, the evaluation of row i does not depend on x_i, x_{i-1}, etc., and so they can be updated as they become available. The equations are shown in expanded form in (2.53).

$$\begin{bmatrix} 1 & 0 & 0 & .. & 0 \\ a_{21} & 1 & 0 & .. & 0 \\ a_{31} & a_{32} & 1 & .. & 0 \\ . & . & . & .... & \\ . & . & . & .... & \\ a_{n1} & a_{n2} & . & .. & 1 \end{bmatrix} \begin{Bmatrix} x_1 \\ x_2 \\ x_3 \\ . \\ . \\ x_n \end{Bmatrix}_{k+1} = \begin{Bmatrix} b_1 \\ b_2 \\ b_3 \\ . \\ . \\ b_n \end{Bmatrix} + \begin{bmatrix} 0 & -a_{12} & -a_{13} & .. & -a_{1n} \\ 0 & 0 & -a_{23} & .. & -a_{2n} \\ 0 & 0 & 0 & .. & -a_{3n} \\ . & . & . & . & .... \\ . & . & . & .... & . \\ 0 & 0 & . & .. & 0 \end{bmatrix} \begin{Bmatrix} x_1 \\ x_2 \\ x_3 \\ . \\ . \\ x_n \end{Bmatrix}_k$$

$$(2.53)$$

It will be obvious by comparing equations (2.52) and (2.51) that for non-sparse systems the right-hand side can be computed again by using a matrix-vector multiply and a vector addition, leading to

$$[[\mathbf{I}] - [\mathbf{L}]]\{\mathbf{x}\}_{k+1} = \{\mathbf{y}\}_k \qquad (2.54)$$

These equations take the form

$$\begin{bmatrix} 1 & 0 & 0 & .. & 0 \\ a_{21} & 1 & 0 & .. & 0 \\ a_{31} & a_{32} & 1 & .. & 0 \\ . & . & . & .... & \\ . & . & . & .... & \\ a_{n1} & a_{n2} & . & .. & 1 \end{bmatrix} \begin{Bmatrix} x_1 \\ x_2 \\ x_3 \\ . \\ . \\ x_n \end{Bmatrix}_{k+1} = \begin{Bmatrix} y_1 \\ y_2 \\ y_3 \\ . \\ . \\ y_n \end{Bmatrix}_k \qquad (2.55)$$

which is just one of the processes we encountered in "direct" methods when we called it "forward-substitution". Library subroutine subfor is available to carry out this task.

Example 2.7

Solve the equations

$$16x_1 + 4x_2 + 8x_3 = 4$$
$$4x_1 + 5x_2 - 4x_3 = 2$$
$$8x_1 - 4x_2 + 22x_3 = 5$$

by Gauss-Seidel iteration.

Solution 2.7

Divide each equation by the diagonal term and rearrange as

$$[[\mathbf{I}] - [\mathbf{L}]]\{\mathbf{x}\}_{k+1} = \{\mathbf{b}\} + [\mathbf{U}]\{\mathbf{x}\}_k$$

hence

$$\begin{bmatrix} 1 & 0 & 0 \\ 0.8 & 1 & 0 \\ 0.3636 & -0.1818 & 1 \end{bmatrix} \begin{Bmatrix} x_1 \\ x_2 \\ x_3 \end{Bmatrix}_{k+1} = \begin{Bmatrix} 0.25 \\ 0.4 \\ 0.2273 \end{Bmatrix} + \begin{bmatrix} 0 & -0.25 & -0.5 \\ 0 & 0 & 0.8 \\ 0 & 0 & 0 \end{bmatrix} \begin{Bmatrix} x_1 \\ x_2 \\ x_3 \end{Bmatrix}_k$$

As a starting guess, let

$$\begin{Bmatrix} x_1 \\ x_2 \\ x_3 \end{Bmatrix}_0 = \begin{Bmatrix} 1 \\ 1 \\ 1 \end{Bmatrix}$$

hence

$$\begin{bmatrix} 1 & 0 & 0 \\ 0.8 & 1 & 0 \\ 0.2626 & -0.1818 & 1 \end{bmatrix} \begin{Bmatrix} x_1 \\ x_2 \\ x_3 \end{Bmatrix}_1 = \begin{Bmatrix} -0.5 \\ 1.2 \\ 0.2273 \end{Bmatrix}$$

which after forward-substitution gives

$$\begin{Bmatrix} x_1 \\ x_2 \\ x_3 \end{Bmatrix}_1 = \begin{Bmatrix} -0.5 \\ 1.6 \\ 0.7 \end{Bmatrix}$$

The second iteration gives

$$\begin{bmatrix} 1 & 0 & 0 \\ 0.8 & 1 & 0 \\ 0.2626 & -0.1818 & 1 \end{bmatrix} \begin{Bmatrix} x_1 \\ x_2 \\ x_3 \end{Bmatrix}_2 = \begin{Bmatrix} -0.5 \\ 0.96 \\ 0.2273 \end{Bmatrix}$$

and again after forward-substitution

$$\begin{Bmatrix} x_1 \\ x_2 \\ x_3 \end{Bmatrix}_2 = \begin{Bmatrix} -0.5 \\ 1.36 \\ 0.6564 \end{Bmatrix} \longrightarrow \begin{Bmatrix} -0.5 \\ 1.0 \\ 0.5 \end{Bmatrix} \quad \textit{(after many iterations)}$$

Convergence is still slow (see Program 2.9), but not as slow as in Jacobi's method.

Program 2.9: Gauss-Seidel iteration for linear simultaneous equations

```
PROGRAM nm29
!---Gauss-Seidel Iteration For Linear Simultaneous Equations---
 USE nm_lib; USE precision; IMPLICIT NONE
 INTEGER::i,iters,limit,n; REAL(iwp)::diag,tol,zero=0.0_iwp
 REAL(iwp),ALLOCATABLE::a(:,:),b(:),u(:,:),x(:),xnew(:)
 OPEN(10,FILE='nm95.dat'); OPEN(11,FILE='nm95.res')
 READ(10,*)n; ALLOCATE(a(n,n),u(n,n),b(n),x(n),xnew(n))
 READ(10,*)a; READ(10,*)b; READ(10,*)x; READ(10,*)tol,limit
 WRITE(11,'(A)')                                                &
   "-Gauss-Seidel Iteration For Linear Simultaneous Equations-"
```

```
WRITE(11,'(/,A)')"Coefficient Matrix"
a=TRANSPOSE(a); DO i=1,n; WRITE(11,'(6E12.4)')a(i,:); END DO
WRITE(11,'(/,A,/,6E12.4)')"Right Hand Side Vector",b
WRITE(11,'(/,A,/,6E12.4)')"Guessed Starting Vector",x
DO i=1,n
   diag=a(i,i); a(i,:)=a(i,:)/diag; b(i)=b(i)/diag
END DO
u=zero; DO i=1,n; u(i,i+1:)=-a(i,i+1:); a(i,i+1:)=zero; END DO
WRITE(11,'(/,A)')"First Few Iterations"; iters=0
DO; iters=iters+1
   xnew=b+MATMUL(u,x); CALL subfor(a,xnew)
   IF(iters<5)WRITE(11,'(6E12.4)')x
   IF(checkit(xnew,x,tol).OR.iters==limit)EXIT; x=xnew
END DO
WRITE(11,'(/,A,/,I5)')"Iterations to Convergence",iters
WRITE(11,'(/,A,/,6E12.4)')"Solution Vector",x
END PROGRAM nm29
```

Number of equations	n		
	3		
Coefficient matrix	(a(i,:),i=1,n)		
	16.	4.	8.
	4.	5.	-4.
	8.	-4.	22.
Right hand side	b		
	4.	2.	5.
Initial guess	x		
	1.	1.	1.
Tolerance and	tol	limit	
iteration limit	1.E-5	100	

Data 2.9: Gauss-Seidel Iteration

-Gauss-Seidel Iteration For Linear Simultaneous Equations-

```
Coefficient Matrix
   0.1600E+02  0.4000E+01   0.8000E+01
   0.4000E+01  0.5000E+01  -0.4000E+01
   0.8000E+01 -0.4000E+01   0.2200E+02
```

Right Hand Side Vector
   0.4000E+01   0.2000E+01   0.5000E+01

Guessed Starting Vector
   0.1000E+01   0.1000E+01   0.1000E+01

First Few Iterations
   0.1000E+01   0.1000E+01   0.1000E+01
  -0.5000E+00   0.1600E+01   0.7000E+00
  -0.5000E+00   0.1360E+01   0.6564E+00
  -0.4182E+00   0.1260E+01   0.6084E+00

Iterations to Convergence
   30

Solution Vector
  -0.2500E+00   0.1000E+01   0.5000E+00

Results 2.9: Gauss-Seidel Iteration

List 2.9:

Scalar integers:
i   simple counter
iters   iteration counter
limit   iteration limit
n   number of equations to be solved

Scalar reals:
diag   temporary store for diagonal components
tol   convergence tolerance
zero   set to 0.0

Dynamic real arrays:
a $n \times n$ matrix of coefficients
b   "right hand side" vector of length n
u   "right hand side" upper triangular matrix
x   approximate solution vector
xnew   improved solution vector

The same scaling of $[A]$ is necessary and matrices $[L]$ and $[U]$ are formed as in equation (2.46). At the same time as $[U]$ is formed from $[A]$, the matrix $[I] - [L]$ overwrites $[A]$. In the iteration loop, the right-hand side of equations

(2.54), called `xnew`, is formed by successive calls to `MATMUL` and vector additions. The forward substitution then takes place using `subfor` leaving the new solution in `xnew`. As before, convergence is checked using `checkit`. The input and output are shown in Data 2.9 and Results 2.9 respectively, illustrating that the number of iterations required to achieve the same tolerance has dropped to 30, compared to Jacobi's 51.

2.9.4 Successive overrelaxation

In this technique, the difference between successive iterations is augmented by a scalar parameter called the "overrelaxation factor" ω where

$$\{\mathbf{x}\}_{k+1} - \{\mathbf{x}\}_k \implies \omega\{\{\mathbf{x}\}_{k+1} - \{\mathbf{x}\}_k\} \tag{2.56}$$

which can be rearranged as

$$\omega\{\mathbf{x}\}_{k+1} = \{\mathbf{x}\}_{k+1} - (1 - \omega)\{\mathbf{x}\}_k \tag{2.57}$$

The Gauss-Seidel method is a special case of this approach where $\omega = 1$, but in general $1 < \omega < 2$.

From the simple Jacobi method (equation 2.51) we have

$$\{\mathbf{x}\}_{k+1} = \{\mathbf{b}\} + [[\mathbf{L}] + [\mathbf{U}]]\{\mathbf{x}\}_k \tag{2.58}$$

hence

$$\omega\{\mathbf{x}\}_{k+1} = \omega\{\mathbf{b}\} + [\omega[\mathbf{L}] + \omega[\mathbf{U}]]\{\mathbf{x}\}_k \tag{2.59}$$

Comparing equations (2.57) and (2.59) we get

$$\{\mathbf{x}\}_{k+1} - (1 - \omega)\{\mathbf{x}\}_k = \omega\{\mathbf{b}\} + [\omega[\mathbf{L}] + \omega[\mathbf{U}]]\{\mathbf{x}\}_k \tag{2.60}$$

With solution updating as in Gauss-Seidel, $\omega[\mathbf{L}]\{\mathbf{x}\}_k$ can be replaced by $\omega[\mathbf{L}]\{\mathbf{x}\}_{k+1}$, hence

$$[[\mathbf{I}] - \omega[\mathbf{L}]]\{\mathbf{x}\}_{k+1} = \omega\{\mathbf{b}\} + [(1 - \omega)[\mathbf{I}] + \omega[\mathbf{U}]]\{\mathbf{x}\}_k \tag{2.61}$$

Equations (2.61) have similar properties to equations (2.52) in that the evaluation of row i on the right-hand side does not depend on x_i, x_{i-1} etc., and so updating is possible.

Program 2.10: Successive overrelaxation for linear simultaneous equations

```
PROGRAM nm210
!--Successive Overrelaxation For Linear Simultaneous Equations--
USE nm_lib; USE precision; IMPLICIT NONE
INTEGER::i,iters,limit,n; REAL(iwp)::diag,omega,one=1.0_iwp,    &
  tol,zero=0.0_iwp
REAL(iwp),ALLOCATABLE::a(:,:),b(:),u(:,:),x(:),xnew(:)
OPEN(10,FILE='nm95.dat'); OPEN(11,FILE='nm95.res')
READ(10,*)n; ALLOCATE(a(n,n),u(n,n),b(n),x(n),xnew(n))
READ(10,*)a; READ(10,*)b; READ(10,*)x,omega,tol,limit
WRITE(11,'(A)')"---Successive Overrelaxation---"
WRITE(11,'(A)')"---For Linear Simultaneous Equations---"
WRITE(11,'(/,A)')"Coefficient Matrix"
a=TRANSPOSE(a); DO i=1,n; WRITE(11,'(6E12.4)')a(i,:); END DO
WRITE(11,'(/,A,/,6E12.4)')"Right Hand Side Vector",b
WRITE(11,'(/,A,/,6E12.4)')"Guessed Starting Vector",x
WRITE(11,'(/,A,/,E12.4)')"Overrelaxation Scalar",omega
DO i=1,n; diag=a(i,i); a(i,:)=a(i,:)/diag; b(i)=omega*b(i)/diag
END DO; u=zero; a=a*omega
DO i=1,n; u(i,i+1:)=-a(i,i+1:); a(i,i+1:)=zero; END DO
DO i=1,n; a(i,i)=one; u(i,i)=one-omega; END DO
WRITE(11,'(/,A)')"First Few Iterations"; iters=0
DO; iters=iters+1
  xnew=b+MATMUL(u,x); CALL subfor(a,xnew)
  IF(iters<5)WRITE(11,'(6E12.4)')x
  IF(checkit(xnew,x,tol).OR.iters==limit)EXIT; x=xnew
END DO
WRITE(11,'(/,A,/,I5)')"Iterations to Convergence",iters
WRITE(11,'(/,A,/,6E12.4)')"Solution Vector",x
END PROGRAM nm210
```

Number of equations	n
	3

Coefficient matrix	(a(i,:),i=1,n)		
	16.	4.	8.
	4.	5.	-4.
	8.	-4.	22.

Right hand side	b		
	4.	2.	5.

```
Initial guess              x
                           1.    1.    1.

Overrelaxation factor      omega
                           1.5

Tolerance and              tol     limit
iteration limit            1.E-5   100
```

Data 2.10: Successive Overrelaxation

```
---Successive Overrelaxation---
---For Linear Simultaneous Equations---

Coefficient Matrix
  0.1600E+02  0.4000E+01  0.8000E+01
  0.4000E+01  0.5000E+01 -0.4000E+01
  0.8000E+01 -0.4000E+01  0.2200E+02

Right Hand Side Vector
  0.4000E+01  0.2000E+01  0.5000E+01

Guessed Starting Vector
  0.1000E+01  0.1000E+01  0.1000E+01

Overrelaxation Scalar
  0.1500E+01

First Few Iterations
  0.1000E+01  0.1000E+01  0.1000E+01
 -0.1250E+01  0.2800E+01  0.1286E+01
 -0.1015E+01  0.1961E+01  0.7862E+00
 -0.4427E+00  0.1094E+01  0.4877E+00

Iterations to Convergence
  18

Solution Vector
 -0.2500E+00  0.1000E+01  0.5000E+00
```

Results 2.10: Successive Overrelaxation

There is one extra input quantity, ω in this case, which is the overrelaxation factor, but no other major changes. First [**A**] and {**b**} are scaled as usual and

---

List 2.10:

Scalar integers:

i	simple counter
iters	iteration counter
limit	iteration limit
n	number of equations to be solved

Scalar reals:

diag	temporary store for diagonal components
omega	overrelaxation factor
one	set to 1.0
tol	convergence tolerance
zero	set to 0.0

Dynamic real arrays:

a	$n \times n$ matrix of coefficients
b	"right hand side" vector of length n
u	"right hand side" upper triangular matrix
x	approximate solution vector
xnew	improved solution vector

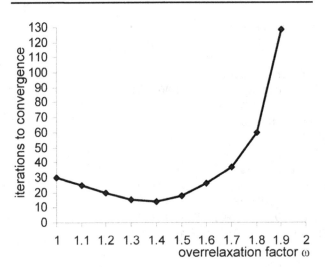

Figure 2.3: Influence of overrelaxation factor on rate of convergence of SOR.

{b} multiplied by ω as required by equation (2.61). [A] is then also multiplied by ω and [U] and [A] organized as before. In order to end up with $[[\mathbf{I}] - \omega[\mathbf{L}]]$ and $[(1 - \omega)[\mathbf{I}] + \omega[\mathbf{U}]]$ on the left- and right-hand sides of equations (2.61) the diagonals of [A] and of [U] are replaced by 1 and $(1 - \omega)$ respectively.

Iteration proceeds using exactly the same loop as for Gauss-Seidel in Program 2.9. For the input shown in Data 2.10 with `omega=1.5` the number of iterations for convergence has dropped to 18 as shown in the output in Results 2.10. The variation of iteration count with the overrelaxation factor for this problem is shown in Figure 2.3 and is seen to be optimal for $\omega \approx 1.4$. In general, however, the optimal value of ω is problem dependent and difficult to predict.

2.10 Gradient methods

The methods described in the previous sections are sometimes called "stationary" methods because there is no attempt made in them to modify the convergence process according to a measure of the error in the trial solution. In gradient methods, by contrast, an error function is repeatedly evaluated, and used to generate new trial solutions. In our first examples we shall confine our interest to equations with symmetrical, positive definite coefficient matrices. Although it will soon be apparent that in this case the gradient methods are very simple to program, the mathematical reasoning behind them is somewhat involved. For example, the reader is referred to Jennings (1977) pp. 212-216 for a reasonably concise exposition in engineering terms.

2.10.1 The method of 'steepest descent'

For any trial solution $\{x\}_k$ the error or "residual" will clearly be expressible as

$$\{r\}_k = \{b\} - [A]\{x\}_k \qquad (2.62)$$

We start with an initial guess of the solution vector $\{x\}_0$. In the method of steepest descent, the error implicit in $\{r\}_k$ is minimized according to the following algorithm:

$$\{r\}_0 = \{b\} - [A]\{x\}_0 \qquad \text{(to start the process)}$$

$$\{u\}_k = [A]\{r\}_k \qquad \text{(a)}$$

$$\alpha_k = \frac{\{r\}_k^T\{r\}_k}{\{r\}_k^T\{u\}_k} \qquad \text{(b)} \qquad\qquad (2.63)$$

$$\{x\}_{k+1} = \{x\}_k + \alpha_k\{r\}_k \qquad \text{(c)}$$

$$\{r\}_{k+1} = \{r\}_k - \alpha_k\{u\}_k \qquad \text{(d)}$$

To implement this algorithm we can make use of the power of Fortran 95 which enables us to multiply a matrix by a vector in step (a), to calculate vector inner or "dot" products in step (b), to multiply a vector by a scalar in steps (c) and (d) and to add and subtract vectors in steps (c) and (d) respectively, all by features intrinsic to the language. We therefore anticipate very efficient implementations of these types of algorithm.

Program 2.11: Steepest descent for linear simultaneous equations

```
PROGRAM nm211
!---Steepest Descent For Linear Simultaneous Equations---
USE nm_lib; USE precision; IMPLICIT NONE
INTEGER::i,iters,limit,n; REAL(iwp)::alpha,tol
REAL(iwp),ALLOCATABLE::a(:,:),b(:),r(:),u(:),x(:),xnew(:)
OPEN(10,FILE='nm95.dat'); OPEN(11,FILE='nm95.res')
READ(10,*)n; ALLOCATE(a(n,n),b(n),x(n),xnew(n),r(n),u(n))
READ(10,*)a; READ(10,*)b; READ(10,*)x,tol,limit
WRITE(11,'(A)')                                                    &
   "---Steepest Descent For Linear Simultaneous Equations---"
WRITE(11,'(/,A)')"Coefficient Matrix"
a=TRANSPOSE(a); DO i=1,n; WRITE(11,'(6E12.4)')a(i,:); END DO
WRITE(11,'(/,A,/,6E12.4)')"Right Hand Side Vector",b
WRITE(11,'(/,A,/,6E12.4)')"Guessed Starting Vector",x
WRITE(11,'(/,A)')"First Few Iterations"; r=b-MATMUL(a,x); iters=0
DO; iters=iters+1
  u=MATMUL(a,r); alpha=DOT_PRODUCT(r,r)/DOT_PRODUCT(r,u)
  xnew=x+alpha*r; r=r-alpha*u
  IF(iters<5)WRITE(11,'(6E12.4)')x
  IF(checkit(xnew,x,tol).OR.iters==limit)EXIT; x=xnew
END DO
WRITE(11,'(/,A,/,I5)')"Iterations to Convergence",iters
WRITE(11,'(/,A,/,6E12.4)')"Solution Vector",x
END PROGRAM nm211
```

Number of equations	n		
	3		

Coefficient matrix	(a(i,:),i=1,n)		
	16.	4.	8.
	4.	5.	-4.
	8.	-4.	22.

Right hand side	b		
	4.	2.	5.

```
Initial guess              x
                           1.    1.    1.

Tolerance and              tol     limit
iteration limit            1.E-5   100
```

Data 2.11: Steepest Descent

List 2.11:

Scalar integers:

i	simple counter
iters	iteration counter
limit	iteration limit
n	number of equations to be solved

Scalar reals:

alpha	see equations (2.63)
tol	convergence tolerance

Dynamic real arrays:

a	$n \times n$ matrix of coefficients
b	"right hand side" vector of length n
r	see equations (2.63)
u	see equations (2.63)
x	approximate solution vector
xnew	improved solution vector

```
---Steepest Descent For Linear Simultaneous Equations---

Coefficient Matrix
   0.1600E+02   0.4000E+01   0.8000E+01
   0.4000E+01   0.5000E+01  -0.4000E+01
   0.8000E+01  -0.4000E+01   0.2200E+02

Right Hand Side Vector
   0.4000E+01   0.2000E+01   0.5000E+01

Guessed Starting Vector
   0.1000E+01   0.1000E+01   0.1000E+01
```

First Few Iterations
```
  0.1000E+01   0.1000E+01   0.1000E+01
  0.9133E-01   0.8864E+00   0.2049E+00
 -0.8584E-01   0.7540E+00   0.4263E+00
 -0.1305E+00   0.7658E+00   0.3976E+00
```

Iterations to Convergence
```
   61
```

Solution Vector
```
 -0.2500E+00   0.9999E+00   0.5000E+00
```

Results 2.11: Steepest Descent

The program uses the same input as the previous iterative programs as shown in Data 2.11, requiring as input the number of equations to be solved (n) followed by the equation left- and right-hand sides (a and b). The iteration tolerance (tol) and limit (limit) complete the data.

The operations involved in equations (2.63) start with matrix-vector multiply $[\mathbf{A}]\{\mathbf{x}\}_0$ and subtraction of the result from $\{\mathbf{b}\}$ to form $\{\mathbf{r}\}_0$. The iterations in equations (2.63) can then proceed by using MATMUL as in step (a) and two dot products giving α in step (b). Steps (c) and (d) involve simple additions of scaled vectors, and it remains only to check the tolerance by comparing the relative change between $\{\mathbf{x}\}_k$ and $\{\mathbf{x}\}_{k+1}$ using checkit, and to continue the iteration or stop if convergence has been achieved. The results are shown in Results 2.11, where it can be seen that 61 iterations are required to achieve the desired accuracy.

2.10.2   The method of 'conjugate gradients'

The results obtained by running Program 2.11 show that the method of steepest descent is not competitive with the other iteration methods tried so far. However, it can be radically improved if the descent vectors are made mutually "conjugate" with respect to $[\mathbf{A}]$. That is, we introduce "descent vectors" $\{\mathbf{p}\}$ which satisfy the relationship

$$\{\mathbf{p}\}_i^T [\mathbf{A}] \{\mathbf{p}\}_j = 0 \quad \text{for } i \neq j \tag{2.64}$$

The equivalent algorithm to equation (2.63) becomes

$$\{\mathbf{p}\}_0 = \{\mathbf{r}\}_0 = \{\mathbf{b}\} - [\mathbf{A}]\{\mathbf{x}\}_0 \quad \text{(to start the process)}$$

$$\{\mathbf{u}\}_k = [\mathbf{A}]\{\mathbf{p}\}_k \qquad \text{(a)}$$

$$\alpha_k = \frac{\{\mathbf{r}\}_k^T \{\mathbf{r}\}_k}{\{\mathbf{p}\}_k^T \{\mathbf{u}\}_k} \qquad \text{(b)}$$

$$\{\mathbf{x}\}_{k+1} = \{\mathbf{x}\}_k + \alpha_k \{\mathbf{p}\}_k \qquad \text{(c)} \qquad\qquad (2.65)$$

$$\{\mathbf{r}\}_{k+1} = \{\mathbf{r}\}_k + \alpha_k \{\mathbf{u}\}_k \qquad \text{(d)}$$

$$\beta_k = \frac{\{\mathbf{r}\}_{k+1}^T \{\mathbf{r}\}_{k+1}}{\{\mathbf{r}\}_k^T \{\mathbf{r}\}_k} \qquad \text{(e)}$$

$$\{\mathbf{p}\}_{k+1} = \{\mathbf{r}\}_{k+1} + \beta_k \{\mathbf{p}\}_k \qquad \text{(f)}$$

Program 2.12: Conjugate gradients for linear simultaneous equations

```
PROGRAM nm212
!---Conjugate Gradients For Linear Simultaneous Equations---
USE nm_lib; USE precision; IMPLICIT NONE
INTEGER::i,iters,limit,n; REAL(iwp)::alpha,beta,tol,up
REAL(iwp),ALLOCATABLE::a(:,:),b(:),p(:),r(:),u(:),x(:),xnew(:)
OPEN(10,FILE='nm95.dat'); OPEN(11,FILE='nm95.res')
READ(10,*)n; ALLOCATE(a(n,n),b(n),x(n),xnew(n),r(n),p(n),u(n))
READ(10,*)a; READ(10,*)b; READ(10,*)x,tol,limit
WRITE(11,'(A)')"---Conjugate Gradients---"
WRITE(11,'(A)')"---For Linear Simultaneous Equations---"
WRITE(11,'(/,A)')"Coefficient Matrix"
a=TRANSPOSE(a); DO i=1,n; WRITE(11,'(6E12.4)')a(i,:); END DO
WRITE(11,'(/,A,/,6E12.4)')"Right Hand Side Vector",b
WRITE(11,'(/,A,/,6E12.4)')"Guessed Starting Vector",x
r=b-MATMUL(a,x); p=r
WRITE(11,'(/,A)')"First Few Iterations"; iters=0
DO; iters=iters+1
   u=MATMUL(a,p); up=DOT_PRODUCT(r,r); alpha=up/DOT_PRODUCT(p,u)
   xnew=x+alpha*p; r=r-alpha*u; beta=DOT_PRODUCT(r,r)/up
```

```
    p=r+beta*p; IF(iters<5)WRITE(11,'(6E12.4)')x
    IF(checkit(xnew,x,tol).OR.iters==limit)EXIT; x=xnew
  END DO
  WRITE(11,'(/,A,/,I5)')"Iterations to Convergence",iters
  WRITE(11,'(/,A,/,6E12.4)')"Solution Vector",x
END PROGRAM nm212
```

```
Number of equations        n
                           3


Coefficient matrix         (a(i,:),i=1,n)
                           16.   4.    8.
                           4.    5.   -4.
                           8.   -4.   22.


Right hand side            b
                           4.    2.    5.


Initial guess              x
                           1.    1.    1.


Tolerance and              tol    limit
iteration limit            1.E-5  100
```

Data 2.12: Conjugate Gradients

```
---Conjugate Gradients---
---For Linear Simultaneous Equations---

Coefficient Matrix
   0.1600E+02  0.4000E+01  0.8000E+01
   0.4000E+01  0.5000E+01 -0.4000E+01
   0.8000E+01 -0.4000E+01  0.2200E+02

Right Hand Side Vector
   0.4000E+01  0.2000E+01  0.5000E+01

Guessed Starting Vector
   0.1000E+01  0.1000E+01  0.1000E+01

First Few Iterations
   0.1000E+01  0.1000E+01  0.1000E+01
   0.9133E-01  0.8864E+00  0.2049E+00
  -0.1283E+00  0.7444E+00  0.4039E+00
  -0.2500E+00  0.1000E+01  0.5000E+00
```

```
Iterations to Convergence
    4

Solution Vector
 -0.2500E+00  0.1000E+01  0.5000E+00
```

Results 2.12: Conjugate Gradients

List 2.12:

Scalar integers:

i	simple counter
iters	iteration counter
limit	iteration limit
n	number of equations to be solved

Scalar reals:

alpha	see equations (2.65)
beta	see equations (2.65)
tol	convergence tolerance
up	see equations (2.65)

Dynamic real arrays:

a	$n \times n$ matrix of coefficients
b	"right hand side" vector of length n
p	see equations (2.65)
r	see equations (2.65)
u	see equations (2.65)
x	approximate solution vector
xnew	improved solution vector

The starting procedure leads to $\{r\}_0$ which is copied into $\{p\}_0$. The iteration process then proceeds. Vector $\{u\}_k$ is computed by MATMUL in step (a) followed by α_k as required by step (b) involving up. Step (c) leads to the updated x, called xnew as before. Step (d) gives the new $\{r\}_{k+1}$, followed by β_k in step (e). It remains to complete step (f) and to check convergence using checkit.

The same input as used previously as shown in Data 2.12 led to the output shown in Results 2.12. The iteration count has dropped to 4, making this the most successful method so far. In fact theoretically, in perfect arithmetic, the method would converge in n (3 in this case) steps.

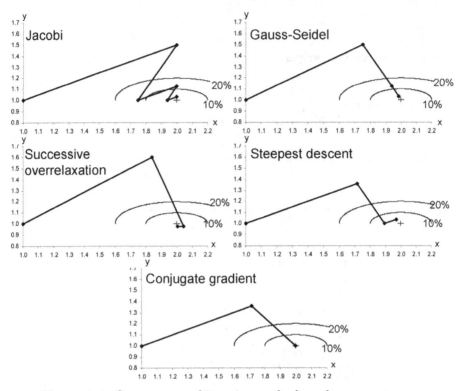

Figure 2.4: Convergence of iterative methods and error contours.

2.10.3   Convergence of iterative methods

Figure 2.4 illustrates how the five iterative methods described so far in this chapter converge on the solution $(x_1, x_2) = (1.0, 2.0)$ of equations (2.41). Also shown are contours of the total error E, defined $\sqrt{e(x_1)^2 + e(x_2)^2}$ where $e(x_1)$ and $e(x_2)$ are the relative errors in x_1 and x_2. Although not easily demonstrated by the small example problems described in this book, gradient methods attempt to steer successive approximations normal to the error contours, and hence towards the true solution by the shortest path in n-dimensional vector space.

2.11   Unsymmetrical systems

The ideas outlined in the previous section can be extended to the case where the equation coefficients are not arranged symmetrically. Again, although the programming of the resulting methods is rather simple, a full mathemati-

cal description is beyond the scope of this book. Readers are referred to
Kelley(1995) and Greenbaum(1997), for example, for more details. They describe a large class of methods generally called "minimum residual" methods
of which "GMRES" or "generalised minimum residual" is a prominent one.
We choose to program a method based on the conjugate gradient approach
of the previous section called "BiCGSTAB", or "stabilized bi-conjugate gradients". The algorithm takes the following steps,

Initial Phase:

$$\{\hat{\mathbf{r}}\}_0 = \{\mathbf{r}\}_0 = \{\mathbf{b}\} - [\mathbf{A}]\{\mathbf{x}\}_0$$
$$\{\mathbf{v}\}_0 = \{\mathbf{p}\}_0 = \{\mathbf{0}\}$$
$$\rho 1_0 = \{\hat{\mathbf{r}}\}_0^T\{\mathbf{r}\}_0$$
$$\rho 0_0 = w_0 = \alpha_0 = 1$$

(2.66)

Iteration Phase: $k = 0, 1, 2, \cdots$

(a)
$$\beta_k = \frac{\rho 1_k \alpha_k}{\rho 0_k w_k}$$
$$\{\mathbf{p}\}_{k+1} = \{\mathbf{r}\}_k + \beta_k (\{\mathbf{p}\}_k - w_k\{\mathbf{v}\}_k)$$
$$\{\mathbf{v}\}_{k+1} = [\mathbf{A}]\{\mathbf{p}\}_{k+1}$$

(2.67)

(b)
$$\alpha_{k+1} = \frac{\rho 1_k}{\{\hat{\mathbf{r}}\}_0^T\{\mathbf{v}\}_{k+1}}$$
$$\{\mathbf{s}\}_{k+1} = \{\mathbf{r}\}_k - \alpha_{k+1}\{\mathbf{v}\}_{k+1}$$
$$\{\mathbf{t}\}_{k+1} = [\mathbf{A}]\{\mathbf{s}\}_{k+1}$$
$$w_{k+1} = \frac{\{\mathbf{t}\}_{k+1}^T\{\mathbf{s}\}_{k+1}}{\{\mathbf{t}\}_{k+1}^T\{\mathbf{t}\}_{k+1}}$$
$$\rho 0_{k+1} = \rho 1_k$$
$$\rho 1_{k+1} = -w_{k+1}\{\hat{\mathbf{r}}\}_0^T\{\mathbf{t}\}_{k+1}$$
$$\{\mathbf{x}\}_{k+1} = \{\mathbf{x}\}_k + \alpha_{k+1}\{\mathbf{p}\}_{k+1}$$
$$\{\mathbf{r}\}_{k+1} = \{\mathbf{s}\}_{k+1} - w_{k+1}\{\mathbf{t}\}_{k+1}$$

(2.68)

Comparing with equations (2.65) we see a similar but more complicated
structure. There are now two (hence "bi") matrix-vector multiplication sections (a) and (b) with associated vector operations. The coding of Program
2.13 is still compact however, and involves only Fortran 95 intrinsic features.

Program 2.13: Stabilized bi-conjugate gradients for linear simultaneous equations

```
PROGRAM nm213
!---Stabilized Bi-Conjugate Gradients---
!---For Linear Simultaneous Equations---
 USE nm_lib; USE precision; IMPLICIT NONE
 INTEGER::i,iters,limit,n; LOGICAL::converged
 REAL(iwp)::alpha,beta,one=1.0_iwp,rho0,rho1,tol,w,zero=0.0_iwp
 REAL(iwp),ALLOCATABLE::a(:,:),b(:),p(:),r(:),r0_hat(:),s(:),    &
  t(:),v(:),x(:)
 OPEN(10,FILE='nm95.dat'); OPEN(11,FILE='nm95.res')
 READ(10,*)n
 ALLOCATE(x(n),v(n),r(n),b(n),r0_hat(n),p(n),s(n),t(n),a(n,n))
 READ(10,*)a; READ(10,*)b; READ(10,*)x,tol,limit
 WRITE(11,'(A)')"---Stabilized Bi-Conjugate Gradients---"
 WRITE(11,'(A)')"---For Linear Simultaneous Equations---"
 WRITE(11,'(/,A)')"Coefficient Matrix"
 a=TRANSPOSE(a); DO i=1,n; WRITE(11,'(6E12.4)')a(i,:); END DO
 WRITE(11,'(/,A,/,6E12.4)')"Right Hand Side Vector",b
 WRITE(11,'(/,A,/,6E12.4)')"Guessed Starting Vector",x
 r=b-MATMUL(a,x); r0_hat=r
 rho0=one; alpha=one; w=one; v=zero; p=zero
 rho1=DOT_PRODUCT(r0_hat,r)
 WRITE(11,'(/,A)')"First Few Iterations"; iters=0
 DO; iters=iters+1; converged=norm(r)<tol*norm(b)
   IF(iters==limit.OR.converged)EXIT
   beta=(rho1/rho0)*(alpha/w); p=r+beta*(p-w*v); v=MATMUL(a,p)
   alpha=rho1/DOT_PRODUCT(r0_hat,v); s=r-alpha*v; t=MATMUL(a,s)
   w=DOT_PRODUCT(t,s)/DOT_PRODUCT(t,t); rho0=rho1
   rho1=-w*DOT_PRODUCT(r0_hat,t); x=x+alpha*p+w*s; r=s-w*t
   IF(iters<5)WRITE(11,'(6E12.4)')x
 END DO
 WRITE(11,'(/,A,/,I5)')"Iterations to Convergence",iters
 WRITE(11,'(/,A,/,6E12.4)')"Solution Vector",x
END PROGRAM nm213
```

List 2.13:

Scalar integers:

i	simple counter
iters	iteration counter
limit	iteration limit
n	number of equations to be solved

Scalar reals:

alpha	see equations (2.68)
beta	see equations (2.67)
one	set to 1.0
rho0	see equations (2.66)
rho1	see equations (2.66)
tol	convergence tolerance
w	see equations (2.66)
zero	set to 0.0

Scalar logical:

converged   set to .TRUE. if converged

Dynamic real arrays:

a	$n \times n$ matrix of coefficients
b	"right hand side" vector of length n
p	see equations (2.66)
r	see equations (2.66)
r0_hat	see equations (2.66)
s	see equations (2.68)
t	see equations (2.67)
v	see equations (2.66)
x	approximate solution vector

Number of equations	n		
	3		
Coefficient matrix	(a(i,:),i=1,n)		
	10.	1.	-5.
	-20.	3.	-20.
	5.	3.	5.
Right hand side	b		
	1.	2.	6.
Initial guess	x		
	1.	1.	1.

```
Tolerance and                 tol    limit
iteration limit               1.E-5   100
```

Data 2.13: Stabilized Bi-Conjugate Gradients

```
---Stabilized Bi-Conjugate Gradients---
---For Linear Simultaneous Equations---

Coefficient Matrix
   0.1000E+02   0.1000E+01  -0.5000E+01
  -0.2000E+02   0.3000E+01   0.2000E+02
   0.5000E+01   0.3000E+01   0.5000E+01

Right Hand Side Vector
   0.1000E+01   0.2000E+01   0.6000E+01

Guessed Starting Vector
   0.1000E+01   0.1000E+01   0.1000E+01

First Few Iterations
   0.1851E+00   0.1113E+01   0.1411E+00
   0.1845E+00   0.1116E+01   0.1399E+00
   0.1000E+01  -0.2000E+01   0.1400E+01

Iterations to Convergence
     4

Solution Vector
   0.1000E+01  -0.2000E+01   0.1400E+01
```

Results 2.13: Stabilized Bi-Conjugate Gradients

The program follows the steps outlined in equations (2.66-2.68). Input is listed as Data 2.13 with output as Results 2.13. For this small system of equations convergence is rapid, taking 4 iterations. For large systems of equations, the number of iterations to convergence can be significantly lower than the number of equations n.

2.12 Preconditioning

The effects of ill-conditioning on coefficient matrices were described in Section 2.7.1 for "direct" solution methods. When iterative methods are used, these effects can translate into very slow convergence rates and indeed to no convergence at all. Therefore it is natural when solving

$$[\mathbf{A}]\{\mathbf{x}\} = \{\mathbf{b}\} \qquad (2.69)$$

to see whether this can be transformed into an equivalent system,

$$[\mathbf{A}^*]\{\mathbf{x}\} = \{\mathbf{b}^*\} \qquad (2.70)$$

for which the convergence properties of the iterative process are better. The transformation process is called "preconditioning" using the "preconditioner" matrix $[\mathbf{P}]$ to give

$$[\mathbf{P}][\mathbf{A}]\{\mathbf{x}\} = [\mathbf{P}]\{\mathbf{b}\} \qquad (2.71)$$

for so-called "left-preconditioning" or

$$[\mathbf{A}][\mathbf{P}]\{\mathbf{x}\} = [\mathbf{P}]\{\mathbf{b}\} \qquad (2.72)$$

for "right-preconditioning".

If $[\mathbf{P}]$ were the inverse of $[\mathbf{A}]$ no iteration would be required but the computational work could be prohibitive. Instead, approximations to $[\mathbf{A}]^{-1}$ are sought which can be obtained cheaply. The simplest of these is a diagonal matrix (stored as a vector) formed by taking the reciprocals of the diagonal terms of $[\mathbf{A}]$. The idea can be extended by including more diagonals of $[\mathbf{A}]$ in the inverse approximation (called "incomplete factorizations"), but in the following programs we illustrate only the pure diagonal form. We begin by applying diagonal preconditioning to the process described in Program 2.12 to form a "preconditioned conjugate gradient" or "PCG" algorithm. The steps in the algorithm are shown in equations (2.73-2.74).

Initial Phase:

$$\{\mathbf{r}\}_0 = \{\mathbf{b}\} - [\mathbf{A}]\{\mathbf{x}\}_0 \qquad (2.73)$$

$$\{\mathbf{precon}\} = \{\frac{1}{\text{diag}[\mathbf{A}]}\}$$

From here on, we don't need $\{\mathbf{b}\}$ so in the actual coding $\{\mathbf{r}\}$ is replaced by $\{\mathbf{b}\}$.

$$\{\mathbf{d}\}_0 = \{\mathbf{precon}\}\{\mathbf{r}\}_0$$

$$\{\mathbf{p}\}_0 = \{\mathbf{d}\}_0$$

Iteration Phase: $k = 0, 1, 2, \cdots$

$$\{\mathbf{u}\}_k = [\mathbf{A}]\{\mathbf{p}\}_k \qquad (2.74)$$

$$\alpha_k = \frac{\{\mathbf{r}\}_k^T \{\mathbf{d}\}_k}{\{\mathbf{p}\}_k^T \{\mathbf{u}\}_k}$$

$$\{\mathbf{x}\}_{k+1} = \{\mathbf{x}\}_k + \alpha_k \{\mathbf{p}\}_k$$

$$\{\mathbf{r}\}_{k+1} = \{\mathbf{r}\}_k - \alpha_k \{\mathbf{u}\}_k$$

$$\{\mathbf{d}\}_{k+1} = \{\mathbf{precon}\}\{\mathbf{r}\}_{k+1}$$

$$\beta_k = \frac{\{\mathbf{r}\}_{k+1}^T \{\mathbf{d}\}_{k+1}}{\{\mathbf{r}\}_k^T \{\mathbf{d}\}_k}$$

$$\{\mathbf{p}\}_{k+1} = \{\mathbf{d}\}_{k+1} + \beta_k \{\mathbf{p}\}_k$$

Program 2.14: Preconditioned conjugate gradients for linear simultaneous equations

```
PROGRAM nm214
!---Preconditioned Conjugate Gradients---
!--- For Linear Simultaneous Equations---
 USE nm_lib; USE precision; IMPLICIT NONE
 INTEGER::i,iters,limit,n
 REAL(iwp)::alpha,beta,one=1.0_iwp,tol,up
 REAL(iwp),ALLOCATABLE::a(:,:),b(:),d(:),p(:),precon(:),u(:),   &
   x(:),xnew(:)
 OPEN(10,FILE='nm95.dat'); OPEN(11,FILE='nm95.res')
 READ(10,*)n
 ALLOCATE(a(n,n),b(n),x(n),xnew(n),d(n),p(n),u(n),precon(n))
 READ(10,*)a; READ(10,*)b; READ(10,*)x,tol,limit
 WRITE(11,'(A)')"---Preconditioned Conjugate Gradients---"
 WRITE(11,'(A)')"---For Linear Simultaneous Equations---"
 WRITE(11,'(/,A)')"Coefficient Matrix"
 a=TRANSPOSE(a); DO i=1,n; WRITE(11,'(6E12.4)')a(i,:); END DO
 WRITE(11,'(/,A,/,6E12.4)')"Right Hand Side Vector",b
 WRITE(11,'(/,A,/,6E12.4)')"Guessed Starting Vector",x
 DO i=1,n; precon(i)=one/a(i,i); END DO
! If x=.0 p and r are just b but in general b=b-a*x
 b=b-MATMUL(a,x); d=precon*b; p=d
 WRITE(11,'(/,A)')"First Few Iterations"; iters=0
 DO; iters=iters+1
   u=MATMUL(a,p); up=DOT_PRODUCT(b,d); alpha=up/DOT_PRODUCT(p,u)
   xnew=x+alpha*p; b=b-alpha*u; d=precon*b
```

```
    beta=DOT_PRODUCT(b,d)/up; p=d+beta*p
    IF(iters<5)WRITE(11,'(6E12.4)')x
    IF(checkit(xnew,x,tol).OR.iters==limit)EXIT; x=xnew
  END DO
  WRITE(11,'(/,A,/,I5)')"Iterations to Convergence",iters
  WRITE(11,'(/,A,/,6E12.4)')"Solution Vector",x
END PROGRAM nm214
```

List 2.14:

Scalar integers:

i	simple counter
iters	iteration counter
limit	iteration limit
n	number of equations to be solved

Scalar reals:

alpha	see equations (2.74)
beta	see equations (2.74)
one	set to 1.0
tol	convergence tolerance
up	numerator of α, see equations (2.74)

Dynamic real arrays:

a	$n \times n$ matrix of coefficients
b	"right hand side" vector of length n
d	see equations (2.74)
p	see equations (2.74)
precon	see equations (2.73)
u	see equations (2.74)
x	approximate solution vector
xnew	improved solution vector

```
Number of equations        n
                           3

Coefficient matrix         (a(i,:),i=1,n)
                           16.    4.    8.
                            4.    5.   -4.
                            8.   -4.   22.

Right hand side            b
                           4.    2.    5.
```

```
Initial guess              x
                           1.    1.    1.

Tolerance and              tol    limit
iteration limit            1.E-5   100
```

Data 2.14: Preconditioned Conjugate Gradients

```
---Preconditioned Conjugate Gradients---
---For Linear Simultaneous Equations---

Coefficient Matrix
   0.1600E+02  0.4000E+01  0.8000E+01
   0.4000E+01  0.5000E+01 -0.4000E+01
   0.8000E+01 -0.4000E+01  0.2200E+02

Right Hand Side Vector
   0.4000E+01  0.2000E+01  0.5000E+01

Guessed Starting Vector
   0.1000E+01  0.1000E+01  0.1000E+01

First Few Iterations
   0.1000E+01  0.1000E+01  0.1000E+01
  -0.4073E-01  0.5837E+00  0.3377E+00
  -0.8833E-01  0.7985E+00  0.3543E+00
  -0.2500E+00  0.1000E+01  0.5000E+00

Iterations to Convergence
     4

Solution Vector
  -0.2500E+00  0.1000E+01  0.5000E+00
```

Results 2.14: Preconditioned Conjugate Gradients

Input and output are shown in Data 2.14 and Results 2.14 respectively. For this small system of equations convergence still takes 4 iterations but for larger systems, convergence can be significantly accelerated.

When preconditioning is applied to the algorithm described by equations (2.66-2.68) the resulting Programs 2.15 (left preconditioning) and 2.16 (right preconditioning) are clearly recognized as derivatives from Program 2.13.

Program 2.15: BiCGSTAB scheme (left preconditioned)

```
PROGRAM nm215
!---BiCGSTAB Scheme(Left Preconditioned)---
!---For Linear Simultaneous Equations---
 USE nm_lib; USE precision; IMPLICIT NONE
 INTEGER::i,iters,limit,n; LOGICAL::converged
 REAL(iwp)::alpha,beta,one=1.0_iwp,rho0,rho1,tol,w,zero=0.0_iwp
 REAL(iwp),ALLOCATABLE::a(:,:),b(:),p(:),precon(:),r(:),        &
   r0_hat(:),s(:),t(:),v(:),x(:)
 OPEN(10,FILE='nm95.dat'); OPEN(11,FILE='nm95.res')
 READ(10,*)n; ALLOCATE(x(n),v(n),r(n),b(n),r0_hat(n),p(n),s(n), &
   t(n),a(n,n),precon(n))
 READ(10,*)a; READ(10,*)b; READ(10,*)x,tol,limit
 WRITE(11,'(A)')"---BiCGSTAB Scheme(Left Preconditioned)---"
 WRITE(11,'(A)')"---For Linear Simultaneous Equations---"
 WRITE(11,'(/,A)')"Coefficient Matrix"
 a=TRANSPOSE(a); DO i=1,n; WRITE(11,'(6E12.4)')a(i,:); END DO
 WRITE(11,'(/,A,/,6E12.4)')"Right Hand Side Vector",b
 WRITE(11,'(/,A,/,6E12.4)')"Guessed Starting Vector",x
!---Simple diagonal preconditioner---
 DO i=1,n; precon(i)=one/a(i,i); END DO
 DO i=1,n; a(i,:)=a(i,:)*precon(i); END DO
!---Apply preconditioner to left hand side---
 b=b*precon; r=b-MATMUL(a,x); r0_hat=r; rho0=one; alpha=one;   &
 w=one; v=zero; p=zero; rho1=DOT_PRODUCT(r0_hat,r)
 WRITE(11,'(/,A)')"First Few Iterations"; iters=0
 DO; iters=iters+1; converged=norm(r)<tol*norm(b)
   IF(iters==limit.OR.converged)EXIT
   beta=(rho1/rho0)*(alpha/w); p=r+beta*(p-w*v); v=MATMUL(a,p)
   alpha=rho1/DOT_PRODUCT(r0_hat,v); s=r-alpha*v; t=MATMUL(a,s)
   w=DOT_PRODUCT(t,s)/DOT_PRODUCT(t,t); rho0=rho1
   rho1=-w*DOT_PRODUCT(r0_hat,t); x=x+alpha*p+w*s; r=s-w*t
   IF(iters<5)WRITE(11,'(6E12.4)')x
 END DO
 WRITE(11,'(/,A,/,I5)')"Iterations to Convergence",iters
 WRITE(11,'(/,A,/,6E12.4)')"Solution Vector",x
END PROGRAM nm215
```

```
Number of equations        n
                           3

Coefficient matrix         (a(i,:),i=1,n)
                           10.   1.  -5.
                          -20.   3. -20.
                            5.   3.   5.

Right hand side            b
                           1.    2.   6.

Initial guess              x
                           1.    1.   1.

Tolerance and              tol     limit
iteration limit            1.E-5   100
```

Data 2.15: BiCGSTAB Scheme (Left Preconditioned)

```
---BiCGSTAB Scheme(Left Preconditioned)---
---For Linear Simultaneous Equations---

Coefficient Matrix
   0.1000E+02  0.1000E+01 -0.5000E+01
  -0.2000E+02  0.3000E+01  0.2000E+02
   0.5000E+01  0.3000E+01  0.5000E+01

Right Hand Side Vector
   0.1000E+01  0.2000E+01  0.6000E+01

Guessed Starting Vector
   0.1000E+01  0.1000E+01  0.1000E+01

First Few Iterations
   0.3725E+00  0.2634E+01  0.6040E-01
  -0.2603E+01  0.1665E+02 -0.4630E+01
   0.1000E+01 -0.2000E+01  0.1400E+01

Iterations to Convergence
   4

Solution Vector
   0.1000E+01 -0.2000E+01  0.1400E+01
```

Results 2.15: BiCGSTAB Scheme (Left Preconditioned)

Input data are as in Data 2.15, with output in Results 2.15. Again due to the small number of equations, no obvious benefits of preconditioning are apparent, but these benefits can be substantial for larger systems.

List 2.15:

Same as List 2.13 with one additional array

Dynamic real arrays:
precon   see equations (2.73)

Program 2.16: BiCGSTAB scheme (right preconditioned)

```
PROGRAM nm216
!---BiCGSTAB Scheme(Right Preconditioned)---
!---For Linear Simultaneous Equations---
 USE nm_lib; USE precision; IMPLICIT NONE
 INTEGER::i,iters,limit,n; LOGICAL::converged
 REAL(iwp)::alpha,beta,one=1.0_iwp,rho0,rho1,tol,w,zero=0.0_iwp
 REAL(iwp),ALLOCATABLE::a(:,:),b(:),p(:),precon(:),p1(:),r(:),  &
   r0_hat(:),s(:),s1(:),t(:),v(:),x(:)
 OPEN(10,FILE='nm95.dat'); OPEN(11,FILE='nm95.res')
 READ(10,*)n
 ALLOCATE(x(n),v(n),r(n),b(n),r0_hat(n),p(n),s(n),t(n),a(n,n),  &
   precon(n),p1(n),s1(n))
 READ(10,*)a; READ(10,*)b; READ(10,*)x,tol,limit
 WRITE(11,'(A)')"---BiCGSTAB Scheme(Right Preconditioned)---"
 WRITE(11,'(A)')"---For Linear Simultaneous Equations---"
 WRITE(11,'(/,A)')"Coefficient Matrix"
 a=TRANSPOSE(a); DO i=1,n; WRITE(11,'(6E12.4)')a(i,:); END DO
 WRITE(11,'(/,A,/,6E12.4)')"Right Hand Side Vector",b
 WRITE(11,'(/,A,/,6E12.4)')"Guessed Starting Vector",x
!---Simple diagonal preconditioner---
 DO i=1,n; precon(i)=one/a(i,i); END DO
 DO i=1,n; a(i,:)=a(i,:)*precon(i); END DO
!---Apply preconditioner to right hand side---
 b=b*precon; r=b-MATMUL(a,x); r0_hat=r; x=x/precon; rho0=one
 alpha=one; w=one; v=zero; p=zero; rho1=DOT_PRODUCT(r0_hat,r)
```

```
WRITE(11,'(/,A)')"First Few Iterations"; iters=0
DO; iters=iters+1; converged=norm(r)<tol*norm(b)
  IF(iters==limit.OR.converged)EXIT
  beta=(rho1/rho0)*(alpha/w); p=r+beta*(p-w*v); p1=p*precon
  v=MATMUL(a,p1); alpha=rho1/DOT_PRODUCT(r0_hat,v); s=r-alpha*v
  s1=s*precon; t=MATMUL(a,s1)
  w=DOT_PRODUCT(t,s)/DOT_PRODUCT(t,t); rho0=rho1
  rho1=-w*DOT_PRODUCT(r0_hat,t); x=x+alpha*p+w*s; r=s-w*t
  IF(iters<5)WRITE(11,'(6E12.4)')x*precon
END DO
x=x*precon
WRITE(11,'(/,A,/,I5)')"Iterations to Convergence",iters
WRITE(11,'(/,A,/,6E12.4)')"Solution Vector",x
END PROGRAM nm216
```

List 2.16:

Same as List 2.13 with three additional arrays

Dynamic real arrays:
precon   see equations (2.73)
p1   see p.81
s1   see p.81

```
Number of equations        n
                           3

Coefficient matrix         (a(i,:),i=1,n)
                           10.   1.   -5.
                          -20.   3.  -20.
                            5.   3.    5.

Right hand side            b
                           1.    2.    6.

Initial guess              x
                           1.    1.    1.

Tolerance and              tol    limit
iteration limit            1.E-5  100
```

Data 2.16: BiCGSTAB Scheme (Right Preconditioned)

```
---BiCGSTAB Scheme(Right Preconditioned)---
---For Linear Simultaneous Equations---

Coefficient Matrix
   0.1000E+02   0.1000E+01  -0.5000E+01
  -0.2000E+02   0.3000E+01   0.2000E+02
   0.5000E+01   0.3000E+01   0.5000E+01

Right Hand Side Vector
   0.1000E+01   0.2000E+01   0.6000E+01

Guessed Starting Vector
   0.1000E+01   0.1000E+01   0.1000E+01

First Few Iterations
   0.7171E+00   0.3563E+01   0.1126E+00
   0.1407E+01  -0.3938E+01   0.2070E+01
   0.1000E+01  -0.2000E+01   0.1400E+01

Iterations to Convergence
      4

Solution Vector
   0.1000E+01  -0.2000E+01   0.1400E+01
```

Results 2.16: BiCGSTAB Scheme (Right Preconditioned)

In this algorithm care has to be taken to precondition intermediate vectors p and s to give p1 and s1 (Kelley,1995). Input and output are as shown in Data 2.16 and Results 2.16 respectively.

2.13   Comparison of direct and iterative methods

In the era of scalar digital computing, say up to 1980, it could be stated that in the majority of cases, direct solution was to be preferred to iterative solution. Exceptions were possible for very sparse systems, and some ill-conditioned systems, but the degree of security offered by direct solution was attractive. It has been shown in the examples of iterative solutions presented in this chapter that a very wide range of efficiencies (as measured by iteration count to convergence and work per iteration) is possible and so the amount of time consumed in the solution of a system of equations is rather unpredictable.

However, the widespread use of "vector" or "parallel" processing computers has led to a revision of previous certainties about equation solution. Comparison of programs like Program 2.14 (preconditioned conjugate gradient technique) with Program 2.2 (LU decomposition) will show that the former consists almost entirely of rather straightforward operations on vectors which can be processed very quickly by nonscalar machines. Even if the coefficient matrix $[\mathbf{A}]$ is sparse or banded, the matrix-vector multiplication operation is vectorizable or parallelizable.

In contrast, the $[\mathbf{L}][\mathbf{U}]$ factorization is seen to contain more complicated code involving conditional statements and variable length loops which present greater difficulties to the programmer attempting to optimize code on a non-scalar machine.

It can therefore be said that algorithm choice for linear equation solution is far from simple for large systems of equations, and depends strongly upon machine architecture. For small systems of equations, direct methods are still attractive.

2.14 Exercises

1. Solve the set of simultaneous equations

$$6x_1 + 3x_2 + 6x_3 = 30$$
$$2x_1 + 3x_2 + 3x_3 = 17$$
$$x_1 + 2x_2 + 2x_3 = 11$$

Answer: $x_1 = 1$, $x_2 = 2$, $x_3 = 3$

2. Solve the system

$$\begin{bmatrix} 1 & 1 & 2 & -4 \\ 2 & -1 & 3 & 1 \\ 3 & 1 & -1 & 2 \\ 1 & -1 & -1 & 1 \end{bmatrix} \begin{Bmatrix} x_1 \\ x_2 \\ x_3 \\ x_4 \end{Bmatrix} = \begin{Bmatrix} 0 \\ 5 \\ 5 \\ 0 \end{Bmatrix}$$

by $[\mathbf{L}][\mathbf{U}]$ factorization.

Answer: $[\mathbf{L}] = \begin{bmatrix} 1 & 0 & 0 & 0 \\ 2 & 1 & 0 & 0 \\ 3 & \frac{2}{3} & 1 & 0 \\ 1 & \frac{2}{3} & \frac{7}{19} & 1 \end{bmatrix}$ $[\mathbf{U}] = \begin{bmatrix} 1 & 1 & 2 & -4 \\ 0 & -3 & -1 & 9 \\ 0 & 0 & -\frac{19}{3} & 8 \\ 0 & 0 & 0 & -\frac{75}{19} \end{bmatrix}$

and $x_1 = x_2 = x_3 = x_4 = 1$

3. Solve the symmetrical equations

$$9.3746x_1 + 3.0416x_2 - 2.4371x_3 = 9.2333$$
$$3.0416x_1 + 6.1832x_2 + 1.2163x_3 = 8.2049$$
$$-2.4371x_1 + 1.2163x_2 + 8.4429x_3 = 3.9339$$

by $[\mathbf{L}][\mathbf{D}][\mathbf{L}]^T$ decomposition.
Answer: $x_1 = 0.8964$, $x_2 = 0.7651$, $x_3 = 0.6145$. The diagonal terms are $D_{11} = 9.3746$, $D_{22} = 5.1964$, $D_{33} = 7.0341$ and the determinant of the coefficient matrix is 342.66.

4. Solve the symmetrical equations

$$5x_1 + 6x_2 - 2x_3 - 2x_4 = 1$$
$$6x_1 - 5x_2 - 2x_3 + 2x_4 = 0$$
$$-2x_1 - 2x_2 + 3x_3 - x_4 = 0$$
$$-2x_1 + 2x_2 - x_3 - 3x_4 = 0$$

Answer: $x_1 = 0.12446$, $x_2 = 0.07725$, $x_3 = 0.11159$, $x_4 = -0.06867$

5. Solve the symmetrical equations

$$x_1 + 2x_2 - 2x_3 + x_4 = 4$$
$$2x_1 + 5x_2 - 2x_3 + 3x_4 = 7$$
$$-2x_1 - 2x_2 + 5x_3 + 3x_4 = -1$$
$$x_1 + 3x_2 + 3x_3 + 2x_4 = 0$$

Answer: $x_1 = 2$, $x_2 = -1$, $x_3 = -1$, $x_4 = 2$

6. Attempt to solve Exercises 4 and 5 by Cholesky's method (Program 2.4).
Answer: Square roots of negative numbers will arise.

7. Solve the symmetrical banded system

$$\begin{bmatrix} 4 & 2 & 0 & 0 \\ 2 & 8 & 2 & 0 \\ 0 & 2 & 8 & 2 \\ 0 & 0 & 2 & 4 \end{bmatrix} = \begin{Bmatrix} 4 \\ 0 \\ 0 \\ 0 \end{Bmatrix}$$

Answer: $x_1 = 1.156$, $x_2 = -0.311$, $x_3 = 0.089$, $x_4 = -0.044$

8. Solve the following system using elimination with pivoting

$$\begin{bmatrix} 1 & 0 & 2 & 3 \\ -1 & 2 & 2 & -3 \\ 0 & 1 & 1 & 4 \\ 6 & 2 & 2 & 4 \end{bmatrix} = \begin{Bmatrix} 1 \\ -1 \\ 2 \\ 1 \end{Bmatrix}$$

Answer: $x_1 = -\frac{13}{70}$, $x_2 = \frac{8}{35}$, $x_3 = -\frac{4}{35}$, $x_4 = \frac{33}{70}$. The interchanged row order is 4, 2, 1, 3.

9. Solve the following equations using elimination with pivoting

$$x_1 + 2x_2 + 3x_3 = 2$$
$$3x_1 + 6x_2 + x_3 = 14$$
$$x_1 + x_2 + x_3 = 2$$

Answer: $x_1 = 1$, $x_2 = 2$, $x_3 = -1$. The interchanged row order is 2, 3, 1.

10. Attempt to solve Exercise 9 without pivoting.
    Answer: Zero pivot found in row 2.

11. Solve the equations

$$20x_1 + 2x_2 - x_3 = 25$$
$$2x_1 + 13x_2 - 2x_3 = 30$$
$$x_1 + x_2 + x_3 = 2$$

using (a) Jacobi and (b) Gauss-Seidel iterations, using a starting guess $x_1 = x_2 = x_3 = 0$.
Answer: $x_1 = 1$, $x_2 = 2$, $x_3 = -1$.

12. Compare the iteration counts, for a tolerance of 1×10^{-5}, in the solution of Exercise 11 by the following methods: (a) Jacobi, (b) Gauss-Seidel, (c) SOR ($\omega = 1.2$), (d) Steepest descent, (e) Conjugate gradients.
    Answer: (a) 16, (b) 10, (c) 30, (d) 69, (e) Does not converge. The last method is suitable only for symmetrical, positive definite systems.

13. Solve Exercises 4 and 5 by the method of Conjugate gradients.
    Answer: For a tolerance of 1×10^{-5}, solution obtained in 5 iterations in both cases.

14. Check that the solution vector $[\, 1.22 \; -1.02 \; 3.04 \,]^T$ is a solution to the system

$$9x_1 + 9x_2 + 8x_3 = 26$$
$$9x_1 + 8x_2 + 7x_3 = 24$$
$$8x_1 + 7x_2 + 6x_3 = 21$$

to within a tolerance of 0.01. Find the true solution. What do these results imply about the system of equations?
Answer: True solution $[1 \; 1 \; 1]^T$. The system is ill-conditioned.

15. Solve the following set of equations by using Gaussian elimination with pivoting to transform the matrix into lower triangular form.

$$\begin{bmatrix} 0 & 2 & 0 & 3 \\ 1 & 0 & 3 & 4 \\ 2 & 3 & 0 & 1 \\ -3 & 5 & 2 & 0 \end{bmatrix} \begin{Bmatrix} x_1 \\ x_2 \\ x_3 \\ x_4 \end{Bmatrix} = \begin{Bmatrix} 0 \\ 7 \\ -9 \\ -12 \end{Bmatrix}$$

Answer: $x_1 = -1$, $x_2 = -3$, $x_3 = 0$, $x_4 = 2$

16. Estimate the solution to the following set of equations using the Gauss-Seidel method with an initial guess of $x_1 = 10$, $x_2 = 0$ and $x_3 = -10$

$$\begin{bmatrix} 4 & -2 & -1 \\ 1 & -6 & 2 \\ 1 & -2 & 12 \end{bmatrix} \begin{Bmatrix} x_1 \\ x_2 \\ x_3 \end{Bmatrix} = \begin{Bmatrix} 40 \\ -28 \\ -86 \end{Bmatrix}$$

Answer: Solution is $[x_1 \ x_2 \ x_3]^T = [10.11 \quad 3.90 \quad -7.36]^T$
After two iterations you should have $[9.45 \quad 3.79 \quad -7.32]^T$

17. Solve the following set of equations using Gaussian elimination.

$$\begin{bmatrix} 4 & 3 & -6 & 1 \\ 4 & 3 & 2 & 2 \\ -6 & -6 & 3 & -1 \\ -1 & 3 & -1 & 2 \end{bmatrix} \begin{Bmatrix} x_1 \\ x_2 \\ x_3 \\ x_4 \end{Bmatrix} = \begin{Bmatrix} -5 \\ 12 \\ -1 \\ -9 \end{Bmatrix}$$

Answer: $[x_1 \ x_2 \ x_3 \ x_4]^T = [3 \ -2 \ 2 \ 1]^T$

18. Decompose the matrix $[\mathbf{A}]$ into the factors $[\mathbf{L}][\mathbf{D}][\mathbf{L}]^T$ and hence compute its determinant

$$[\mathbf{A}] = \begin{bmatrix} 2 & 1 & 0 \\ 1 & 3 & 1 \\ 0 & 1 & 2 \end{bmatrix}$$

Answer: $[\mathbf{L}][\mathbf{D}][\mathbf{L}]^T = \begin{bmatrix} 1 & 0 & 0 \\ 0.5 & 1 & 0 \\ 0 & 0.4 & 1 \end{bmatrix} \begin{bmatrix} 2 & 0 & 0 \\ 0 & 2.5 & 0 \\ 0 & 0 & 1.6 \end{bmatrix} \begin{bmatrix} 1 & 0.5 & 0 \\ 0 & 1 & 0.4 \\ 0 & 0 & 1 \end{bmatrix}$,

$\det[\mathbf{A}] = 8$

19. In a less stringent approach to avoiding zero-diagonal terms known as "Partial pivoting", at each stage of the elimination process, the terms in the column below the active diagonal are scanned, and the row containing the largest (absolute) term is interchanged with the row containing the diagonal.

Use this method to solve the following set of equations:

$$\begin{bmatrix} 4 & -4 & -6 \\ 4 & 3 & 2 \\ -6 & -6 & 3 \end{bmatrix} \begin{Bmatrix} x_1 \\ x_2 \\ x_3 \end{Bmatrix} = \begin{Bmatrix} 2.5 \\ -4.6 \\ 4.95 \end{Bmatrix}$$

Answer: $x_1 = -0.55$, $x_2 = -0.50$, $x_3 = -0.45$

20. Use Cholesky's method to solve the following two systems of simultaneous equations

$$\begin{bmatrix} 1 & 2 & -2 \\ 2 & 5 & -4 \\ -2 & -4 & 5 \end{bmatrix} \begin{Bmatrix} x_1 \\ x_2 \\ x_3 \end{Bmatrix} = \begin{Bmatrix} -4 \\ 2 \\ 4 \end{Bmatrix}, \quad \begin{bmatrix} 1 & 2 & -2 \\ 2 & 5 & -4 \\ -2 & -4 & 5 \end{bmatrix} \begin{Bmatrix} x_1 \\ x_2 \\ x_3 \end{Bmatrix} = \begin{Bmatrix} 2 \\ -4 \\ 4 \end{Bmatrix}$$

Briefly describe any labor saving procedures that may have helped you to solve these two problems.

Answer: $[-32 \quad 10 \quad -4]^T$, $[34 \quad -8 \quad 8]^T$. Factorize once, followed by forward and back-substitution for each problem.

21. Attempt to solve the following system of equations using Cholesky's Method.

$$\begin{bmatrix} 3 & -2 & 1 \\ -2 & 3 & 2 \\ 1 & 2 & 2 \end{bmatrix} \begin{Bmatrix} x_1 \\ x_2 \\ x_3 \end{Bmatrix} = \begin{Bmatrix} 3 \\ -3 \\ 2 \end{Bmatrix}$$

What do your calculations indicate about the nature of the coefficient matrix?

Answer: Cholesky factorization not possible, hence symmetrical coefficient matrix is not positive definite.

22. Perform a couple of Gauss-Seidel iterations on the following system of equations

$$\begin{bmatrix} 1.00 & -0.62 & 0.37 \\ -0.62 & 1.00 & -0.51 \\ 0.37 & -0.51 & 1.00 \end{bmatrix} \begin{Bmatrix} x_1 \\ x_2 \\ x_3 \end{Bmatrix} = \begin{Bmatrix} 3 \\ -3 \\ 2 \end{Bmatrix}$$

Answer: $[2.5 \quad -1.1668 \quad 0.4799]^T$ after 2 iteration with an initial guess of $[1 \ 1 \ 1]^T$. Solution $[1.808 \quad -1.622 \quad 0.504]^T$

23. An engineering analysis involves repeated solution ($i = 1, 2 \cdots$ etc.) of a set of equations of the form: $[\mathbf{A}]\{\mathbf{x}\}_i = \{\mathbf{b}\}_i$ where

$$[\mathbf{A}] = \begin{bmatrix} 4 & 3 & -6 \\ 3 & 3 & 2 \\ -6 & -2 & 3 \end{bmatrix}$$

Choose an appropriate method for solving these equations, and use it to solve for the case when $\{\mathbf{b}\}_1 = \begin{Bmatrix} -4 \\ 3 \\ -6 \end{Bmatrix}$

Work to 4 decimal places of accuracy.

Answer: The appropriate method involves factorization of $[\mathbf{A}]$.
$x_1 = 2.2530$, $x_2 = -2.0241$, $x_3 = 1.1566$

24. The coefficient matrix of a symmetrical, positive definite system of equations is stored in "skyline" form as:

$$\begin{Bmatrix} 1080.0 \\ -180.0 \\ 120.0 \\ 60.0 \\ 20.0 \\ 40.0 \end{Bmatrix}$$

with diagonal terms in the first, third and sixth locations. Retrieve the original coefficient matrix and solve the equations to an accuracy of four decimal places using a suitable method with a right hand side vector of:

$$\begin{Bmatrix} 10.0 \\ 0.0 \\ 0.0 \end{Bmatrix}$$

Answer: $[\,0.0170 \quad 0.0324 \quad -0.0417\,]^T$

25. Prove that the BiCGStab algorithm (Program 2.13) leads to a solution of the unsymmetrical equation systems of Excercises 1,2,8,9 and 11 in n or $n+1$ iterations in every case. For these small systems, preconditioning (Programs 2.15 or 2.16) is not necessary.

Chapter 3

Nonlinear Equations

3.1 Introduction

In the previous chapter we dealt with "linear" equations which did not involve powers or products of the unknowns. A common form of "nonlinear" equations which frequently arises in practice does contain such powers, for example,

$$x_1^3 - 2x_1x_2 + x_2^2 = 4$$
$$x_1 + 4x_1x_2 + 3x_2^2 = 7 \tag{3.1}$$

would be a pair of nonlinear equations satisfied by various combinations of x_1 and x_2.

In the simplest situation, we might have a single nonlinear equation such as

$$y = f(x) = x^3 - x - 1 = 0 \tag{3.2}$$

A graphical interpretation helps us to understand the nature of the solutions for x which satisfy equation (3.2). A plot of $f(x)$ versus x is shown in Figure 3.1. Where $f(x)$ intersects the line $y = 0$ is clearly a solution of the equation, often called a "root", which in this case has the value $x \approx 1.3247$.

Note that we could also write equation (3.2) as

$$y = g(x) = x^3 - x = 1 \tag{3.3}$$

and look for the intersection of $g(x)$ with the line $y = 1$ as a solution of the equation, also shown in the figure.

Since we now know that $x \approx 1.3247$ is a solution, we can factorize $f(x)$ to yield

$$f(x) \approx (x - 1.3247)(x^2 + 1.3247x + 0.7549) = 0 \tag{3.4}$$

Taking the roots of the quadratic equation (3.4), we arrive at solutions

$$x \approx \frac{-1.3247 \pm \sqrt{-1.2648}}{2} \tag{3.5}$$

showing that the remaining two roots are imaginary ones.

It is immediately apparent that finding solutions to general sets of nonlinear equations will be quite a formidable numerical task. As in the last chapter, where we said that many physical systems produced diagonally dominant and/or symmetric systems of linear equations, it is fortunate that in many physical situations which give rise to nonlinear sets of equations, the nature of the problem limits the possible values that the roots may have. For example, we may know in advance that all the roots must be real, or even that they must all be real and positive. In this chapter we shall concentrate on these limited problems.

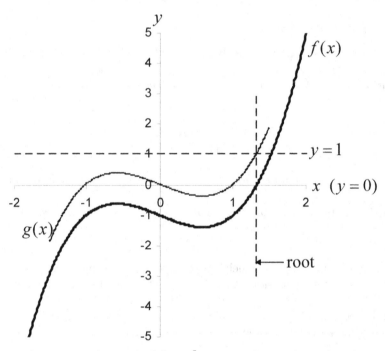

Figure 3.1: Root of $f(x) = x^3 - x - 1$ obtained graphically.

We shall also see that the methods for solving nonlinear equations are intrinsically iterative in character. Referring to Figure 3.1, it should be clear that such an iterative process might depend strongly on the quality of an initial guessed solution. For example, a guess of $x = 1$ or $x = 2$ in that case might have much more chance of success than a guess of $x = -100$ or $x = 100$.

3.2 Iterative substitution

A simple iterative process is to replace an equation like equation (3.2) which had the form

$$f(x) = x^3 - x - 1 = 0$$

by an equivalent equation

$$x = F(x) \tag{3.6}$$

The iterative process proceeds by a guess being made for x, substitution of this guess on the right-hand side of equation (3.6) and comparison of the result with the guessed x. In the unlikely event that equality results, the solution has been found. If not, the new $F(x)$ is assumed to be a better estimate of x and the process is repeated.

An immediate dilemma is that there is no single way of determining $F(x)$. In the case of equation (3.2), we could write

$$x = F_1(x) = x^3 - 1 \qquad \text{(a)}$$
$$x = F_2(x) = \frac{1}{x^2 - 1} \qquad \text{(b)} \tag{3.7}$$
$$x = F_3(x) = \sqrt[3]{x + 1} \qquad \text{(c)}$$

When these are plotted, as in Figure 3.2, further difficulties are apparent. In each case the root is correctly given by the intersection of $y = F_i(x)$ with $y = x$. The function $F_3(x)$, however, has no real value for $x < -1.0$ and the function $F_2(x)$ has singular points at $x \pm 1.0$. It can also be seen that $F_1(x)$ and $F_2(x)$ are changing very rapidly in the region of the root ($x = 1.3247$) in contrast to $F_3(x)$ which is changing very slowly. The following simple analysis gives insight into the convergence properties of the method.

Consider a typical step of the method "close" to the required solution, then

$$x_{i+1} = F(x_i) \tag{3.8}$$

If the exact solution is α, then

$$\alpha = F(\alpha) \tag{3.9}$$

hence from equations (3.8) and (3.9)

$$\alpha - x_{i+1} = F(\alpha) - F(x_i)$$
$$\approx (\alpha - x_i)\frac{dF}{dx} \tag{3.10}$$

If the method is to converge, $|\alpha - x_{i+1}| < |\alpha - x_i|$, hence

$$\left|\frac{dF}{dx}\right| < 1 \tag{3.11}$$

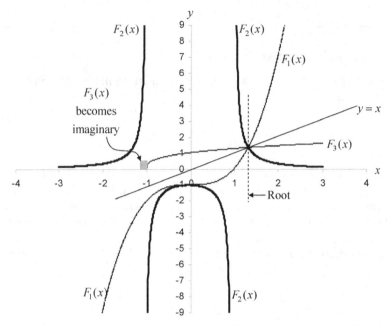

Figure 3.2: Different choices of x=F(x) in equation (3.7).

It is clear from Figure 3.2 that only $F_3(x)$ satisfies this convergence criterion in the vicinity of the root. Furthermore, $F_3(x)$ from equation 3.7(c) will only yield a correct result if the starting guess is $x > -1.0$.

Program 3.1: Iterative substitution for a single root

```
PROGRAM nm31
!---Iterative Substitution for a Single Root---
 USE nm_lib; USE precision; IMPLICIT NONE
 INTEGER::iters,limit; REAL(iwp)::tol,x0,x1
 OPEN(10,FILE='nm95.dat'); OPEN(11,FILE='nm95.res')
 WRITE(11,'(A)')"---Iterative Substitution for a Single Root---"
 READ(10,*)x0,tol,limit
 WRITE(11,'(/,A,/,E12.4)')"Guessed Starting Value",x0
 WRITE(11,'(/,A)')"First Few Iterations"; iters=0
 DO; iters=iters+1
   x1=f31(x0)
   IF(check(x1,x0,tol).OR.iters==limit)EXIT; x0=x1
   IF(iters<5)WRITE(11,'(3E12.4)')x1
```

```
END DO
WRITE(11,'(/,A,/,I5)')"Iterations to Convergence",iters
WRITE(11,'(/,A,/,E12.4)')"Solution",x1
CONTAINS

FUNCTION f31(x)
 IMPLICIT NONE
 REAL(iwp),INTENT(IN)::x; REAL(iwp)::f31
 f31=(x+1.0_iwp)**(1.0_iwp/3.0_iwp)
RETURN
END FUNCTION f31

END PROGRAM nm31
```

List 3.1:

Scalar integers:
iters   iteration counter
limit   iteration limit

Scalar reals:
tol   convergence tolerance
x0   approximate solution
x1   improved solution

```
Initial value            x0
                         1.2

Tolerance and            tol    limit
iteration limit          1.E-5  100
```

Data 3.1: Iterative Substitution

```
---Iterative Substitution for a Single Root---

Guessed Starting Value
  0.1200E+01

First Few Iterations
  0.1301E+01
  0.1320E+01
```

```
0.1324E+01
0.1325E+01
```

```
Iterations to Convergence
    7
```

```
Solution
    0.1325E+01
```

Results 3.1: Iterative Substitution

The program expects $F(x)$ to be provided in the user-supplied function f31 at the end of the main program. In this case, the function $F_3(x) = \sqrt[3]{x+1}$ is generated as

$$f31=(x+1.0\_iwp)**(1.0\_iwp/3.0\_iwp)$$

The program merely reads a starting value x_0, and calculates a new value $x_1 = F(x_0)$. If x_1 is close enough to x_0, checked by library subroutine check (This is done in the same way as checkit does for arrays.), the process terminates. To guard against divergent solutions, a maximum number of iterations limit is prescribed as data. Input and output are shown in Data and Results 3.1 respectively.

One might question whether the starting guess influences the computation very much (as long as, of course, $x_0 > -1.0$). The table below shows that with a tolerance of 1×10^{-5} the influence is small in this case.

Starting value x_0	Number of iterations to convergence
0.0	8
1.2	7
1.3	8
10.0	9
100.0	10

3.3   Multiple roots and other difficulties

It can readily be shown that the nonlinear equation

$$f(x) = x^4 - 6x^3 + 12x^2 - 10x + 3 = 0 \qquad (3.12)$$

can be factorized into

$$f(x) = (x-3)(x-1)(x-1)(x-1) = 0 \qquad (3.13)$$

so that of the four roots, three are coincident. This function is illustrated in Figure 3.3. Following our experience with equation (3.2), we may think of expressing the equation for iterative purposes as

$$x = F(x) = \sqrt[4]{6x^3 - 12x^2 + 10x - 3} \qquad (3.14)$$

This function is also shown in Figure 3.3 where it can be seen that for $x < 1.0$, $F(x)$ becomes imaginary. For x in the range $1 < x < 2.16$, $\dfrac{dF}{dx} > 1$, although in the vicinity of the root at $x = 3$, $\dfrac{dF}{dx} < 1$.

When Program 3.1 is used to attempt to solve this problem, a starting guess $x_0 = 1.1$ converges to the root $x = 3$ in 311 iterations while starting guesses of $x_0 = 2.0$ and $x_0 = 4.0$ converge to the root $x = 3$ in close to 100 iterations. We can see that convergence is slow, and that it is impossible to converge on the root $x = 1$ at all.

Example 3.1

Use Iterative Substitution to find a root close to 0.5 of the function

$$f(x) = x^3 - 3x + 1 = 0$$

Solution 3.1

Possible rearrangements take the form

$$x = F_1(x) = \frac{x^3 + 1}{3} \qquad (a)$$

$$x = F_2(x) = \frac{1}{3 - x^2} \qquad (b)$$

$$x = F_3(x) = \sqrt[3]{3x - 1} \qquad (c)$$

Taking arrangement (a) with initial guess $x_0 = 0.5$ leads to

$$x_1 = \frac{0.5^3 + 1}{3} = 0.3750$$

$$x_2 = \frac{0.3750^3 + 1}{3} = 0.3509$$

$$x_1 = \frac{0.3509^3 + 1}{3} = 0.3477$$

which is almost converged on the exact solution $x \approx 0.3473$ to 4 decimal places.

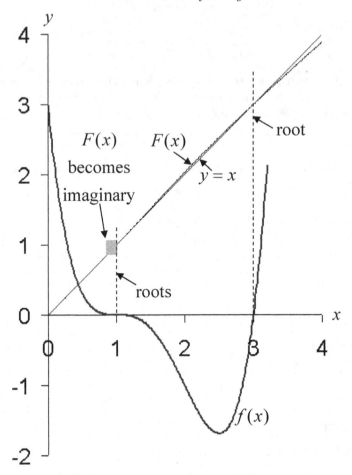

Figure 3.3: Functions $f(x)$ and $F(x)$ from equations (3.13) and (3.14).

In this example it is easily shown that the three roots of the cubic in this
example are to 4 decimal places given by 1.5321, 0.3473 and -1.8794. Analysis of rearrangements (a), (b) and (c) above are summarized in the table below.

Root	$\dfrac{dF_1}{dx} = x^2$	$\dfrac{dF_2}{dx} = \dfrac{2x}{(3 - x^2)^2}$	$\dfrac{dF_3}{dx} = \dfrac{1}{(3x - 1)^{2/3}}$
-1.8794	3.53	-13.35	imaginary
0.3473	0.12	0.08	8.41
1.5321	2.35	7.19	0.43

The Table indicates the values of the derivatives $\dfrac{dF_i}{dx}$ in the vicinity of each of the roots. Derivatives with absolute values less than one (in bold) indicate that some arrangements of the original function can converge on certain roots but not others. For example $F_1(x)$ and $F_2(x)$ can converge on the root $x = 0.3473$, while $F_3(x)$ can converge on the root $x = 1.5321$. None of the arrangements can converge on the root $x = -1.8794$ since the absolute values of the derivatives that are real are all greater than one. In the next section, we turn to other methods, beginning with those based on interpolation between two estimates to find a root.

3.4 Interpolation methods

This class of methods is based on the assumption that the function changes sign in the vicinity of a root. In fact if, for the type of function shown in equation (3.13), there had been a double root rather than a triple one, no sign change would have occurred at the double root, but this occurrence is relatively rare. Let us begin by assuming that the vicinity of a root, involving a change in sign of the function, has been located - perhaps graphically. The next two sections describe methods whereby the root can be accurately evaluated.

3.4.1 Bisection method

To take a typical example, we know that the function given by equation (3.2), namely

$$y = f(x) = x^3 - x - 1 = 0 \tag{3.15}$$

has a root close to 1.3, and that $f(x)$ changes sign at that root (see Figure 3.1). To carry out the bisection process, we would begin with an underestimate of the root, say $x = 1.0$, and proceed to evaluate $f(x)$ at equal increments of x until a sign change occurs. This can be checked, for example, by noting that the product of $f(x_{i+1})$ and $f(x_i)$ is negative. We avoid for the moment the obvious difficulties in this process, namely that we may choose steps which are too small and involve excessive work or that for more complicated functions we may choose steps which are too big and miss two or more roots which have involved a double change of sign of the function.

Having established a change of sign of $f(x)$, we have two estimates of the root which bracket it. The value of the function half way between the estimates is then found, i.e., $f(\frac{x_{i+1} + x_i}{2})$ or $f(x_{mid})$. If the sign of the function at this midpoint is the same as that of $f(x_i)$, the root is "closer" to x_{i+1} and x_{mid} replaces x_i for the next bisection. Alternatively, of course, it replaces x_{i+1}. When successive values of x_{mid} are "close enough", i.e., within a certain

Numerical Methods for Engineers

tolerance, iterations can be stopped.

Example 3.2

Use the Bisection method to find a root of the function,

$$f(x) = x^3 - x - 1 = 0$$

which lies in the range $1.3 < x < 1.4$.

Solution 3.2

A tabular approach is useful, i.e.,

x	$f(x)$
1.3	-0.1030
1.4	0.3440
1.35	0.1104
1.325	0.0012
1.3125	-0.0515
1.31875	-0.0253
1.32188	-0.0121
1.32344	-0.0055
1.32422	-0.0021 etc.

Hence at this stage the root lies in the range

$$1.32422 < x < 1.325$$

Program 3.2: Bisection method for a single root

```
PROGRAM nm32
!---Bisection Method for a Single Root---
 USE nm_lib; USE precision; IMPLICIT NONE
 INTEGER::iters,limit
 REAL(iwp)::half=0.5_iwp,tol,xmid,xi,xip1,xold,zero=0.0_iwp
 OPEN(10,FILE='nm95.dat'); OPEN(11,FILE='nm95.res')
 WRITE(11,'(A)')"---Bisection Method for a Single Root---"
 READ(10,*)xi,xip1,tol,limit; WRITE(11,'(/,A,/,E12.4,A,E12.4)') &
   "Starting Range",xi," to", xip1
 WRITE(11,'(/,A)')"First Few Iterations"; iters=0; xold=xi
 DO; iters=iters+1; xmid=half*(xi+xip1)
  IF(f32(xi)*f32(xmid)<zero)THEN
    xip1=xmid; ELSE; xi=xmid; END IF
```

```
 IF(iters<5)WRITE(11,'(3E12.4)')xmid
 IF(check(xmid,xold,tol).OR.iters==limit)EXIT; xold=xmid
END DO
WRITE(11,'(/,A,/,I5)')"Iterations to Convergence",iters
WRITE(11,'(/,A,/,E12.4)')"Solution",xmid
CONTAINS

FUNCTION f32(x)
 IMPLICIT NONE
 REAL(iwp),INTENT(IN)::x; REAL(iwp)::f32
 f32=x**3-x-1.0_iwp
RETURN
END FUNCTION f32

END PROGRAM nm32
```

List 3.2:

Scalar integers:

iters	iteration counter
limit	iteration limit

Scalar reals:

half	set to 0.5
tol	convergence tolerance
xmid	average of x_i and x_{i+1}
xi	underestimate of root x_i
xip1	overestimate of root x_{i+1}
xold	average from last iteration
zero	set to 0.0

Initial values	xi	xip1
	1.0	2.0

Tolerance and iteration limit	tol	limit
	1.E-5	100

Data 3.2: Bisection Method

```
---Bisection Method for a Single Root---

Starting Range
   0.1000E+01 to  0.2000E+01

First Few Iterations
   0.1500E+01
   0.1250E+01
   0.1375E+01
   0.1312E+01

Iterations to Convergence
   17

Solution
   0.1325E+01
```

Results 3.2: Bisection Method

The function is again equation (3.2) and this has been programmed into function f32 at the end of the main program. The input and output are as shown in Data and Results 3.2 respectively. Numbers to be input are the first underestimate of the root, the first overestimate, the iteration tolerance and the maximum number of iterations allowed.

As long as convergence has not been achieved, or the iteration limit reached, the estimates of the root are bisected to give xmid, and the lower or upper estimate are updated as required. A library routine check checks if convergence has been obtained.

When the tolerance has been achieved, or the iterations exhausted, the current estimate of the root is printed together with the number of iterations to achieve it.

As shown in Results 3.2, for the tolerance of 1×10^{-5}, the Bisection method takes 17 iterations to converge to the root $x = 1.325$ from lower and upper starting limits of 1.0 and 2.0.

3.4.2   False position method

Again this is based on finding roots of opposite sign and interpolating between them, but by a method which is generally more efficient than bisection.

Figure 3.4 shows a plot of a function $f(x)$ with a root in the region of $x = 2$. Let our initial under- and overestimates be $x = 1$ and $x = 3$. The False Position method interpolates linearly between $(x_i, f(x_i))$ and $(x_{i+1}, f(x_{i+1}))$, taking the intersection of the interpolating line with the x-axis as an improved guess at the root. Using the same procedure as in the Bisection method, the improved guess replaces either the previous lower or upper bound as the case may be. The interpolation can be written

$$x_{new} = x_i - f(x_i) \left[\frac{x_{i+1} - x_i}{f(x_{i+1}) - f(x_i)} \right]$$ (3.16)

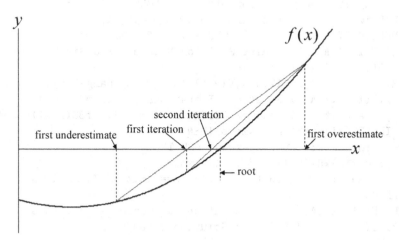

Figure 3.4: False Position method.

Example 3.3

Use the False Position method to find a root of the function

$$f(x) = x^3 - x - 1 = 0$$

which lies in the range $1.3 < x < 1.4$.

Solution 3.3

A tabular approach is again useful, and using equation (3.16) we get

x	$f(x)$
1.3	-0.1030
1.4	0.3440
1.32304	-0.0071
1.32460	-0.0005
1.32471	-0.0000 etc.

Convergence of this method can also be observed by $f(x) \rightarrow 0$.

Program 3.3: False position method for a single root

```
PROGRAM nm33
!---False Position Method for a Single Root---
 USE nm_lib; USE precision; IMPLICIT NONE; INTEGER::iters,limit
 REAL(iwp)::tol,xi,xip1,xnew,xold,zero=0.0_iwp
 OPEN(10,FILE='nm95.dat'); OPEN(11,FILE='nm95.res')
 WRITE(11,'(A)')"---False Position Method for a Single Root---"
 READ(10,*)xi,xip1,tol,limit
 WRITE(11,'(/,A,/,E12.4,A,E12.4)')"Starting Range",xi," to", xip1
 WRITE(11,'(/,A)')"First Few Iterations"; iters=0; xold=xi
 DO; iters=iters+1; xnew=xi-f33(xi)*(xip1-xi)/(f33(xip1)-f33(xi))
   IF(f33(xi)*f33(xnew)<zero)THEN
     xip1=xnew; ELSE; xi=xnew; END IF
 IF(iters<5)WRITE(11,'(3E12.4)')xnew
   IF(check(xnew,xold,tol).OR.iters==limit)EXIT; xold=xnew
 END DO
 WRITE(11,'(/,A,/,I5)')"Iterations to Convergence",iters
 WRITE(11,'(/,A,/,E12.4)')"Solution",xnew
 CONTAINS

 FUNCTION f33(x)
  IMPLICIT NONE; REAL(iwp),INTENT(IN)::x; REAL(iwp)::f33
  f33=x**3-x-1.0_iwp
  RETURN
 END FUNCTION f33

END PROGRAM nm33
```

List 3.3:

Scalar integers:

iters	iteration counter
limit	iteration limit

Scalar reals:

tol	convergence tolerance
xi	underestimate of root x_i
xip1	overestimate of root x_{i+1}
xnew	interpolated value
xold	interpolated value from last iteration
zero	set to 0.0

```
Initial values            xi     xip1
                          1.0    2.0

Tolerance and             tol    limit
iteration limit           1.E-5  100
```

Data 3.3: False Position Method

```
---False Position Method for a Single Root---

Starting Range
  0.1000E+01 to  0.2000E+01

First Few Iterations
   0.1167E+01
   0.1253E+01
   0.1293E+01
   0.1311E+01

Iterations to Convergence
   13

Solution
   0.1325E+01
```

Results 3.3: False Position Method

The program is very similar to the previous one described for the Bisection method. The same problem is solved with data and output shown in Data and Results 3.3. It can be seen that the False Position method gives convergence in 13 iterations as opposed to 17 iterations with the same tolerance by the Bisection method. This increased efficiency will generally be observed except in cases where the function varies rapidly close to one of the guessed roots.

3.5   Extrapolation methods

A disadvantage of interpolation methods is the need to find sign changes in the function before carrying out the calculation. Extrapolation methods do not suffer from this problem, but this is not to say that convergence difficulties are avoided, as we shall see. Probably the most widely used of all extrapolation methods is often called the Newton-Raphson method. It bases its extrapolation procedure on the slope of the function at the guessed root.

3.5.1 Newton-Raphson method

Suppose we expand a Taylor series about a single guess at a root, x_i. For a "small" step h in the x direction the Taylor expansion is

$$f(x_{i+1}) = f(x_i + h) = f(x_i) + hf'(x_i) + \frac{h^2}{2!}f''(x_i) + \cdots \qquad (3.17)$$

Dropping terms in the expansion higher than $f'(x)$, and assuming that x_{i+1} is a root, i.e., $f(x_i + h) = 0$, we can write

$$f(x_i) + hf'(x_i) = 0 \qquad (3.18)$$

or

$$f(x_i) + (x_{i+1} - x_i)f'(x_i) = 0 \qquad (3.19)$$

After rearrangement, we can write the recursive formula

$$x_{i+1} = x_i - \frac{f(x_i)}{f'(x_i)} \qquad (3.20)$$

Obviously this has a simple graphical interpretation as shown in Figure 3.5. The extrapolation is just tangential to the function at the guessed point x_i. The new feature of the method is the need to calculate the derivative of the function, and this can present difficulties (see Chapter 5). However, for simple algebraic expressions like equation (3.2) the differentiation is easily done, and the derivative returned as a function at the end of the main program as will be shown in Program 3.4.

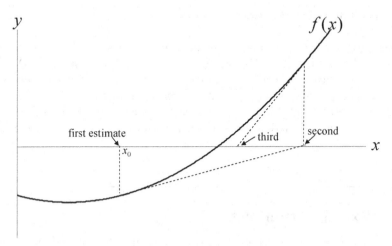

Figure 3.5: The Newton-Raphson process.

This method often has good convergence properties; however, by comparing equations (3.20) and (3.8) we can conclude that the Newton-Raphson method will converge only if

$$\left| \frac{d}{dx} \left(x - \frac{f(x)}{f'(x)} \right) \right| < 1 \tag{3.21}$$

or

$$\left| \frac{f(x)f''(x)}{(f'(x))^2} \right| < 1 \tag{3.22}$$

in the vicinity of the root.

Analysis of this expression is beyond the scope of this text, but it can be seen that the method may run into difficulties if $f(x)$ becomes large or $f'(x)$ becomes very small. With reference to Figure 3.5 if an initial guess happens to give $f'(x) \approx 0$, the tangent will run virtually parallel to the x-axis. In this case the next estimate of x would be far removed from the root and convergence slow.

Example 3.4

Use the Newton-Raphson method to find a root close to $x = 2$ of the function

$$f(x) = x^3 - x - 1 = 0$$

Solution 3.4

The Newton-Raphson formula is

$$x_{i+1} = x_i - \frac{f(x_i)}{f'(x_i)}$$

hence in this case

$$x_{i+1} = x_i - \frac{x_i^3 - x_i - 1}{3x_i^2 - 1}$$

We can use a tabular approach as follows

x	$f(x)$
2.0	5.0
1.54545	1.1458
1.35961	0.1537
1.32580	0.0046
1.32472	0.0000 etc.

displaying rapid convergence in this case. It may be noted that the column marked $f(x)$ was not strictly necessary in this case, since the recursive formula was derived explicitly in terms of x_i only.

Program 3.4: Newton-Raphson method for a single root

```
PROGRAM nm34
!---Newton-Raphson Method for a Single Root---
 USE nm_lib; USE precision; IMPLICIT NONE
 INTEGER::iters,limit; REAL(iwp)::tol,x0,x1
 OPEN(10,FILE='nm95.dat'); OPEN(11,FILE='nm95.res')
 WRITE(11,'(A)')"---Newton-Raphson Method for a Single Root---"
 READ(10,*)x0,tol,limit
 WRITE(11,'(/,A,/,E12.4)')"Guessed Starting Value",x0
 WRITE(11,'(/,A)')"First Few Iterations"; iters=0
 DO; iters=iters+1
   x1=x0-f34(x0)/f34dash(x0)
   IF(check(x1,x0,tol).OR.iters==limit)EXIT; x0=x1
   IF(iters<5)WRITE(11,'(3E12.4)')x1
 END DO
 WRITE(11,'(/,A,/,I5)')"Iterations to Convergence",iters
 WRITE(11,'(/,A,/,E12.4)')"Solution",x1
 CONTAINS

 FUNCTION f34(x)
  IMPLICIT NONE
  REAL(iwp),INTENT(IN)::x; REAL(iwp)::f34
  f34=x**3-x-1.0_iwp
  RETURN
 END FUNCTION f34

 FUNCTION f34dash(x)
  IMPLICIT NONE
  REAL(iwp),INTENT(IN)::x; REAL(iwp)::f34dash
  f34dash=3.0_iwp*x**2-1.0_iwp
  END FUNCTION f34dash

END PROGRAM nm34
```

Initial value	x0	
	1.2	
Tolerance and	tol	limit
iteration limit	1.E-5	100

Data 3.4: Newton-Raphson Method

```
---Newton-Raphson Method for a Single Root---

Guessed Starting Value
  0.1200E+01

First Few Iterations
  0.1342E+01
  0.1325E+01
  0.1325E+01

Iterations to Convergence
     4

Solution
  0.1325E+01
```

Results 3.4: Newton-Raphson Method

List 3.4:

Scalar integers:

iters	iteration counter
limit	iteration limit

Scalar reals:

tol	convergence tolerance
x0	approximate solution
x1	improved solution

Program 3.4 differs from the simple Iterative Substitution Program 3.1 in only one line, where x_i (called x0 in the program) is updated to x_{i+1} (called x1) according to equation (3.20). The function and its derivative are contained in the user-supplied functions f34 and f34dash.

When the program is run with the data from Data 3.4, convergence is now achieved in 4 iterations as shown in Results 3.4. This can be compared with the 7 iterations needed for convergence for the same problem and data using Iterative Substitution (Results 3.1).

3.5.2 A modified Newton-Raphson method

For large systems of equations, the need to be forever calculating derivatives can make the iteration process expensive. A modified method can be used (there are many other "modified" methods) in which the first evaluation of $f'(x_0)$ is used for all further extrapolations. This is shown graphically in Figure 3.6, where it can be seen that the extrapolation lines are now parallel with a gradient dictated by the derivative corresponding to the initial guessed value of x_0.

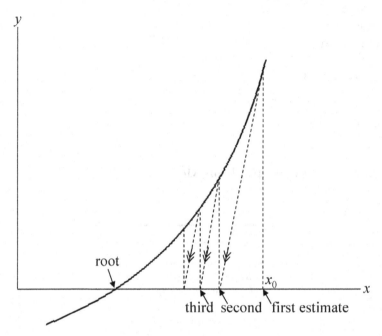

Figure 3.6: A Modified Newton-Raphson process with a constant gradient.

Example 3.5

Use the Modified Newton-Raphson method to find a root close to $x = 2$ of the function,

$$f(x) = x^3 - x - 1 = 0$$

Solution 3.5

The Modified Newton-Raphson formula is

$$x_{i+1} = x_i - \frac{f(x_i)}{f'(x_0)}$$

hence in this case

$$x_{i+1} = x_i - \frac{x_i^3 - x_i - 1}{3(2)^2 - 1}$$

$$= x_i - \frac{x_i^3 - x_i - 1}{11}$$

We can use a tabular approach as follows

x	$f(x)$
2.0	5.0
1.54545	1.14576
1.44129	0.55275
1.39104	0.30063
1.36371	0.17241
1.34801	0.10164
1.33880	0.06085
1.33327	0.03676 etc.

The convergence in this case is rather slow because the gradient at $x = 2$, which remains unchanged, is a poor approximation to the "correct" gradient at the root $x \approx 1.32472$.

Program 3.5: Modified Newton-Raphson method for a single root

```
PROGRAM nm35
!---Modified Newton-Raphson Method for a Single Root---
 USE nm_lib; USE precision; IMPLICIT NONE
 INTEGER::iters,limit; REAL(iwp)::fd,tol,x0,x1
 OPEN(10,FILE='nm95.dat'); OPEN(11,FILE='nm95.res')
 WRITE(11,'(A)')"---Newton-Raphson Method for a Single Root---"
 READ(10,*)x0,tol,limit
 WRITE(11,'(/,A,/,E12.4)')"Guessed Starting Value",x0
 WRITE(11,'(/,A)')"First Few Iterations"; iters=0; fd=f35dash(x0)
 DO; iters=iters+1
   x1=x0-f35(x0)/fd
   IF(check(x1,x0,tol).OR.iters==limit)EXIT; x0=x1
   IF(iters<5)WRITE(11,'(3E12.4)')x1
 END DO
 WRITE(11,'(/,A,/,I5)')"Iterations to Convergence",iters
 WRITE(11,'(/,A,/,E12.4)')"Solution",x1
 CONTAINS
```

```
FUNCTION f35(x)
 IMPLICIT NONE
 REAL(iwp),INTENT(IN)::x; REAL(iwp)::f35
 f35=x**3-x-1.0_iwp
 RETURN
END FUNCTION f35

FUNCTION f35dash(x)
 IMPLICIT NONE
 REAL(iwp),INTENT(IN)::x; REAL(iwp)::f35dash
 f35dash=3.0_iwp*x**2-1.0_iwp
 END FUNCTION f35dash

END PROGRAM nm35
```

List 3.5:

Scalar integers:
iters   iteration counter
limit   iteration limit

Scalar reals:
fd   holds initial gradient $f'(x_0)$
tol   convergence tolerance
x0   approximate solution
x1   improved solution

| Initial value | x0 |
| | 1.0   (or 1.2) |

| Tolerance and | tol | limit |
| iteration limit | 1.E-5 | 100 |

Data 3.5: Modified Newton-Raphson Method

```
---Newton-Raphson Method for a Single Root---

Guessed Starting Value
  0.1000E+01

First Few Iterations
  0.1500E+01
```

```
0.1062E+01
0.1494E+01
0.1074E+01
0.1492E+01
  .
  .
  .
0.1490E+01
0.1081E+01
0.1490E+01
0.1081E+01
0.1490E+01
```

```
Iterations to Convergence
  100
```

```
Solution
  0.1081E+01
```

Results 3.5a: Modified Newton-Raphson Method (first example)

```
---Newton-Raphson Method for a Single Root---
```

```
Guessed Starting Value
  0.1200E+01
```

```
First Few Iterations
  0.1342E+01
  0.1319E+01
  0.1326E+01
  0.1324E+01
```

```
Iterations to Convergence
    8
```

```
Solution
  0.1325E+01
```

Results 3.5b: Modified Newton-Raphson Method (second example)

The program is almost identical to the previous one. The function and its derivative are contained in the user-supplied functions f35 and f35dash. Unlike the full Newton-Raphson algorithm, this "modified" version calls the function f35dash once only. The initial derivative is called fd and used thereafter as a constant. The program is used to solve the familiar cubic equation, but with two different starting guesses of $x_0 = 1.0$ and $x_0 = 1.2$ as shown in Data 3.5. The results in each case are given in Results 3.5a and 3.5b respectively, illustrating one of the typical problems of Newton-Raphson type methods. It can be seen that with a starting value of 1.0 the estimated root oscillates between two constant values, and will never converge to the correct solution. On the other hand, if the starting guess is changed to 1.2, convergence occurs in 8 iterations, which can be compared with the 4 iterations required with the "full" Newton-Raphson algorithm from Program 3.4.

We can see that the process is far from automatic and that the closer to the solution the initial guess is, the better.

3.6 Acceleration of convergence

For slowly convergent iterative calculations, "Aitken's δ^2 acceleration" process is sometimes effective in extrapolating from the converging solutions to the converged result.

The method can be simply stated. If we have three estimates of a root, x_i, x_{i+1} and x_{i+2} the "δ^2" process extrapolates to the new solution

$$x_{new} = \frac{x_i x_{i+2} - x_{i+1}^2}{x_i - 2x_{i+1} + x_{i+2}} \tag{3.23}$$

3.7 Systems of nonlinear equations

For simplicity of presentation, we have so far concentrated on a single nonlinear equation. In practice we may have systems of nonlinear equations to solve simultaneously of the general form

$$f_1(x_1, x_2, \cdots, x_n) = 0$$
$$f_2(x_1, x_2, \cdots, x_n) = 0$$
$$\vdots \tag{3.24}$$
$$f_n(x_1, x_2, \cdots, x_n) = 0$$

A solution or root of such a system will consist of a set of variables of the form $(x_1, x_2, \cdots, x_n)$ and there will typically be multiple solutions that satisfy the equations.

The methods already described, namely Iterative Substitution and Newton-Raphson, can be extended to tackle systems of nonlinear equations, and these will be revisited in subsequent sections.

3.7.1 Iterative substitution for systems

The standard equations (3.24) are rearranged in the form

$$x_1 = F_1(x_1, x_2, \cdots, x_n)$$
$$x_2 = F_2(x_1, x_2, \cdots, x_n)$$
$$\vdots \qquad\qquad\qquad (3.25)$$
$$x_n = F_n(x_1, x_2, \cdots, x_n)$$

and the method proceeds by making an initial guess for all the variables $(x_1, x_2, \cdots, x_n)_0$. These values are substituted into the right hand sides of equation (3.25) yielding updated values of the variables $(x_1, x_2, \cdots, x_n)_1$ and so on. Iterations continue until none of the variables is changing by more than a specified tolerance.

A disadvantage of this method is that for most systems of nonlinear equations, several rearrangements of the type shown in equations (3.25) are possible. Some rearrangements will have convergent properties and others may not. In the case of a single equation, convergence was conditional on the magnitude of the gradient of the function $F(x)$ in the vicinity of the root as shown by equation (3.11). From similar arguments, two nonlinear equations, for example, will only be convergent if

$$\left|\frac{\partial F_1}{\partial x_1}\right| + \left|\frac{\partial F_1}{\partial x_2}\right| < 1$$
$$\left|\frac{\partial F_2}{\partial x_1}\right| + \left|\frac{\partial F_2}{\partial x_2}\right| < 1 \qquad (3.26)$$

As the number of equations increases, the restrictions on convergence become ever more stringent, limiting the value of the method in practice.

Example 3.6

Use Iterative Substitution to find a root close to $x_1 = 1$, $x_2 = 1$ of the functions

$$2x_1^2 + x_2^2 = 4.32$$
$$x_1^2 - x_2^2 = 0$$

Solution 3.6

It is recommended that the equations are first arranged in "standard form" whereby the right hand sides equals zero, thus

$$f_1(x_1, x_2) = 2x_1^2 + x_2^2 - 4.32 = 0$$
$$f_2(x_1, x_2) = x_1^2 - x_2^2 = 0$$

We now rearrange the equations such that each variable appears on the left hand side. There are several ways that this could be done, with one possible arrangement being as follows

$$x_1 = F_1(x_1, x_2) = \sqrt{2.16 - 0.5x_2^2}$$

$$x_2 = F_2(x_1, x_2) = \sqrt{x_1^2} = x_1$$

With the recommended starting guess of $x_1 = x_2 = 1.0$ we can set up a table

Iteration number	x_1	x_2
0	1.00	1.00
1	1.29	1.00
2	1.29	1.29
3	1.15	1.29
4	1.15	1.15
5	1.22	1.15
6	1.22	1.22
7	1.19	1.22

which is clearly converging on the result $x_1 = x_2 = 1.2$. In the manner of Gauss-Seidel iteration, the convergence could be improved by using updated values as soon as they become available.

Program 3.6: Iterative substitution for systems of equations

```
PROGRAM nm36
!---Iterative Substitution for Systems of Equations---
 USE nm_lib; USE precision; IMPLICIT NONE
 INTEGER::iters,limit,n; REAL(iwp)::tol
 REAL(iwp),ALLOCATABLE::x(:)
 OPEN(10,FILE='nm95.dat'); OPEN(11,FILE='nm95.res')
 WRITE(11,'(A)')                                              &
    "---Iterative Substitution for Systems of Equations---"
 READ(10,*)n; ALLOCATE(x(n)); READ(10,*)x,tol,limit
 WRITE(11,'(/,A,/,6E12.4)')"Guessed Starting Vector",x
 WRITE(11,'(/,A)')"First Few Iterations"; iters=0
 DO; iters=iters+1
   IF(checkit(f36(x),x,tol).OR.iters==limit)EXIT
   x=f36(x); IF(iters<5)WRITE(11,'(6E12.4)')x
```

```
END DO
WRITE(11,'(/,A,/,I5)')"Iterations to Convergence",iters
WRITE(11,'(/,A,/,6E12.4)')"Solution Vector",x
CONTAINS

FUNCTION f36(x)
 IMPLICIT NONE
 REAL(iwp),INTENT(IN)::x(:); REAL(iwp)::f36(UBOUND(x,1))
 f36(1)=SQRT(2.16_iwp-0.5_iwp*x(2)*x(2))
 f36(2)=x(1)
END FUNCTION f36

END PROGRAM nm36
```

List 3.6:

Scalar integers:
iters   iteration counter
limit   iteration limit
n   number of equations

Scalar reals:
tol   convergence tolerance

Dynamic real arrays:
x   vector holding old and updated variables

Number of equations	n	
	2	
Initial values	x	
	1.0	1.0
Tolerance and	tol	limit
iteration limit	1.E-5	100

Data 3.6: Iterative Substitution for Systems of Equations

```
---Iterative Substitution for Systems of Equations---

Guesssed Starting Vector
   0.1000E+01   0.1000E+01


First Few Iterations
   0.1288E+01   0.1000E+01
   0.1288E+01   0.1288E+01
   0.1153E+01   0.1288E+01
   0.1153E+01   0.1153E+01

Iterations to Convergence
   31

Solution Vector
   0.1200E+01   0.1200E+01
```

Results 3.6: Iterative Substitution for Systems of Equations

The program is essentially the same as Program 3.1 with arrays in place of simple variables. Functions $F_1, F_2, \cdots, F_n$ from equations (3.25) are provided in the user-supplied function called f36. Data and results for a problem with two equations are shown in Data and Results 3.6. Additional data n indicates the number of nonlinear equations to be solved. The library function checkit is used to check convergence of successive solution vectors. With a tolerance of tol=0.00001 convergence is achieved in 31 iterations for the starting guess $x_1 = x_2 = 1.0$.

3.7.2   Newton-Raphson for systems

This is based on Taylor expansions in several variables. For example, in the case of two equations, such as equations (3.26), suppose we expand a Taylor series about a guess at a root (x_1^i, x_2^i) [1]. For "small" steps Δx_1 in the x_1 direction and Δx_2 in the x_2 direction, the first order Taylor expansion becomes

$$f_1^{i+1} = f_1^i + \Delta x_1 \left(\frac{\partial f_1}{\partial x_1}\right)^i + \Delta x_2 \left(\frac{\partial f_1}{\partial x_2}\right)^i$$

$$f_2^{i+1} = f_2^i + \Delta x_1 \left(\frac{\partial f_2}{\partial x_1}\right)^i + \Delta x_2 \left(\frac{\partial f_2}{\partial x_2}\right)^i$$

---

[1] We have placed the iteration counter as a superscript in this section to avoid conflict with the variable number subscripts.

where

$$x_k^{i+1} = x_k^i + \Delta x_k$$

$$f_k^i = f_k(x_1^i, x_2^i)$$

$$\left.\begin{aligned}\left(\frac{\partial f_k}{\partial x_1}\right)^i &= \frac{\partial f_k(x_1^i, x_2^i)}{\partial x_1}\\[2mm]\left(\frac{\partial f_k}{\partial x_2}\right)^i &= \frac{\partial f_k(x_1^i, x_2^i)}{\partial x_2}\end{aligned}\right\} \quad k = 1, 2$$

If we assume (x_1^{i+1}, x_2^{i+1}) is a root, then $f_1^{i+1} = f_2^{i+1} = 0$ and we can write

$$f_1 + \Delta x_1 \frac{\partial f_1}{\partial x_1} + \Delta x_2 \frac{\partial f_1}{\partial x_2} = 0$$

$$f_2 + \Delta x_1 \frac{\partial f_2}{\partial x_1} + \Delta x_2 \frac{\partial f_2}{\partial x_2} = 0$$

or in matrix form

$$\begin{bmatrix} \dfrac{\partial f_1}{\partial x_1} & \dfrac{\partial f_1}{\partial x_2} \\[4mm] \dfrac{\partial f_2}{\partial x_1} & \dfrac{\partial f_2}{\partial x_2} \end{bmatrix} \begin{Bmatrix} \Delta x_1 \\ \Delta x_2 \end{Bmatrix} = \begin{Bmatrix} -f_1 \\ -f_2 \end{Bmatrix} \tag{3.27}$$

where all functions and derivatives are evaluated at (x_1^i, x_2^i).

Thus for a system of n simultaneous nonlinear equations, we must solve n simultaneous *linear* equations to find the vector of variable changes

$$\begin{Bmatrix} \Delta x_1 \\ \Delta x_2 \\ \vdots \\ \Delta x_n \end{Bmatrix} \tag{3.28}$$

which are then used to update all the variables

$$\begin{Bmatrix} x_1 \\ x_2 \\ \vdots \\ x_n \end{Bmatrix}^{i+1} = \begin{Bmatrix} x_1 \\ x_2 \\ \vdots \\ x_n \end{Bmatrix}^{i} + \begin{Bmatrix} \Delta x_1 \\ \Delta x_2 \\ \vdots \\ \Delta x_n \end{Bmatrix} \tag{3.29}$$

The process repeats itself iteratively until successive solution vectors from equation (3.19) are hardly changing as dictated by a convergence criterion.

The matrix on the left-hand side of equation (3.27) is called the "Jacobian matrix" [J] and its determinant simply "the Jacobian" . Clearly, if the Jacobian is zero, the process fails, and if it is close to zero, slow convergence may be anticipated.

For a starting guess of $x_1 = x_2 = 1.0$, the iteration matrix for the first iteration on the problem shown in Example 3.6 is

$$\begin{bmatrix} 4 & 2 \\ 2 & -2 \end{bmatrix} \begin{Bmatrix} \Delta x_1 \\ \Delta x_2 \end{Bmatrix} = \begin{Bmatrix} 1.32 \\ 0.0 \end{Bmatrix}$$

which yields $\Delta x_1 = \Delta x_2 = 0.22$ and hence an improved solution of $x_1 = x_2 = 1.22$

Example 3.7

Use the Newton-Raphson method to find a root of the equations

$$xy = 1$$
$$x^2 + y^2 = 4$$

close to $x = 1.8$ and $y = 0.5$.

Solution 3.7

First we arrange the equations in standard form with subscript notation as

$$f_1(x_1, x_2) = x_1 x_2 - 1 = 0$$
$$f_2(x_1, x_2) = x_1^2 + x_2^2 - 4 = 0$$

Form the Jacobian matrix by differentiation according to equation (3.17)

$$[J] = \begin{bmatrix} x_2 & x_1 \\ 2x_1 & 2x_2 \end{bmatrix}$$

In a small system of equations such as this it is convenient to simply invert the Jacobian matrix, thus

$$[J]^{-1} = \frac{1}{2x_2^2 - 2x_1^2} \begin{bmatrix} 2x_2 & -x_1 \\ -2x_1 & x_2 \end{bmatrix}$$

1st iteration

Initial guess, $x_1 = 1.8$, $x_2 = 0.5$, hence

$$[J]^{-1} = \begin{bmatrix} -0.1672 & 0.3010 \\ 0.6020 & -0.0836 \end{bmatrix} \quad and \quad \begin{Bmatrix} -f_1 \\ -f_2 \end{Bmatrix} = \begin{Bmatrix} 0.1 \\ 0.51 \end{Bmatrix}$$

Thus,

$$\begin{Bmatrix} \Delta x_1 \\ \Delta x_2 \end{Bmatrix} = \begin{bmatrix} -0.1672 & 0.3010 \\ 0.6020 & -0.0836 \end{bmatrix} \begin{Bmatrix} 0.1 \\ 0.51 \end{Bmatrix} = \begin{Bmatrix} 0.1368 \\ 0.0176 \end{Bmatrix}$$

2nd iteration

Updated values

$$x_1 = 1.8 + 0.1368 = 1.9368$$
$$x_2 = 0.5 + 0.0176 = 0.5176$$

hence

$$[\mathbf{J}]^{-1} = \begin{bmatrix} -0.1486 & 0.2780 \\ 0.5560 & -0.0743 \end{bmatrix} \quad and \quad \begin{Bmatrix} -f_1 \\ -f_2 \end{Bmatrix} = \begin{Bmatrix} -0.0025 \\ -0.0191 \end{Bmatrix}$$

Thus,

$$\begin{Bmatrix} \Delta x_1 \\ \Delta x_2 \end{Bmatrix} = \begin{bmatrix} -0.1486 & 0.2780 \\ 0.5560 & -0.0743 \end{bmatrix} \begin{Bmatrix} -0.0025 \\ -0.0191 \end{Bmatrix} = \begin{Bmatrix} -0.0049 \\ 0.0000 \end{Bmatrix}$$

3rd iteration

Updated values

$$x_1 = 1.9368 - 0.0049 = 1.9319$$
$$x_2 = 0.5176 + 0.0000 = 0.5176 \quad etc.$$

Note how $\Delta x_1, \Delta x_2 \longrightarrow 0$ as iterations proceed and the root is approached.

Program 3.7: Newton-Raphson for systems of equations

```
PROGRAM nm37
!---Newton-Raphson for Systems of Equations---
USE nm_lib; USE precision; IMPLICIT NONE
INTEGER::iters,limit,n; REAL(iwp)::tol
REAL(iwp),ALLOCATABLE::x0(:),x1(:)
OPEN(10,FILE='nm95.dat'); OPEN(11,FILE='nm95.res')
WRITE(11,'(A)')"---Newton-Raphson for Systems of Equations---"
READ(10,*)n; ALLOCATE(x0(n),x1(n)); READ(10,*)x0,tol,limit
WRITE(11,'(/,A,/,6E12.4)')"Guessed Starting Vector",x0
WRITE(11,'(/,A)')"First Few Iterations"; iters=0
DO; iters=iters+1
  x1=x0-MATMUL(inverse(f37dash(x0)),f37(x0))
  IF(checkit(x1,x0,tol).OR.iters==limit)EXIT; x0=x1
  IF(iters<5)WRITE(11,'(6E12.4)')x0
```

```
END DO
WRITE(11,'(/,A,/,I5)')"Iterations to Convergence",iters
WRITE(11,'(/,A,/,6E12.4)')"Solution Vector",x0
CONTAINS

FUNCTION f37(x)
 IMPLICIT NONE
 REAL(iwp),INTENT(IN)::x(:); REAL(iwp)::f37(UBOUND(x,1))
 f37(1)=2.0_iwp*x(1)*x(1)+x(2)*x(2)-4.32_iwp
 f37(2)=x(1)*x(1)-x(2)*x(2)
END FUNCTION f37

FUNCTION f37dash(x)
 IMPLICIT NONE
 REAL(iwp),INTENT(IN)::x(:)
 REAL(iwp)::f37dash(UBOUND(x,1),UBOUND(x,1))
 f37dash(1,1) = 4.0_iwp*x(1); f37dash(2,1) = 2.0_iwp*x(1)
 f37dash(1,2) = 2.0_iwp*x(2); f37dash(2,2) =-2.0_iwp*x(2)
END FUNCTION f37dash

END PROGRAM nm37
```

List 3.7:

Scalar integers:
iters   iteration counter
limit   iteration limit
n   number of equations

Scalar reals:
tol   convergence tolerance

Dynamic real arrays:
x0   vector holding old variables
x1   vector holding updated variables

```
Number of equations        n
                           2

Initial values             x0
                           1.0  1.0
```

```
Tolerance and              tol    limit
iteration limit            1.E-5  100
```

Data 3.7: Newton-Raphson for Systems of Equations

```
---Newton-Raphson for Systems of Equations---

Guessed Starting Vector
  0.1000E+01  0.1000E+01

First Few Iterations
  0.1220E+01  0.1220E+01
  0.1200E+01  0.1200E+01
  0.1200E+01  0.1200E+01

Iterations to Convergence
     4

Solution Vector
  0.1200E+01  0.1200E+01
```

Results 3.7: Newton-Raphson for Systems of Equations

The input data and output results are given in Data and Results 3.7 respectively. The number of equations n is provided followed by an initial guess of the solution vector x0 and the tolerance and iteration limit. The actual functions and their derivatives are contained in the user-supplied functions f37 and f37dash. The program assumes that the Jacobian matrix is small enough to be inverted by the library function inverse. Note that for large systems, inverse should be replaced by an appropriate solution algorithm from Chapter 2. Results 3.7 shows that convergence was achieved in this simple case in 4 iterations.

3.7.3 Modified Newton-Raphson method for systems

In the full Newton-Raphson algorithm, the presence of function inverse, or an alternative equation solver if n becomes large, means that substantial computation is required at each iteration. Furthermore, since the coefficient matrix is typically nonsymmetric and changes from one iteration to the next, no advantage can be taken of factorization strategies. Modified Newton-Raphson strategies seek to reduce the amount of work performed at each iteration, but at the expense of needing more iterations for convergence. For example, instead of updating and inverting the Jacobian matrix every iteration, we could update it periodically, every m^{th} iteration say.

An extreme, but very simple form of modification is where we compute and invert the Jacobian matrix once only, corresponding to the initial guess.

This was the method described in Program 3.5 for a single equation, and is presented again here as Program 3.8 for systems of equations. In general, it will require more iterations than the full Newton-Raphson procedure. It will be a question in practice of whether the increase in iterations in the modified method more than compensates for the reduction in equation solving.

Example 3.8

Use the Modified Newton-Raphson method to find a root of the equations

$$e^x + y = 0$$
$$\cosh y - x = 3.5$$

close to $x = -2.4$ and $y = -0.1$.

Solution 3.8

First we arrange the equations in standard form with subscript notation as

$$f_1(x_1, x_2) = e^{x_1} + x_2 = 0$$
$$f_2(x_1, x_2) = -x_1 + \cosh x_2 - 3.5 = 0$$

Form the Jacobian matrix by differentiation according to equation (3.27)

$$[\mathbf{J}] = \begin{bmatrix} e^{x_1} & 1 \\ -1 & \sinh x_2 \end{bmatrix}$$

In a small system of equations such as this it is convenient to simply invert the Jacobian matrix, thus

$$[\mathbf{J}]^{-1} = \frac{1}{1 + e^{x_1} \sinh x_2} \begin{bmatrix} \sinh x_2 & -1 \\ 1 & e^{x_1} \end{bmatrix}$$

Initial guess, $x_1 = -2.4$, $x_2 = -0.1$, hence

$$[\mathbf{J}]^{-1} = \begin{bmatrix} -0.1011 & -1.0092 \\ 1.0092 & 0.0915 \end{bmatrix}$$

In the modified version of Newton-Raphson described in this text, $[\mathbf{J}]^{-1}$ remains constant.

1st iteration

Initial guess, $x_1 = -2.4$, $x_2 = -0.1$, hence

$$\begin{Bmatrix} -f_1 \\ -f_2 \end{Bmatrix} = \begin{Bmatrix} -e^{x_1} - x_2 \\ x_1 - \cosh x_2 \end{Bmatrix} = \begin{Bmatrix} 0.0093 \\ 0.0950 \end{Bmatrix}$$

Thus,

$$\left\{ \begin{array}{c} \Delta x_1 \\ \Delta x_2 \end{array} \right\} = \left[\begin{array}{cc} -0.1011 & -1.0092 \\ 1.0092 & 0.0915 \end{array} \right] \left\{ \begin{array}{c} 0.0093 \\ 0.0950 \end{array} \right\} = \left\{ \begin{array}{c} -0.0968 \\ 0.0181 \end{array} \right\}$$

2nd *iteration*

Updated values

$$x_1 = -2.4 - 0.0968 = -2.4968$$
$$x_2 = 0.1 + 0.0181 = -0.0819$$

hence

$$\left\{ \begin{array}{c} -f_1 \\ -f_2 \end{array} \right\} = \left\{ \begin{array}{c} -e^{x_1} - x_2 \\ x_1 - \cosh x_2 \end{array} \right\} = \left\{ \begin{array}{c} -0.0004 \\ -0.0002 \end{array} \right\}$$

Thus,

$$\left\{ \begin{array}{c} \Delta x_1 \\ \Delta x_2 \end{array} \right\} = \left[\begin{array}{cc} -0.1011 & -1.0092 \\ 1.0092 & 0.0915 \end{array} \right] \left\{ \begin{array}{c} -0.0004 \\ -0.0002 \end{array} \right\} = \left\{ \begin{array}{c} 0.0002 \\ -0.0004 \end{array} \right\}$$

3rd *iteration*

Updated values

$$x_1 = -2.4968 + 0.0002 = 1.9319$$
$$x_2 = -0.0819 - 0.0004 = 0.5176 \quad etc.$$

Program 3.8: Modified Newton-Raphson for systems of equations

```
PROGRAM nm38
!---Modified Newton-Raphson for Systems of Equations---
USE nm_lib; USE precision; IMPLICIT NONE
INTEGER::iters,limit,n; REAL(iwp)::tol
REAL(iwp),ALLOCATABLE::inv(:,:),x0(:),x1(:)
OPEN(10,FILE='nm95.dat'); OPEN(11,FILE='nm95.res')
WRITE(11,'(A)')                                                    &
   "---Modified Newton-Raphson for Systems of Equations---"
READ(10,*)n; ALLOCATE(x0(n),x1(n),inv(n,n))
READ(10,*)x0,tol,limit
```

```
WRITE(11,'(/,A,/,6E12.4)')"Guessed Starting Vector",x0
WRITE(11,'(/,A)')"First Few Iterations"; iters=0
inv=inverse(f38dash(x0))
DO; iters=iters+1
  x1=x0-MATMUL(inv,f38(x0))
  IF(checkit(x1,x0,tol).OR.iters==limit)EXIT; x0=x1
  IF(iters<5)WRITE(11,'(6E12.4)')x0
END DO
WRITE(11,'(/,A,/,I5)')"Iterations to Convergence",iters
WRITE(11,'(/,A,/,6E12.4)')"Solution Vector",x0
CONTAINS

FUNCTION f38(x)
 IMPLICIT NONE
 REAL(iwp),INTENT(IN)::x(:); REAL(iwp)::f38(UBOUND(x,1))
 f38(1)=2.0_iwp*x(1)*x(1)+x(2)*x(2)-4.32_iwp
 f38(2)=x(1)*x(1)-x(2)*x(2)
END FUNCTION f38

FUNCTION f38dash(x)
 IMPLICIT NONE
 REAL(iwp),INTENT(IN)::x(:)
 REAL(iwp)::f38dash(UBOUND(x,1),UBOUND(x,1))
 f38dash(1,1)=4.0_iwp*x(1); f38dash(1,2)= 2.0_iwp*x(2)
 f38dash(2,1)=2.0_iwp*x(1); f38dash(2,2)=-2.0_iwp*x(2)
END FUNCTION f38dash

END PROGRAM nm38
```

List 3.8:

Scalar integers:
iters   iteration counter
limit   iteration limit
n   number of equations

Scalar reals:
tol   convergence tolerance

Dynamic real arrays:
inv   holds inverse of the Jacobian matrix evaluated with initial guess
x0   vector holding old variables
x1   vector holding updated variables

Number of equations	n
	2

Initial values	x0
	1.0 1.0

Tolerance and	tol	limit
iteration limit	1.E-5	100

Data 3.8: Modified Newton-Raphson for Systems of Equations

---Modified Newton-Raphson for Systems of Equations---

Guessed Starting Vector
  0.1000E+01 0.1000E+01

First Few Iterations
  0.1220E+01 0.1220E+01
  0.1196E+01 0.1196E+01
  0.1201E+01 0.1201E+01
  0.1200E+01 0.1200E+01

Iterations to Convergence
  7

Solution Vector
  0.1200E+01 0.1200E+01

Results 3.8: Modified Newton-Raphson for Systems of Equations

The only difference from Program 3.7 is that **inverse** is moved outside the iteration loop. When run with the data given in Data 3.8 it produces the results shown in Results 3.8. In this simple case, convergence is reached in 7 iterations using the modified method which can be compared with 4 iterations using the full Newton-Raphson approach with Program 3.7

3.8   Exercises

1. Find a root of the equation $x^4 - 8x^3 + 23x^2 + 16x - 50 = 0$ in the vicinity of 1.0 by Iterative Substitution.
   Answer: 1.4142

2. Find the root of the equation $x^3 - 3x^2 + 2x - 0.375 = 0$ in the vicinity of 1.0 by Iterative Substitution in the form $x = [(x^3 + 2x - 0.375)/3]^{1/2}$.
   Answer: 0.5 in 95 iterations for a tolerance of 1×10^{-5}.

3. Find a root of the equation in Exercise 1 in the range $1.0 < x < 2.0$ by the Bisection Method.
   Answer: 1.4142 in 17 iterations for a tolerance of 1×10^{-5}.

4. Find a root of the equation in Exercise 1 in the range $1.0 < x < 2.0$ by the False Position Method.
   Answer: 1.4142 in 3 iterations for a tolerance of 1×10^{-5}.

5. Find a root of the equation in Exercise 2 in the range $0.35 < x < 1.0$ by the Bisection Method.
   Answer: 0.5 in 17 iterations for a tolerance of 1×10^{-5}.

6. Find a root of the equation in Exercise 2 in the range $0.4 < x < 0.6$ by the False Position Method.
   Answer: 0.5 in 12 iterations for a tolerance of 1×10^{-5}.

7. Find a root of the equation in Exercise 1 in the vicinity of 1.0 by the Newton-Raphson Method.
   Answer: 1.4142 in 3 iterations for a tolerance of 1×10^{-5}.

8. Find a root of the equation in Exercise 2 in the vicinity of 1.0 by the Newton-Raphson Method.
   Answer: 0.5 in 6 iterations for a tolerance of 1×10^{-5}.

9. Find a root of the equation in Exercise 1 in the vicinity of 1.0 by the Modified Newton-Raphson Method.
   Answer: 1.4142 in 5 iterations for a tolerance of 1×10^{-5}.

10. Find a root of the equation in Exercise 2 in the vicinity of 1.0 by the Modified Newton-Raphson Method.
    Answer: 0.5 in 30 iterations for a tolerance of 1×10^{-5}.

11. Solve the equations

$$x_1 + x_2 - x_2^{1/2} - 0.25 = 0$$
$$8x_1^2 + 16x_2 - 8x_1x_2 - 5 = 0$$

by Iterative Substitution from a starting guess $x_1 = x_2 = 1$.
Answer: $x_1 = 0.5$ and $x_2 = 0.25$ in 11 iterations for a tolerance of 1×10^{-5}.

12. Solve the equations

$$2x_1^2 - 4x_1x_2 - x_2^2 = 0$$
$$2x_2^2 + 10x_1 - x_1^2 - 4x_1x_2 - 5 = 0$$

by the Newton-Raphson procedure starting from $x_1 = x_2 = 1$.
Answer: $x_1 = 0.58$ and $x_2 = 0.26$ in 5 iterations for a tolerance of 1×10^{-5}.

13. Solve the equations in Exercise 12 by the Modified Newton-Raphson procedure starting from (a) $x_1 = x_2 = 1$ and (b) $x_1 = 0.5$, $x_2 = 0.25$. *Answer:* $x_1 = 0.58$, $x_2 = 0.26$ in (a) 1470 iterations and (b) 6 iterations for a tolerance of 1×10^{-5}.

14. Solve the equations in Exercise 11 starting from $x_1 = 1.0$, $x_2 = 0.1$ by
    (a) the Newton-Raphson procedure and
    (b) the Modified Newton-Raphson procedure
    Answer: $x_1 = 0.5$, $x_2 = 0.25$ in (a) 5 iterations and (b) 20 iterations for a tolerance of 1×10^{-5}.

15. Estimate a root of the following equations close to $\theta = 0.5$ and $\alpha = 0.5$ using the Modified Newton-Raphson Method.

$$3\cos\theta + \sin\alpha = 3.1097$$

$$\alpha\tan\theta = 0.2537$$

Answer: $\theta = 0.488$, $\alpha = 0.478$

16. Estimate a root of the following equations close to $x = 0.5$ and $y = 2.5$ using the Modified Newton-Raphson Method.

$$x\sin y + x^2\cos y = 0$$

$$e^x + xy - 2 = 0$$

Answer: Solution is $x = 0.248$, $y = 2.899$
After two iterations you should have $x = 0.249$, $y = 2.831$

17. The depth of embedment d of a sheet-pile wall is governed by the equation:

$$d = (d^3 + 2.87d^2 - 10.28)/4.62$$

An engineer has estimated the correct depth to be $d = 2.5$. Use the Modified Newton Raphson Method to improve on this estimate.

Answer: 2.002

18. An engineer has estimated the critical depth of flow down an open channel, y_c, to be in the range 4-5 ft. If the governing equation is given by:

$$\frac{Q^2 b}{g A^3} = 1$$

where:

Q	=	Flow rate	=	$520 \text{ ft}^3/\text{s}$
g	=	Gravity	=	32.2 ft/s^2
A	=	Area of flow	=	$(5 + 2y_c)y_c \text{ ft}^2$
b	=	Surface width	=	$5 + 4y_c \text{ ft}$

Improve on this estimate, correct to one decimal place, using the False Position method.

Answer: 4.2

19. The rate of fluid flow down a sloping channel of rectangular cross-section is given by the equation:

$$Q = \frac{S^{1/2}(BH)^{5/3}}{n(B + 2H)^{2/3}}$$

where:

Q	=	Flow rate $(5 \text{ m}^3/\text{s})$
B	=	Channel width (20 m)
H	=	Depth of flow
S	=	Slope gradient (0.0002)
n	=	Roughness (0.03)

An engineer has estimated that the depth of flow must lie in the range $0.6 \text{ m} < \text{H} < 0.9 \text{ m}$. Use any suitable method to estimate the actual depth of flow.

Answer: $H = 0.7023$m

20. The Hazen-Williams equation governing gravity flow down circular pipes is of the form:

$$Q = 0.281CD^{2.63}S^{0.54}$$

where, assuming use of a consistent system of units

Q	=	flow rate
C	=	pipe roughness coefficient
D	=	pipe diameter
S	=	pipe gradient

A particular pipe has a flow rate of 273, a roughness coefficient of 110 and a gradient of 0.0023. The diameter of the pipe is estimated to be about 10. Perform a few iterations of the Newton-Raphson method to improve this estimate.

Answer: 7.969 after 4 iterations with a tolerance of 1×10^{-5}

21. The vertical stress σ_z generated at point X in an elastic continuum under the edge of a strip footing supporting a uniform pressure q shown in Figure 3.7 is given by Boussinesq's formula to be:

$$\sigma_z = \frac{q}{\pi}\{\alpha + \sin\alpha\cos\alpha\}$$

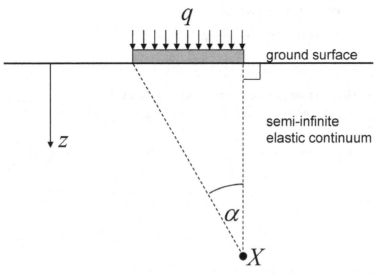

Figure 3.7

Use the Modified Newton-Raphson Method to estimate the value of α at which the vertical stress σ_z will be 25% of the footing stress q. Use an initial guess of $\alpha = 0.4$. Note that α must be expressed in radians.

Answer: 0.4159

22. The displacement/time function for free vibration of a simple damped oscillator is given by:

$$u(t) = e^{-4.95t}[7.48\sin 13.23t + 0.05\cos 13.23t]$$

Use the False Position method to estimate the time in the range $0.2 < t < 0.3$ at which the oscillator first passes through its equilibrium position at $u = 0.0$.

Answer: 0.237 secs

23. The depth d of submergence of a solid pine sphere of radius $r = 100$, floating in water, is given by the equation:

$$\frac{\pi(2552 - 30d^2 + d^3)}{3} = 0$$

Within what range must the required root lie? Use the Bisection method with appropriate initial guesses to compute d to the nearest whole number.

Answer: 12 (exact 11.8615)

24. Find the root of the equation:

$$2x(1 - x^2 + x)\ln x = x^2 - 1$$

in the interval $[0.1, 0.5]$ correct to two decimal places.

Answer: 0.33

Chapter 4

Eigenvalue Equations

4.1 Introduction

Equations of the type

$$[\mathbf{A}]\{\mathbf{x}\} = \lambda\{\mathbf{x}\} \tag{4.1}$$

often occur in practice, for example in the analysis of structural stability or the natural frequencies of vibrating systems. We have to find a vector $\{\mathbf{x}\}$ which, when multiplied by $[\mathbf{A}]$, yields a scalar multiple of itself. This multiple λ is called an "eigenvalue" or "characteristic value" of $[\mathbf{A}]$ and we shall see there are n of these for a matrix of order n. Physically they might represent frequencies of oscillation. There are also n vectors $\{\mathbf{x}\}$, one associated with each of the eigenvalues λ. These are called "eigenvectors" or "characteristic vectors" . Physically they might represent the mode shapes of oscillation.

A specific example of equation (4.1) might be

$$\begin{bmatrix} 16 & -24 & 18 \\ 3 & -2 & 0 \\ -9 & 18 & -17 \end{bmatrix} \begin{Bmatrix} x_1 \\ x_2 \\ x_3 \end{Bmatrix} = \lambda \begin{Bmatrix} x_1 \\ x_2 \\ x_3 \end{Bmatrix} \tag{4.2}$$

which can be rewritten

$$\begin{bmatrix} 16 - \lambda & -24 & 18 \\ 3 & -2 - \lambda & 0 \\ -9 & 18 & -17 - \lambda \end{bmatrix} \begin{Bmatrix} x_1 \\ x_2 \\ x_3 \end{Bmatrix} = \begin{Bmatrix} 0 \\ 0 \\ 0 \end{Bmatrix} \tag{4.3}$$

A nontrivial solution of this set of linear simultaneous equations is only possible if the determinant of the coefficients is zero

$$\begin{vmatrix} 16 - \lambda & -24 & 18 \\ 3 & -2 - \lambda & 0 \\ -9 & 18 & -17 - \lambda \end{vmatrix} = 0 \tag{4.4}$$

Expanding the determinant gives

$$\lambda^3 + 3\lambda^2 - 36\lambda + 32 = 0 \tag{4.5}$$

which is called the "characteristic polynomial". Clearly one way of solving eigenvalue equations would be to reduce them to an nth degree characteristic polynomial and use the methods of the previous chapter to find its roots. This is sometimes done, perhaps as part of a total solution process, but on its own is not usually the best means of solving eigenvalue equations.

In the case of equation (4.5) the characteristic polynomial has simple factors

$$(\lambda - 4)(\lambda - 1)(\lambda + 8) = 0 \qquad (4.6)$$

and so the eigenvalues of our matrix are 4, 1 and -8.

Note that for arbitrary matrices $[\mathbf{A}]$ the characteristic polynomial is likely to yield imaginary as well as real roots. We shall restrict our discussion to matrices with real eigenvalues and indeed physical constraints will often mean that matrices of interest are symmetrical and "positive definite" (see Chapter 2) in which case all the eigenvalues are real and positive.

Having found an eigenvalue, its associated eigenvector can, in principle, be found by solving a set of linear simultaneous equations. For example, for the case of $\lambda = 1$ substituted into equation (4.3) we get

$$\begin{bmatrix} 15 & -24 & 18 \\ 3 & -3 & 0 \\ -9 & 18 & -18 \end{bmatrix} \begin{Bmatrix} x_1 \\ x_2 \\ x_3 \end{Bmatrix} = \begin{Bmatrix} 0 \\ 0 \\ 0 \end{Bmatrix} \qquad (4.7)$$

Carrying out the first stage of a Gaussian elimination gives

$$\begin{bmatrix} 15.0 & -24.0 & 18.0 \\ 0.0 & 1.8 & -3.6 \\ 0.0 & 3.6 & -7.2 \end{bmatrix} \begin{Bmatrix} x_1 \\ x_2 \\ x_3 \end{Bmatrix} = \begin{Bmatrix} 0 \\ 0 \\ 0 \end{Bmatrix} \qquad (4.8)$$

As we knew already from the zero determinant, the system of equations exhibits linear dependence and we have fewer equations than unknowns. We can however see from the second and third equations of (4.8) that the ratio of x_2 to x_3 is $2:1$ and by substitution in the first equation of (4.8) that the ratio of x_1 to x_2 is $1:1$. So any vector with the ratios $x_1 : x_2 : x_3 = 2:2:1$ is an eigenvector associated with the eigenvalue $\lambda = 1$. Similar operations could be used to find the eigenvectors corresponding to the other two eigenvalues.

4.1.1   Orthogonality and normalization of eigenvectors

The previous section showed that eigenvectors have unique directions, but their magnitudes are arbitrary. This is easily shown by multiplying both sides of equation (4.1) by an arbitrary constant α to give

$$[\mathbf{A}]\alpha\{\mathbf{x}\} = \lambda\alpha\{\mathbf{x}\} \qquad (4.9)$$

In this modified equation, $\alpha\{\mathbf{x}\}$ and $\{\mathbf{x}\}$ are equally valid eigenvectors corresponding to eigenvalue λ, of array $[\mathbf{A}]$. Two methods of eigenvector normalization are commonly used. Consider the symmetrical matrix

$$[\mathbf{A}] = \begin{bmatrix} 4 & \frac{1}{2} & 0 \\ \frac{1}{2} & 4 & \frac{1}{2} \\ 0 & \frac{1}{2} & 4 \end{bmatrix} \tag{4.10}$$

which has eigenvalues

$$\lambda_1 = 4 + \frac{1}{\sqrt{2}}, \quad \lambda_2 = 4, \quad \lambda_3 = 4 - \frac{1}{\sqrt{2}} \tag{4.11}$$

with corresponding eigenvectors

$$\{\mathbf{x}_1\} = \left\{ \begin{array}{c} 1 \\ \sqrt{2} \\ 1 \end{array} \right\}, \quad \{\mathbf{x}_2\} = \left\{ \begin{array}{c} 1 \\ 0 \\ -1 \end{array} \right\}, \quad \{\mathbf{x}_3\} = \left\{ \begin{array}{c} 1 \\ -\sqrt{2} \\ 1 \end{array} \right\} \tag{4.12}$$

A convenient form of eigenvector normalization involves scaling the length, otherwise known as the "L2 norm" or "Euclidean norm" to unity. Thus for an eigenvector of n components, we scale the vector such that

$$\left(\sum_{i=1}^{n} x_i^2 \right)^{1/2} = 1 \tag{4.13}$$

which is easily achieved by dividing all the components of an unnormalized eigenvector by the square root of the sum of the squares of all the components.

Thus, for the first eigenvector from (4.12) we would divide each component by a factor given by

$$\sqrt{1^2 + (\sqrt{2})^2 + 1^2} = 2 \tag{4.14}$$

This, together with similar operations on the other two eigenvectors, leads to the normalized eigenvectors

$$\{\mathbf{x}_1\} = \left\{ \begin{array}{c} \frac{1}{2} \\ \frac{1}{\sqrt{2}} \\ \frac{1}{2} \end{array} \right\}, \quad \{\mathbf{x}_2\} = \left\{ \begin{array}{c} \frac{1}{\sqrt{2}} \\ 0 \\ -\frac{1}{\sqrt{2}} \end{array} \right\}, \quad \{\mathbf{x}_3\} = \left\{ \begin{array}{c} \frac{1}{2} \\ -\frac{1}{\sqrt{2}} \\ \frac{1}{2} \end{array} \right\} \tag{4.15}$$

Eigenvectors of symmetrical matrices are said to exhibit "orthogonality" one to the other. That is, the dot product of any two different eigenvectors from the same matrix will equal zero, while the dot product of an eigenvector with itself gives its length squared. An attractive feature of setting the Euclidean norm to unity, therefore, is that the eigenvector dot products either equal zero or unity, and can be summarized as

$$\{\mathbf{x}_i\}^T\{\mathbf{x}_j\} = \begin{cases} 1 \text{ for } i = j \\ 0 \text{ for } i \neq j \end{cases} \tag{4.16}$$

A simple alternative form of eigenvector normalization is to divide through by the component with the largest (absolute) magnitude. This results in a normalized eigenvector whose *largest* component equals unity. Returning again to the eigenvectors of equations (4.12), the first eigenvector would be divided through by $\sqrt{2}$, the second eigenvector would be unaltered because its largest term is already equal to unity and the third eigenvector would be divided through by $-\sqrt{2}$ leading to the normalized vectors

$$\{\mathbf{x}_1\} = \left\{ \begin{array}{c} \frac{1}{\sqrt{2}} \\ 1 \\ \frac{1}{\sqrt{2}} \end{array} \right\}, \quad \{\mathbf{x}_2\} = \left\{ \begin{array}{c} 1 \\ 0 \\ -1 \end{array} \right\}, \quad \{\mathbf{x}_3\} = \left\{ \begin{array}{c} -\frac{1}{\sqrt{2}} \\ 1 \\ -\frac{1}{\sqrt{2}} \end{array} \right\} \tag{4.17}$$

4.1.2   Properties of eigenvalues and eigenvectors

The numerical methods described in this chapter make use of theorems that describe the influence of various matrix transformation on the eigenvalues and eigenvectors. Some of the most useful of these relationships are summarized here as a reference.

- If an $n \times n$ matrix $[\mathbf{A}]$ has real eigenvalues $\lambda_1, \lambda_2, \cdots, \lambda_n$, $[\mathbf{I}]$ is the unit matrix and p is a scalar shift, then

$[\mathbf{A}][\mathbf{A}]$ ⠀⠀⠀⠀⠀ has eigenvalues ⠀ $\lambda_1^2, \lambda_2^2, \cdots, \lambda_n^2$

$[\mathbf{A}] + p[\mathbf{I}]$ ⠀⠀⠀ has eigenvalues ⠀ $\lambda_1 + p, \lambda_2 + p, \cdots, \lambda_n + p$

$[\mathbf{A}]^{-1}$ ⠀⠀⠀⠀⠀ has eigenvalues ⠀ $\dfrac{1}{\lambda_1}, \dfrac{1}{\lambda_2}, \cdots, \dfrac{1}{\lambda_n}$

$[[\mathbf{A}] - p[\mathbf{I}]]^{-1}$ ⠀ has eigenvalues ⠀ $\dfrac{1}{\lambda_1 - p}, \dfrac{1}{\lambda_2 - p}, \cdots, \dfrac{1}{\lambda_n - p}$

All these cases have the same eigenvectors.

- The sum of the diagonals of an $n \times n$ matrix is called the "trace" and is also equal to the sum of the eigenvalues, thus

$$\mathrm{tr}[\mathbf{A}] = \sum_{i=1}^{n} a_{ii} = \sum_{i=1}^{n} \lambda_i \tag{4.18}$$

- The product of the eigenvalues of an $n \times n$ matrix equals the determinant of the matrix, thus

$$\det[\mathbf{A}] = \lambda_1 \lambda_2 \cdots \lambda_n \tag{4.19}$$

- The diagonals of an $n \times n$ triangular matrix are the eigenvalues.

- Given two square matrices $[\mathbf{A}]$ and $[\mathbf{B}]$.
  (a) $[\mathbf{A}][\mathbf{B}]$ and $[\mathbf{B}][\mathbf{A}]$ will have the same eigenvalues.
  (b) If $\{\mathbf{x}\}$ is an eigenvector of $[\mathbf{A}][\mathbf{B}]$, $[\mathbf{B}]\{\mathbf{x}\}$ is an eigenvector of $[\mathbf{B}][\mathbf{A}]$.
  This is shown by writing the equation

$$[\mathbf{A}][\mathbf{B}]\{\mathbf{x}\} = \lambda\{\mathbf{x}\} \tag{4.20}$$

and then premultiplying both sides by $[\mathbf{B}]$ to give

$$[\mathbf{B}][\mathbf{A}][\mathbf{B}]\{\mathbf{x}\} = \lambda[\mathbf{B}]\{\mathbf{x}\} \tag{4.21}$$

- Given two square matrices $[\mathbf{A}]$ and $[\mathbf{P}]$, where $[\mathbf{P}]$ is nonsingular.
  (a) $[\mathbf{A}]$ and $[\mathbf{P}]^{-1}[\mathbf{A}][\mathbf{P}]$ have the same eigenvalues.
  (b) If $\{\mathbf{x}\}$ is an eigenvector of $[\mathbf{A}]$, $[\mathbf{P}]^{-1}\{\mathbf{x}\}$ is an eigenvector of $[\mathbf{P}]^{-1}[\mathbf{A}][\mathbf{P}]$.
  This is shown by writing the equation

$$[\mathbf{A}]\{\mathbf{x}\} = \lambda\{\mathbf{x}\} \tag{4.22}$$

and then premultiplying both sides by $[\mathbf{P}]^{-1}$ to give

$$[\mathbf{P}]^{-1}[\mathbf{A}]\{\mathbf{x}\} = \lambda[\mathbf{P}]^{-1}\{\mathbf{x}\} \tag{4.23}$$

Insert the identity matrix in the form $[\mathbf{P}][\mathbf{P}]^{-1} = [\mathbf{I}]$ to give

$$[\mathbf{P}]^{-1}[\mathbf{A}][\mathbf{P}][\mathbf{P}]^{-1}\{\mathbf{x}\} = \lambda[\mathbf{P}]^{-1}\{\mathbf{x}\} \tag{4.24}$$

- If $[\mathbf{A}]$ is an $n \times n$ matrix with eigenvalues and eigenvectors $\lambda_1\{\mathbf{x}_1\}$, $\lambda_2\{\mathbf{x}_2\}$, $\cdots$, $\lambda_n\{\mathbf{x}_n\}$ and $[\mathbf{P}]$ is a matrix whose columns are the eigenvectors of $[\mathbf{A}]$ scaled so that their Euclidean norms equal unity, then

$$[\mathbf{P}]^{-1}[\mathbf{A}][\mathbf{P}] = \begin{bmatrix} \lambda_1 & 0 & 0 & \cdots & 0 \\ 0 & \lambda_2 & 0 & \cdots & 0 \\ \vdots & \vdots & \vdots & \vdots & \vdots \\ 0 & \cdots & 0 & \lambda_{n-1} & 0 \\ 0 & \cdots & 0 & 0 & \lambda_n \end{bmatrix} \tag{4.25}$$

4.1.3   Solution methods for eigenvalue equations

Because of the presence of the unknown vector $\{\mathbf{x}\}$ on both sides of equation (4.1) we can see that solution methods for eigenvalue problems will be essentially iterative in character. We have already seen one such method, which involved finding the roots of the characteristic polynomial. A second class of methods comprises "transformation methods" in which the matrix $[\mathbf{A}]$ in equation (4.1) is iteratively transformed into a new matrix, say $[\mathbf{A}^*]$, which has the same eigenvalues as $[\mathbf{A}]$. However, these eigenvalues are easier to compute than the eigenvalues of the original matrix. A third class of methods comprises "vector iteration" methods which are perhaps the most obvious of all. Just as we did in Chapter 3 for iterative substitution in the solution of nonlinear equations, a guess is made for $\{\mathbf{x}\}$ on the left-hand side of equation (4.1), the product $[\mathbf{A}]\{\mathbf{x}\}$ is formed, and compared with the right-hand side. The guess is then iteratively adjusted until agreement is reached. In the following sections we shall deal with these classes of methods in reverse order, beginning with vector iteration.

4.2   Vector iteration

This procedure is sometimes called the "power" method and is only able to find the largest (absolute) eigenvalue and corresponding eigenvector of the matrix it operates on. As will be shown however, this does not turn out to be quite the restriction it seems, since modifications to the initial matrix will enable other eigensolutions to be found. The rate of convergence of vector iteration depends upon the nature of the eigenvalues.

Example 4.1

Find the "largest" eigenvalue and corresponding eigenvector of the matrix

$$\begin{bmatrix} 16 & -24 & 18 \\ 3 & -2 & 0 \\ -9 & 18 & -17 \end{bmatrix}$$

Solution 4.1

The eigenvalue equation is of the form

$$\begin{bmatrix} 16 & -24 & 18 \\ 3 & -2 & 0 \\ -9 & 18 & -17 \end{bmatrix} \begin{Bmatrix} x_1 \\ x_2 \\ x_3 \end{Bmatrix} = \lambda \begin{Bmatrix} x_1 \\ x_2 \\ x_3 \end{Bmatrix}$$

and let us guess the solution $\{\mathbf{x}\} = [\, 1 \;\; 1 \;\; 1 \,]^T$.

Matrix-by-vector multiplication on the left-hand side yields

$$\begin{bmatrix} 16 & -24 & 18 \\ 3 & -2 & 0 \\ -9 & 18 & -17 \end{bmatrix} \begin{Bmatrix} 1 \\ 1 \\ 1 \end{Bmatrix} = \begin{Bmatrix} 10 \\ 1 \\ -8 \end{Bmatrix} = 10 \begin{Bmatrix} 1.0 \\ 0.1 \\ -0.8 \end{Bmatrix}$$

where we have normalized the resulting vector (see Section 4.1.1) by dividing by the largest absolute term in it to give $|x_i|_{max} = 1$. The normalized $\{\mathbf{x}\}$ is then used for the next round, so the second iteration becomes

$$\begin{bmatrix} 16 & -24 & 18 \\ 3 & -2 & 0 \\ -9 & 18 & -17 \end{bmatrix} \begin{Bmatrix} 1.0 \\ 0.1 \\ -0.8 \end{Bmatrix} = \begin{Bmatrix} -0.8 \\ 2.8 \\ 6.4 \end{Bmatrix} = 6.4 \begin{Bmatrix} -0.125 \\ 0.4375 \\ 1.0 \end{Bmatrix}$$

and the third

$$\begin{bmatrix} 16 & -24 & 18 \\ 3 & -2 & 0 \\ -9 & 18 & -17 \end{bmatrix} \begin{Bmatrix} -0.125 \\ 0.4375 \\ 1.0 \end{Bmatrix} = \begin{Bmatrix} 5.5 \\ -1.25 \\ -8.0 \end{Bmatrix} = -8.0 \begin{Bmatrix} -0.6875 \\ 0.15625 \\ 1.0 \end{Bmatrix}$$

and the fourth

$$\begin{bmatrix} 16 & -24 & 18 \\ 3 & -2 & 0 \\ -9 & 18 & -17 \end{bmatrix} \begin{Bmatrix} -0.6875 \\ 0.15625 \\ 1.0 \end{Bmatrix} = \begin{Bmatrix} 3.17 \\ -2.39 \\ -7.865 \end{Bmatrix} = -7.865 \begin{Bmatrix} -0.404 \\ 0.304 \\ 1.0 \end{Bmatrix}$$

illustrating convergence towards the eigenvalue $\lambda = -8$, which is the eigenvalue of largest absolute value (see the solution to the characteristic polynomial from equation 4.6).

Program 4.1: Vector iteration for "largest" eigenvalue and its eigenvector

```
PROGRAM nm41
!---Vector Iteration for 'Largest' Eigenvalue and Eigenvector---
USE nm_lib; USE precision; IMPLICIT NONE
INTEGER::i,iters,limit,n; REAL(iwp)::big,l2,tol,zero=0.0_iwp
REAL(iwp),ALLOCATABLE::a(:,:),x(:),x1(:)
OPEN(10,FILE='nm95.dat'); OPEN(11,FILE='nm95.res')
READ(10,*)n; ALLOCATE(a(n,n),x(n),x1(n))
READ(10,*)a; READ(10,*)x; READ(10,*)tol,limit
WRITE(11,'(A)')"---Vector Iteration for 'Largest' Eigenvalue &
   &and Eigenvector---"; WRITE(11,'(/,A)')"Coefficient Matrix"
a=TRANSPOSE(a); DO i=1,n; WRITE(11,'(6E12.4)')a(i,:); END DO
```

```
WRITE(11,'(/,A)')"Guessed Starting Vector"
WRITE(11,'(6E12.4)')x
WRITE(11,'(/,A)')"First Few Iterations" ; iters=0
DO; iters=iters+1; x1=MATMUL(a,x); big=zero
  DO i=1,n; IF(ABS(x1(i))>ABS(big))big=x1(i); END DO; x1=x1/big
  IF(checkit(x1,x,tol).OR.iters==limit)EXIT; x=x1
  IF(iters<5)WRITE(11,'(6E12.4)')x
END DO
12=norm(x1); x1=x1/12
WRITE(11,'(/,A,/,I5)')"Iterations to Convergence",iters
WRITE(11,'(/,A,/,E12.4)')"Largest Eigenvalue",big
WRITE(11,'(/,A,/,6E12.4)')"Corresponding Eigenvector",x1
END PROGRAM nm41
```

List 4.1:

Scalar integers:

i	simple counter
iters	iteration counter
limit	iteration limit
n	number of equations

Scalar reals:

big	term of largest absolute magnitude in a vector
12	L2 norm of a vector
tol	convergence tolerance
zero	set to 0.0

Dynamic real arrays:

a	matrix of coefficients
x	old estimate of eigenvector
x1	new estimate of eigenvector

```
Number of equations      n
                         3

Coefficient matrix       (a(i,:),i=1,n)
                         10.    5.    6.
                          5.   20.    4.
                          6.    4.   30.
```

```
Initial guess              x
                           1.    1.    1.

Tolerance and              tol    limit
iteration limit            1.E-5  100
```

Data 4.1: Vector Iteration

```
---Vector Iteration for 'Largest' Eigenvalue and Eigenvector---

Coefficient Matrix
   0.1000E+02   0.5000E+01   0.6000E+01
   0.5000E+01   0.2000E+02   0.4000E+01
   0.6000E+01   0.4000E+01   0.3000E+02

Guessed Starting Vector
   0.1000E+01   0.1000E+01   0.1000E+01

First Few Iterations
   0.5250E+00   0.7250E+00   0.1000E+01
   0.4126E+00   0.5860E+00   0.1000E+01
   0.3750E+00   0.5107E+00   0.1000E+01
   0.3588E+00   0.4692E+00   0.1000E+01

Iterations to Convergence
   19

Largest Eigenvalue
   0.3371E+02

Corresponding Eigenvector
   0.3002E+00   0.3664E+00   0.8807E+00
```

Results 4.1: Vector Iteration

Input and output for the program are shown in Data 4.1 and Results 4.1 respectively. The program reads in the number of equations to be solved followed by the coefficient matrix and the initial guessed eigenvector. The tolerance required of the iteration process and the maximum number of iterations complete the data set. After a matrix-vector multiplication using intrinsic routine MATMUL, the new eigenvector is normalized so that its largest component is 1.0. The checkit function then checks if convergence has been achieved and updates the eigenvector. If more iterations are required, the program returns to MATMUL and the process is repeated. After convergence, the eigenvector is normalized to have a length of unity. Results 4.1 shows that convergence to the "largest" eigenvalue $\lambda = 33.71$ is achieved in 19 iterations to the tolerance requested of 1×10^{-5}.

4.2.1 Shifted vector iteration

The rate of convergence of the vector iteration method depends upon the nature of the eigenvalues. For closely spaced eigenvalues, convergence can be slow. For example, taking the $[\mathbf{A}]$ matrix of equation (4.10) and using Program 4.1 with an initial guess of $\{\mathbf{x}\} = [1 \ 1 \ 1]^T$, convergence to 4.7071 is only achieved after 26 iterations for a tolerance of 1×10^{-5}. The device of "shifting" can improve convergence rates and is based on the solution of the modified problem,

$$[[\mathbf{A}] - p[\mathbf{I}]] \{\mathbf{x}\} = (\lambda - p)\{\mathbf{x}\} \tag{4.26}$$

For example, again solving the $[\mathbf{A}]$ matrix of equation (4.10) with the same initial guess, but using a shift of 3.5 (see Program 4.2) leads to convergence to 4.7071 in 7 iterations.

A useful feature of shifted iteration is that it enables the ready calculation of the *smallest* eigenvalue of $[\mathbf{A}]$ for systems whose eigenvalues are all positive. Vector iteration can first be used to compute the largest eigenvalue of $[\mathbf{A}]$, say p, which is then used as the "shift" in the modified problem of equation (4.26). Vector iteration can then be used to find the "largest" eigenvalue of the modified matrix $[[\mathbf{A}] - p[\mathbf{I}]]$. Finally p is added to this value leading to the eigenvalue of $[\mathbf{A}]$ closest to zero.

Program 4.2: Shifted vector iteration for eigenvalue and its eigenvector

```
PROGRAM nm42
!--Shifted Vector Iteration for Eigenvalue and its Eigenvector--
USE nm_lib; USE precision; IMPLICIT NONE
INTEGER::i,iters,limit,n; REAL(iwp)::big,l2,shift,tol,          &
   zero=0.0_iwp; REAL(iwp),ALLOCATABLE::a(:,:),x(:),x1(:)
OPEN(10,FILE='nm95.dat'); OPEN(11,FILE='nm95.res')
READ(10,*)n; ALLOCATE(a(n,n),x(n),x1(n))
READ(10,*)a; READ(10,*)shift; READ(10,*)x; READ(10,*)tol,limit
WRITE(11,'(A)')                                                 &
   "--Shifted Vector Iteration Eigenvalue and its Eigenvector--"
WRITE(11,'(/,A)')"Coefficient Matrix"
a=TRANSPOSE(a); DO i=1,n; WRITE(11,'(6E12.4)')a(i,:); END DO
WRITE(11,'(/,A,/,E12.4)')"Shift",shift
WRITE(11,'(/,A)')"Guessed Starting Vector"
```

```
WRITE(11,'(3E12.4)')x; DO i=1,n; a(i,i)=a(i,i)-shift; END DO
WRITE(11,'(/,A)')"First Few Iterations"; iters=0
DO; iters=iters+1; x1=MATMUL(a,x); big=zero
  DO i=1,n; IF(ABS(x1(i))>ABS(big))big=x1(i); END DO; x1=x1/big
  IF(checkit(x1,x,tol).OR.iters==limit)EXIT; x=x1
  IF(iters<5)WRITE(11,'(6E12.4)')x
END DO; 12=norm(x1); x1=x1/12
WRITE(11,'(/,A,/,I5)')"Iterations to Convergence",iters
WRITE(11,'(/,A,/,E12.4)')"Eigenvalue",big+shift
WRITE(11,'(/,A,/,6E12.4)')"Corresponding Eigenvector",x1
END PROGRAM nm42
```

List 4.2:

Scalar integers:

i	simple counter
iters	iteration counter
limit	iteration limit
n	number of equations

Scalar reals:

big	term of largest absolute magnitude in a vector
12	L2 norm of a vector
shift	shift
tol	convergence tolerance
zero	set to 0.0

Dynamic real arrays:

a	matrix of coefficients
x	old estimate of eigenvector
x1	new estimate of eigenvector

Number of equations	n		
	3		
Coefficient matrix	(a(i,:),i=1,n)		
	10.	5.	6.
	5.	20.	4.
	6.	4.	30.
Shift	shift		
	33.71		

```
Initial guess              x
                           1.     1.      1.

Tolerance and              tol    limit
iteration limit            1.E-5  100
```

Data 4.2: Shifted Vector Iteration

This program only differs from the previous one in that `shift` is read, a line is added to perform the subtraction of $p\,[\mathbf{I}]$ from $[\mathbf{A}]$ and `shift` is added to the eigenvalue before printing. The input and output are shown in Data and Results 4.2 respectively. Using the largest eigenvalue of $\lambda = 33.71$ found in the Program 4.1 example as the shift, it can be seen that Program 4.2 leads to the smallest eigenvalue $\lambda = 7.142$.

```
---Shifted Vector Iteration for Eigenvalue and Eigenvector---

Coefficient Matrix
  0.1000E+02  0.5000E+01  0.6000E+01
  0.5000E+01  0.2000E+02  0.4000E+01
  0.6000E+01  0.4000E+01  0.3000E+02

Shift
  0.3371E+02

Guessed Starting Vector
  0.1000E+01  0.1000E+01  0.1000E+01

First Few Iterations
  0.1000E+01  0.3706E+00 -0.4949E+00
  0.1000E+01  0.8298E-01 -0.3753E+00

  0.1000E+01 -0.9242E-01 -0.3024E+00
  0.1000E+01 -0.1946E+00 -0.2598E+00

Iterations to Convergence
  20

Eigenvalue
  0.7142E+01

Corresponding Eigenvector
  0.9334E+00 -0.3032E+00 -0.1919E+00
```

Results 4.2: Shifted Vector Iteration

4.2.2 Shifted inverse iteration

A more direct way of achieving convergence of the vector iteration method on eigenvalues other than the "largest" is to recast equation (4.1) in the form

$$[[\mathbf{A}] - p[\mathbf{I}]]^{-1}\{\mathbf{x}\} = \frac{1}{\lambda - p}\{\mathbf{x}\} \tag{4.27}$$

where p is a scalar "shift", $[\mathbf{I}]$ is the unit matrix and λ is an eigenvalue of $[\mathbf{A}]$. As shown in Section 4.1.2, the eigenvectors of $[[\mathbf{A}] - p[\mathbf{I}]]^{-1}$ are the same as those of $[\mathbf{A}]$, but it can be shown that the eigenvalues of $[[\mathbf{A}] - p[\mathbf{I}]]^{-1}$ are $1/(\lambda - p)$. Hence, the largest eigenvalue of $[[\mathbf{A}] - p[\mathbf{I}]]^{-1}$ leads to the eigenvalue of $[\mathbf{A}]$ that is *closest* to p. Thus, if the largest eigenvalue of $[[\mathbf{A}] - p[\mathbf{I}]]^{-1}$ is μ, then the eigenvalue of $[\mathbf{A}]$ closest to p is given by

$$\lambda = \frac{1}{\mu} + p \tag{4.28}$$

For small matrices it would be possible just to invert $[[\mathbf{A}] - p[\mathbf{I}]]$ and to use the inverse to solve equation (4.27) iteratively in exactly the same algorithm as was used in Program 4.1. For larger matrices however, we saw in Chapter 2 that factorization methods are the most applicable for equation solution, especially when dealing with multiple right-hand sides and a constant coefficient matrix. Thus, whereas in the normal shifted iteration method we would have to compute in every iteration

$$[[\mathbf{A}] - p[\mathbf{I}]]\{\mathbf{x}\}_0 = \{\mathbf{x}\}_1^* \tag{4.29}$$

in the inverse iteration we have to compute

$$[[\mathbf{A}] - p[\mathbf{I}]]^{-1}\{\mathbf{x}\}_0 = \{\mathbf{x}\}_1^* \tag{4.30}$$

or

$$[[\mathbf{A}] - p[\mathbf{I}]]\{\mathbf{x}\}_1^* = \{\mathbf{x}\}_0 \tag{4.31}$$

By factorizing $[[\mathbf{A}] - p[\mathbf{I}]]$ (see Chapter 2) using the appropriate library subroutine `lufac` we can write

$$[[\mathbf{A}] - p[\mathbf{I}]] = [\mathbf{L}][\mathbf{U}] \tag{4.32}$$

and so

$$[\mathbf{L}][\mathbf{U}]\{\mathbf{x}\}_1^* = \{\mathbf{x}\}_0 \tag{4.33}$$

If we now let

$$[\mathbf{U}]\{\mathbf{x}\}_1^* = \{\mathbf{y}\}_0 \tag{4.34}$$

and

$$[\mathbf{L}]\{\mathbf{y}\}_0 = \{\mathbf{x}\}_0 \tag{4.35}$$

we can see that equation (4.33) is solved for $\{\mathbf{x}\}_1^*$ by solving in succession equation (4.35) for $\{\mathbf{y}\}_0$ and equation (4.34) for $\{\mathbf{x}\}_1^*$. These processes are just the forward- and back-substitution processes we saw in Chapter 2, for which subroutines subfor and subbac were developed.

By altering p in a systematic way, all the eigenvalues of $[\mathbf{A}]$ can be found by this method.

Example 4.2

Use shifted inverse iteration to find the eigenvalue of the matrix

$$[\mathbf{A}] = \begin{bmatrix} 3 & 2 \\ 3 & 4 \end{bmatrix}$$

that is closest to 2.

Solution 4.2

For such a small problem by hand we can use vector iteration operating directly on the matrix, so let

$$[\mathbf{B}] = [[\mathbf{A} - p\,[\mathbf{I}]]^{-1}$$

where $p = 2$ in this case.

$$[\mathbf{B}] = \left[\begin{bmatrix} 3 & 2 \\ 3 & 4 \end{bmatrix} - 2 \begin{bmatrix} 1 & 0 \\ 0 & 1 \end{bmatrix} \right]^{-1}$$

hence

$$[\mathbf{B}] = \begin{bmatrix} 1 & 2 \\ 3 & 2 \end{bmatrix}^{-1} = -0.25 \begin{bmatrix} 2 & -2 \\ -3 & 1 \end{bmatrix} = \begin{bmatrix} -0.5 & 0.5 \\ 0.75 & -0.25 \end{bmatrix}$$

Let

$$\{\mathbf{x}\}_0 = \begin{Bmatrix} 1 \\ 1 \end{Bmatrix} \quad \text{and} \quad \{\mathbf{x}\}_k^* \text{ be the value of } \{\mathbf{x}\}_k \text{ before normalization.}$$

First iteration ($k = 1$)

$$\{\mathbf{x}\}_1^* = \begin{bmatrix} -0.5 & 0.5 \\ 0.75 & -0.25 \end{bmatrix} \begin{Bmatrix} 1 \\ 1 \end{Bmatrix} = \begin{Bmatrix} 0.0 \\ 0.5 \end{Bmatrix}, \{\mathbf{x}\}_1 = \begin{Bmatrix} 0.0 \\ 1.0 \end{Bmatrix} \quad \lambda_1 = 0.5$$

Second iteration ($k = 2$)

$$\{\mathbf{x}\}_2^* = \begin{bmatrix} -0.5 & 0.5 \\ 0.75 & -0.25 \end{bmatrix} \begin{Bmatrix} 0 \\ 1 \end{Bmatrix} = \begin{Bmatrix} 0.5 \\ -0.25 \end{Bmatrix}, \{\mathbf{x}\}_2 = \begin{Bmatrix} 1.0 \\ -0.5 \end{Bmatrix} \quad \lambda_2 = 0.5$$

Third iteration ($k = 3$)

$$\{\mathbf{x}\}_3^* = \begin{bmatrix} -0.5 & 0.5 \\ 0.75 & -0.25 \end{bmatrix} \begin{Bmatrix} 1.0 \\ -0.5 \end{Bmatrix} = \begin{Bmatrix} -0.75 \\ 0.875 \end{Bmatrix}$$

$$\{\mathbf{x}\}_3 = \left\{ \begin{array}{c} -0.8571 \\ 1.0 \end{array} \right\} \quad \lambda_3 = 0.875$$

Fourth iteration $(k = 4)$

$$\{\mathbf{x}\}_4^* = \left[\begin{array}{cc} -0.5 & 0.5 \\ 0.75 & -0.25 \end{array} \right] \left\{ \begin{array}{c} -0.8571 \\ 1.0 \end{array} \right\} = \left\{ \begin{array}{c} 0.9286 \\ -0.8929 \end{array} \right\}$$

$$\{\mathbf{x}\}_4 = \left\{ \begin{array}{c} 1.0 \\ -0.9616 \end{array} \right\} \quad \lambda_4 = 0.9286$$

and after many iterations $\{\mathbf{x}\} = \left\{ \begin{array}{c} 1.0 \\ -1.0 \end{array} \right\} \quad \lambda = 1.0000$

Program 4.3: Shifted inverse iteration for nearest eigenvalue and eigenvector

```
PROGRAM nm43
!---Shifted Inverse iteration for Nearest Eigenvalue
!and Eigenvector---
USE nm_lib; USE precision; IMPLICIT NONE
INTEGER::i,iters,limit,n
REAL(iwp)::big,l2,one=1.0_iwp,shift,tol,zero=0.0_iwp
REAL(iwp),ALLOCATABLE::a(:,:),lower(:,:),upper(:,:),x(:),x1(:)
OPEN(10,FILE='nm95.dat'); OPEN(11,FILE='nm95.res')
READ(10,*)n; ALLOCATE(a(n,n),lower(n,n),upper(n,n),x(n),x1(n))
READ(10,*)a; READ(10,*)shift; READ(10,*)x; READ(10,*)tol,limit
WRITE(11,'(A)')"---Shifted Inverse Iteration---"
WRITE(11,'(A)')"---for Nearest Eigenvalue and Eigenvector---"
WRITE(11,'(/,A)')"Coefficient Matrix"
a=TRANSPOSE(a); DO i=1,n; WRITE(11,'(6E12.4)')a(i,:); END DO
WRITE(11,'(/,A)')"Guessed Starting Vector"
WRITE(11,'(6E12.4)')x; WRITE(11,'(/,A,/,E12.4)')"Shift",shift
DO i=1,n; a(i,i)=a(i,i)-shift; END DO; CALL lufac(a,lower,upper)
WRITE(11,'(/,A)')"First Few Iterations"; x1=x; iters=0
DO; iters=iters+1
   CALL subfor(lower,x1); CALL subbac(upper,x1); big=zero
   DO i=1,n; IF(ABS(x1(i))>ABS(big))big=x1(i); END DO; x1=x1/big
   IF(checkit(x1,x,tol).OR.iters==limit)EXIT; x=x1
   IF(iters<5)WRITE(11,'(6E12.4)')x
END DO
l2=norm(x1); x1=x1/l2
WRITE(11,'(/,A,/,I5)')"Iterations to Convergence",iters
WRITE(11,'(/,A,/,E12.4)')"Nearest Eigenvalue",one/big+shift
WRITE(11,'(/,A,/,6E12.4)')"Corresponding Eigenvector",x1
END PROGRAM nm43
```

List 4.3:

Scalar integers:
i	simple counter
iters	iteration counter
limit	iteration limit
n	number of equations

Scalar reals:
big	term of largest absolute magnitude in a vector
12	L2 norm of a vector
one	set to 1.0
shift	shift
tol	convergence tolerance
zero	set to 0.0

Dynamic real arrays:
a	matrix of coefficients
lower	lower triangular factor of $[\mathbf{A}] - p\,[\mathbf{I}]$
upper	upper triangular factor of $[\mathbf{A}] - p\,[\mathbf{I}]$
x	old estimate of eigenvector
x1	new estimate of eigenvector

```
Number of equations          n
                             3

Coefficient matrix           (a(i,:),i=1,n)
                             10.    5.    6.
                              5.   20.    4.
                              6.    4.   30.

Shift                        shift
                             20.0

Initial guess                x
                             1.    1.    1.

Tolerance and                tol     limit
iteration limit              1.E-5   100
```

Data 4.3: Shifted Inverse iteration

```
---Shifted Inverse Iteration---
---for Nearest Eigenvalue and its Eigenvector---

Coefficient Matrix
  0.1000E+02  0.5000E+01  0.6000E+01
  0.5000E+01  0.2000E+02  0.4000E+01
  0.6000E+01  0.4000E+01  0.3000E+02

Guessed Starting Vector
  0.1000E+01  0.1000E+01  0.1000E+01

Shift
  0.2000E+02

First Few Iterations
  0.2391E+00  0.1000E+01 -0.7065E+00
  0.2293E+00  0.1000E+01 -0.4823E+00
  0.2237E+00  0.1000E+01 -0.4931E+00
  0.2237E+00  0.1000E+01 -0.4923E+00

Iterations to Convergence
     6

Nearest Eigenvalue
  0.1915E+02

Corresponding Eigenvector
  0.1967E+00  0.8796E+00 -0.4330E+00
```

Results 4.3: Shifted Inverse iteration

Typical input data and output results are shown in Data and Results 4.3 respectively. The number of equations and the coefficients of $[\mathbf{A}]$ are read in, followed by the scalar shift, the first guess of vector $\{\mathbf{x}\}$, the tolerance and the maximum number of iterations. The initial guess $\{\mathbf{x}\}$ is copied into x1 for future reference, and the iteration loop entered. Matrix $[\mathbf{A}] - p\,[\mathbf{I}]$ is formed (still called $[\mathbf{A}]$) followed by factorization using lufac, and calls to subfor and subbac to complete the determination of $\{\mathbf{x}\}_1^*$ following equations (4.34) and (4.35). Vector $\{\mathbf{x}\}_1^*$ is then normalized to $\{\mathbf{x}\}_1$ and the convergence check invoked. When convergence is complete the converged vector $\{\mathbf{x}\}_1$ is normalized so that $\left(\sum_{i=1}^{n} x_i^2\right)^{1/2} = 1$ (see Section 4.1.1) and the eigenvector, the eigenvalue of the original $[\mathbf{A}]$ closest to p and the number of iterations to convergence printed. In the example shown, the intermediate eigenvalue of the matrix considered by Programs 4.1 and Program 4.2 is found with a shift of $p = 20$, which is the average of the largest and smallest eigenvalues previously found. As shown in Results 4.3, convergence to the eigenvalue, $\lambda = 19.15$, is achieved in 6 iterations.

4.3   Intermediate eigenvalues by deflation

In the previous section we have shown that by using vector iteration, or simple variations of it, convergence to the numerically largest eigenvalue, the numerically smallest or the eigenvalue closest to a given quantity can usually be obtained. Suppose, however, that the second largest eigenvalue of a system is to be investigated. One means of doing this is called "deflation" and it consists essentially in removing the largest eigenvalue from the system of equations, once it has been computed by, for example, vector iteration.

Eigenvectors obey the orthogonality rules described in Section 4.1.1, namely that

$$\{\mathbf{x}_i\}^T\{\mathbf{x}_j\} = \begin{cases} 1 \text{ for } i = j \\ 0 \text{ for } i \neq j \end{cases} \tag{4.36}$$

We can use this property to establish a modified matrix $[\mathbf{A}^*]$ such that

$$[\mathbf{A}^*] = [\mathbf{A}] - \lambda_1\{\mathbf{x}_1\}\{\mathbf{x}_1\}^T \tag{4.37}$$

where λ_1 is the largest eigenvalue of $[\mathbf{A}]$ and $\{\mathbf{x}_1\}$ its corresponding eigenvector.

We now multiply this equation by any eigenvector $\{\mathbf{x}_i\}$ to give

$$[\mathbf{A}^*]\{\mathbf{x}_i\} = [\mathbf{A}]\{\mathbf{x}_i\} - \lambda_1\{\mathbf{x}_1\}\{\mathbf{x}_1\}^T\{\mathbf{x}_i\} \tag{4.38}$$

so when $i = 1$ equation (4.38) can be written as

$$\begin{aligned} [\mathbf{A}^*]\{\mathbf{x}_1\} &= [\mathbf{A}]\{\mathbf{x}_1\} - \lambda_1\{\mathbf{x}_1\}\{\mathbf{x}_1\}^T\{\mathbf{x}_1\} \\ &= \lambda_1\{\mathbf{x}_1\} - \lambda_1\{\mathbf{x}_1\} \\ &= 0 \end{aligned} \tag{4.39}$$

and when $i > 1$ equation (4.38) can be written as

$$[\mathbf{A}^*]\{\mathbf{x}_i\} = \lambda_i\{\mathbf{x}_i\} \tag{4.40}$$

Thus the first eigenvalue of $[\mathbf{A}^*]$ is zero, and all other eigenvalues of $[\mathbf{A}^*]$ are the same as those of $[\mathbf{A}]$.

Having "deflated" $[\mathbf{A}]$ to $[\mathbf{A}^*]$, the largest eigenvalue of $[\mathbf{A}^*]$ will thus equal the second largest eigenvalue of $[\mathbf{A}]$ and can be found, for example, by vector iteration.

Example 4.3

Show that the second largest eigenvalue of

$$\begin{bmatrix} 2 & 1 \\ 1 & 2 \end{bmatrix}$$

becomes the largest eigenvalue of the modified problem following deflation.

Solution 4.3

The characteristic polynomial of the original problem is readily shown to be

$$(\lambda - 3)(\lambda - 1) = 0$$

and so the eigenvalues are $\lambda_1 = 3$ and $\lambda_2 = 1$ with corresponding normalized eigenvectors $[1/\sqrt{2} \;\; 1/\sqrt{2}]^T$ and $[1/\sqrt{2} \;\; -1/\sqrt{2}]^T$ respectively.

From equation (4.37),

$$[\mathbf{A}^*] = \begin{bmatrix} 2 & 1 \\ 1 & 2 \end{bmatrix} - 3 \begin{Bmatrix} 1/\sqrt{2} \\ 1/\sqrt{2} \end{Bmatrix} [1/\sqrt{2} \;\; 1/\sqrt{2}]$$

$$= \begin{bmatrix} 2 & 1 \\ 1 & 2 \end{bmatrix} - 3 \begin{bmatrix} 0.5 & 0.5 \\ 0.5 & 0.5 \end{bmatrix}$$

$$= \begin{bmatrix} 0.5 & -0.5 \\ -0.5 & 0.5 \end{bmatrix}$$

The eigenvalues of $[\mathbf{A}^*]$ are then given by the characteristic polynomial

$$\lambda(\lambda - 1) = 0$$

illustrating that the remaining nonzero eigenvalue of $[\mathbf{A}^*]$ is the second largest eigenvalue of $[\mathbf{A}]$, namely $\lambda = 1$.

Example 4.4

The matrix

$$[\mathbf{A}] = \begin{bmatrix} 4 & \frac{1}{2} & 0 \\ \frac{1}{2} & 4 & \frac{1}{2} \\ 0 & \frac{1}{2} & 4 \end{bmatrix}$$

has eigensolutions given by

$$\lambda_1 = 4 + \frac{1}{\sqrt{2}}, \{\mathbf{x}_1\} = \begin{Bmatrix} \frac{1}{2} \\ \frac{1}{\sqrt{2}} \\ \frac{1}{2} \end{Bmatrix}; \quad \lambda_2 = 4, \{\mathbf{x}_2\} = \begin{Bmatrix} \frac{1}{\sqrt{2}} \\ 0 \\ -\frac{1}{\sqrt{2}} \end{Bmatrix}$$

$$\text{and} \quad \lambda_3 = 4 - \frac{1}{\sqrt{2}}, \{\mathbf{x}_3\} = \begin{Bmatrix} \frac{1}{2} \\ -\frac{1}{\sqrt{2}} \\ \frac{1}{2} \end{Bmatrix}$$

Obtain the deflated matrix and use Program 4.1 is find its largest eigenvalue.

Solution 4.4

The deflation process leads to the modified matrix

$$[\mathbf{A}^*] = \begin{bmatrix} 3 - \frac{1}{4\sqrt{2}} & \frac{1}{4} - \sqrt{2} & -1 - \frac{1}{4\sqrt{2}} \\ \frac{1}{4} - \sqrt{2} & 2 - \frac{1}{2\sqrt{2}} & \frac{1}{4} - \sqrt{2} \\ -1 - \frac{1}{4\sqrt{2}} & \frac{1}{4} - \sqrt{2} & 3 - \frac{1}{4\sqrt{2}} \end{bmatrix}$$

When Program 4.1 is applied to this matrix with a starting guess of $\{\mathbf{x}\} = [1 \ 1 \ 1]^T$ convergence occurs not to the second eigenvalue of $[\mathbf{A}]$ $\lambda_2 = 4$, but rather to the third eigenvalue $\lambda_3 = 4 - \frac{1}{\sqrt{2}} = 3.2929$ with associated eigenvector $[\frac{1}{2} \ -\frac{1}{\sqrt{2}} \ \frac{1}{2}]^T = [0.5 \ -0.7071 \ 0.5]^T$ in 2 iterations.

A change in the guessed starting eigenvector, to $\{\mathbf{x}\} = [1 \ 0 \ -\frac{1}{2}]^T$ for example, leads to convergence on the expected second eigenvalue and eigenvector in 46 iterations for the given tolerance.

This example shows that care must be taken with vector iteration methods when applied to deflated matrices or matrices with closely spaced eigenvalues if one is to be sure that convergence to a desired eigenvalue has been attained.

4.4   The generalized eigenvalue problem $[\mathbf{K}]\{\mathbf{x}\} = \lambda[\mathbf{M}]\{\mathbf{x}\}$

Frequently in engineering practice there is an extra matrix on the right-hand side of the eigenvalue equation leading to the form

$$[\mathbf{K}]\{\mathbf{x}\} = \lambda[\mathbf{M}]\{\mathbf{x}\} \tag{4.41}$$

For example in natural frequency or buckling analysis, $[\mathbf{K}]$ would be the "stiffness matrix" and $[\mathbf{M}]$ the "mass" or "geometric" matrix of the system.

By rearrangement of equation (4.41) we could write either of the equivalent eigenvalue equations,

$$[\mathbf{M}]^{-1}[\mathbf{K}]\{\mathbf{x}\} = \lambda\{\mathbf{x}\} \tag{4.42}$$

or

$$[\mathbf{K}]^{-1}[\mathbf{M}]\{\mathbf{x}\} = \frac{1}{\lambda}\{\mathbf{x}\} \tag{4.43}$$

The present implementation yields the largest eigenvalue $1/\lambda$ of equation (4.43), the reciprocal of which corresponds to the smallest eigenvalue λ of equation (4.42) (see Section 4.1.2).

Performing vector iteration on equation (4.41) we can let $\lambda = 1$ and make a guess at $\{\mathbf{x}\}_0$ on the right-hand side. A matrix-by-vector multiplication then yields

$$[\mathbf{M}]\{\mathbf{x}\}_0 = \{\mathbf{y}\}_0 \tag{4.44}$$

allowing a new estimate of $\{\mathbf{x}\}_1^*$ to be established by solving the set of linear equations

$$[\mathbf{K}]\{\mathbf{x}\}_1^* = \{\mathbf{y}\}_0 \tag{4.45}$$

When the new $\{\mathbf{x}\}_1^*$ has been computed, it may be normalized by dividing through by the "largest" component to give $\{\mathbf{x}\}_1$ and the procedure repeated from equation (4.44) onwards to convergence.

Since the $[\mathbf{K}]$ matrix is unchanged throughout the iterative process, the repeated solution of equation (4.45) is efficiently achieved by obtaining the $[\mathbf{L}][\mathbf{U}]$ factors of $[\mathbf{K}]$ (see Program 2.2). Once this is done, all that remains is forward- and back-substitution to compute $\{\mathbf{x}\}_i^*$ at each iteration.

Program 4.4: Vector iteration for $[\mathbf{K}]\{\mathbf{x}\} = \lambda[\mathbf{M}]\{\mathbf{x}\}$

```
PROGRAM nm44
!---Vector Iteration for Kx=lambda*Mx---
 USE nm_lib; USE precision; IMPLICIT NONE
 INTEGER::i,iters,limit,n; REAL(iwp)::big,l2,one=1.0_iwp,tol,    &
   zero=0.0_iwp
 REAL(iwp),ALLOCATABLE::k(:,:),m(:,:),lower(:,:),upper(:,:),    &
   x(:),x1(:)
 OPEN(10,FILE='nm95.dat'); OPEN(11,FILE='nm95.res')
 READ(10,*)n
 ALLOCATE(k(n,n),lower(n,n),m(n,n),upper(n,n),x(n),x1(n))
 DO i=1,n; READ(10,*)k(i,i:n); k(i:n,i)=k(i,i:n); END DO;
 DO i=1,n; READ(10,*)m(i,i:n); m(i:n,i)=m(i,i:n); END DO;
```

```
READ(10,*)x; READ(10,*)tol,limit
WRITE(11,'(A)')"---Vector Iteration for Kx=lambda*Mx---"
WRITE(11,'(/,A)')"Matrix K"
DO i=1,n; WRITE(11,'(6E12.4)')k(i,:); END DO
WRITE(11,'(/,A)')"Matrix M"
DO i=1,n; WRITE(11,'(6E12.4)')m(i,:); END DO
WRITE(11,'(/,A)')"Guessed Starting Vector"
WRITE(11,'(6E12.4)')x; CALL lufac(k,lower,upper)
WRITE(11,'(/,A)')"First Few Iterations"; iters=0
DO; iters=iters+1; x1=MATMUL(m,x)
   CALL subfor(lower,x1); CALL subbac(upper,x1)
   big=zero; DO i=1,n; IF(ABS(x1(i))>ABS(big))big=x1(i); END DO
   x1=x1/big; IF(checkit(x1,x,tol).OR.iters==limit)EXIT; x=x1
   IF(iters<5)WRITE(11,'(6E12.4)')x
END DO
l2=norm(x1); x1=x1/l2
WRITE(11,'(/,A,/,I5)')"Iterations to Convergence",iters
WRITE(11,'(/,A,/,E12.4)')                                          &
   "'Smallest' Eigenvalue of Inv(M)*K",one/big
WRITE(11,'(/,A,/,6E12.4)')"Corresponding Eigenvector",x1
END PROGRAM nm44
```

List 4.4:

Scalar integers:
i	simple counter
iters	iteration counter
limit	iteration limit
n	number of equations

Scalar reals:
big	term of largest absolute magnitude in a vector
l2	L2 norm of a vector
one	set to 1.0
tol	convergence tolerance
zero	set to 0.0

Dynamic real arrays:
k	coefficients of matrix $[\mathbf{K}]$
m	coefficients of matrix $[\mathbf{M}]$
lower	lower triangular factor of $[\mathbf{K}]$
upper	upper triangular factor of $[\mathbf{K}]$
x	old estimate of eigenvector
x1	new estimate of eigenvector

```
Number of equations        n
                           4

Coefficient matrix         (k(i,:),i=1,n)
                           8.0         4.0      -24.0       0.0
                                      16.0        0.0       4.0
                                                192.0      24.0
                                                            8.0

Coefficient matrix         (m(i,:),i=1,n)
                           0.06667  -0.01667    -0.1        0.0
                                     0.13333     0.0     -0.01667
                                                 4.8        0.1
                                                         0.06667

Initial guess              x
                           1.          1.          1.        1.

Tolerance and              tol    limit
iteration limit            1.E-5   100
```

Data 4.4: **Vector Iteration for** $[K]\{x\} = \lambda[M]\{x\}$

```
---Vector Iteration for Kx=lambda*Mx---

Matrix K
  0.8000E+01   0.4000E+01  -0.2400E+02   0.0000E+00
  0.4000E+01   0.1600E+02   0.0000E+00   0.4000E+01
 -0.2400E+02   0.0000E+00   0.1920E+03   0.2400E+02
  0.0000E+00   0.4000E+01   0.2400E+02   0.8000E+01

Matrix M
  0.6667E-01  -0.1667E-01  -0.1000E+00   0.0000E+00
 -0.1667E-01   0.1333E+00   0.0000E+00  -0.1667E-01
 -0.1000E+00   0.0000E+00   0.4800E+01   0.1000E+00
  0.0000E+00  -0.1667E-01   0.1000E+00   0.6667E-01

Guessed Starting Vector
  0.1000E+01   0.1000E+01   0.1000E+01   0.1000E+01

First Few Iterations
  0.1000E+01   0.1639E-01   0.3443E+00  -0.9672E+00
  0.1000E+01   0.6400E-03   0.3209E+00  -0.9987E+00
```

```
0.1000E+01   0.2628E-04   0.3191E+00  -0.9999E+00
0.1000E+01   0.1068E-05   0.3189E+00  -0.1000E+01
```

Iterations to Convergence
   6

'Smallest' Eigenvalue of Inv(M)*K
   0.9944E+01

Corresponding Eigenvector
   0.6898E+00   0.5358E-09   0.2200E+00 -0.6898E+00

Results 4.4: Vector Iteration for $[K]\{x\} = \lambda[M]\{x\}$

Data 4.4 shows data from a typical engineering analysis which leads to the generalized problem of equation (4.41). In this case the $[K]$ represents the stiffness matrix of a compressed strut and $[M]$ the geometric matrix relating to the destabilizing effect of the compressive force (see, e.g., Smith and Griffiths 2004). There are four equations to be solved to a tolerance of 10^{-5} with an iteration limit of 100. The guessed starting vector $\{x\}_0$ is $[1.0\ 1.0\ 1.0\ 1.0]^T$.

The data present the number of equations n followed by the upper triangles of symmetrical matrices $[K]$ and $[M]$. In preparation for the equation solution of equation (4.45), $[K]$ is factorized using subroutine lufac. The iteration loop begins by the multiplication of $[M]$ by $\{x\}_0$ as required by equation (4.44). Forward- and back-substitution by subroutines subfor and subbac, respectively, complete the equation solution and the resulting vector is normalized. The convergence check is invoked and iteration continues if convergence is incomplete, unless the iteration limit has been reached. The final normalization involving the sum of the squares of the components of the eigenvector is then carried out and the normalized eigenvector and number of iterations printed. In this case the reciprocal of the "largest" eigenvalue is the "buckling" load of the strut, which is also printed. The output is shown in Results 4.4 where the estimate of the buckling load of 9.94 after 6 iterations can be compared with the exact solution of $\pi^2 = 9.8696$.

4.4.1   Conversion of generalized problem to symmetrical standard form

Several solution techniques for eigenvalue problems require the equation to be cast in the "standard form"

$$[A]\{x\} = \lambda\{x\} \tag{4.46}$$

where $[A]$ is also symmetrical.

If the generalized eigenvalue equation is encountered as

$$[\mathbf{K}]\{\mathbf{x}\} = \lambda[\mathbf{M}]\{\mathbf{x}\} \qquad (4.47)$$

it is always possible to convert it to the standard form of equations (4.42) or (4.43); however even for symmetrical $[\mathbf{K}]$ and $[\mathbf{M}]$ the products $[\mathbf{K}]^{-1}[\mathbf{M}]$ and $[\mathbf{M}]^{-1}[\mathbf{K}]$ are not in general symmetrical. In order to preserve symmetry, the following strategy can be used.

Starting from equation (4.47) we can factorize $[\mathbf{M}]$ by Cholesky's method (see Chapter 2) to give

$$[\mathbf{M}] = [\mathbf{L}][\mathbf{L}]^{T} \qquad (4.48)$$

hence

$$[\mathbf{K}]\{\mathbf{x}\} = \lambda[\mathbf{L}][\mathbf{L}]^{T}\{\mathbf{x}\} \qquad (4.49)$$

Now let

$$[\mathbf{L}]^{T}\{\mathbf{x}\} = \{\mathbf{z}\} \qquad (4.50)$$

or

$$\{\mathbf{x}\} = [\mathbf{L}]^{-T}\{\mathbf{z}\} \qquad (4.51)$$

which after substitution in (4.49) gives

$$[\mathbf{K}][\mathbf{L}]^{-T}\{\mathbf{z}\} = \lambda[\mathbf{L}]\{\mathbf{z}\} \qquad (4.52)$$

Finally

$$[\mathbf{L}]^{-1}[\mathbf{K}][\mathbf{L}]^{-T}\{\mathbf{z}\} = \lambda\{\mathbf{z}\} \qquad (4.53)$$

which is a standard eigenvalue equation in which the left-hand side matrix $[\mathbf{L}]^{-1}[\mathbf{K}][\mathbf{L}]^{-\mathbf{T}}$ is also symmetrical.

The transformed equation (4.53) has the same eigenvalues λ as the generalized equation (4.47) but different eigenvectors $\{\mathbf{z}\}$. Once the transformed eigenvectors $\{\mathbf{z}\}$ have been found they can easily be converted back to $\{\mathbf{x}\}$ from equation (4.51).

Program 4.5: Conversion of $[\mathrm{K}]\{\mathrm{x}\} = \lambda[\mathrm{M}]\{\mathrm{x}\}$ to symmetrical standard form

```
PROGRAM nm45
!---Conversion of Kx=lambda*Mx to Symmetrical Standard Form---
USE nm_lib; USE precision; IMPLICIT NONE
INTEGER::i,j,n
REAL(iwp),ALLOCATABLE::c(:,:),d(:),e(:),k(:,:),m(:,:),s(:,:)
OPEN(10,FILE='nm95.dat'); OPEN(11,FILE='nm95.res')
READ(10,*)n; ALLOCATE(c(n,n),d(n),e(n),k(n,n),m(n,n),s(n,n))
DO i=1,n; READ(10,*)k(i,i:n); k(i:n,i)=k(i,i:n); END DO;
```

```
DO i=1,n; READ(10,*)m(i,i:n); m(i:n,i)=m(i,i:n); END DO;
WRITE(11,'(A)')"---Conversion of Kx=lambda*Mx to Symmetrical &
   &Standard Form---"
WRITE(11,'(/,A)')"Matrix K"
DO i=1,n; WRITE(11,'(6E12.4)')k(i,:); END DO
WRITE(11,'(/,A)')"Matrix M"
DO i=1,n; WRITE(11,'(6E12.4)')m(i,:); END DO
CALL ldlt(m,d); d=SQRT(d)
DO j=1,n; DO i=j,n; m(i,j)=m(i,j)/d(j); END DO; END DO
DO j=1,n; e=k(:,j); CALL subfor(m,e); c(:,j)=e; END DO
WRITE(11,'(/,A)')"Matrix C"
DO i=1,n; WRITE(11,'(6E12.4)')c(i,:); END DO
DO j=1,n; e=c(j,:); CALL subfor(m,e); s(:,j)=e; END DO
WRITE(11,'(/,A)')"Final symmetrical matrix S"
DO i=1,n; WRITE(11,'(6E12.4)')s(i,:); END DO
END PROGRAM nm45
```

List 4.5:

Scalar integers:
i   simple counter
j   simple counter
n   number of equations

Dynamic real arrays:
c   temporary storage of matrix $[C] = [L]^{-1}[K]$
d   diagonal matrix (vector) in $[L][D][L]^T$
e   temporary storage vector
k   symmetrical array $[K]$ in equation (4.47)
m   symmetrical array $[M]$ in equation (4.47)
s   final symmetrical matrix $[S] = [L]^{-1}[K][L]^{-T}$

```
Number of equations        n
                           4

Coefficient matrix         (k(i,:),i=1,n)
                           8.0        4.0      -24.0        0.0
                                     16.0        0.0        4.0
                                               192.0       24.0
                                                            8.0
```

```
Coefficient matrix        (m(i,:),i=1,n)
                          0.06667 -0.01667 -0.1      0.0
                                   0.13333  0.0     -0.01667
                                            4.8      0.1
                                                     0.06667
```

Data 4.5: Conversion of $[K]\{x\} = \lambda[M]\{x\}$ to Symmetrical Standard Form

```
---Conversion of Kx=lambda*Mx to Symmetrical Standard Form---

Matrix K
  0.8000E+01  0.4000E+01 -0.2400E+02  0.0000E+00
  0.4000E+01  0.1600E+02  0.0000E+00  0.4000E+01
 -0.2400E+02  0.0000E+00  0.1920E+03  0.2400E+02
  0.0000E+00  0.4000E+01  0.2400E+02  0.8000E+01

Matrix M
  0.6667E-01 -0.1667E-01 -0.1000E+00  0.0000E+00
 -0.1667E-01  0.1333E+00  0.0000E+00 -0.1667E-01
 -0.1000E+00  0.0000E+00  0.4800E+01  0.1000E+00
  0.0000E+00 -0.1667E-01  0.1000E+00  0.6667E-01

Matrix C
  0.3098E+02  0.1549E+02 -0.9295E+02  0.0000E+00
  0.1670E+02  0.4730E+02 -0.1670E+02  0.1113E+02
 -0.5029E+01  0.4311E+01  0.7184E+02  0.1149E+02
  0.4001E+01  0.2400E+02  0.8000E+02  0.3200E+02

Final symmetrical matrix S
  0.1200E+03  0.6466E+02 -0.1948E+02  0.1549E+02
  0.6466E+02  0.1432E+03  0.8496E+01  0.6957E+02
 -0.1948E+02  0.8496E+01  0.3011E+02  0.4215E+02
  0.1549E+02  0.6957E+02  0.4215E+02  0.1333E+03
```

Results 4.5: Conversion of $[K]\{x\} = \lambda[M]\{x\}$ to Symmetrical Standard Form

The program achieves the transformations described in equations (4.47)-(4.53) by introducing the temporary array $[C]$ defined as

$$[C] = [L]^{-1}[K] \tag{4.54}$$

which is solved by repeated forward substitutions of

$$[L][C] = [K] \tag{4.55}$$

using the columns of $[\mathbf{K}]$.

The final symmetrical matrix from equation (4.53) is stored in array $[\mathbf{S}]$ given as

$$[\mathbf{S}] = [\mathbf{L}]^{-1}[\mathbf{K}][\mathbf{L}]^{-T} = [\mathbf{C}][\mathbf{L}]^{-T} \qquad (4.56)$$

hence

$$[\mathbf{S}]^T = [\mathbf{S}] = [\mathbf{L}]^{-1}[\mathbf{C}]^T \qquad (4.57)$$

Thus, $[\mathbf{S}]$ is found by repeated forward substitutions of

$$[\mathbf{L}][\mathbf{S}] = [\mathbf{C}]^T \qquad (4.58)$$

using the columns of $[\mathbf{C}]^T$.

The data shown in Data 4.5 indicate that the program begins by reading in the number of equations n and the upper triangles of symmetrical matrices $[\mathbf{K}]$ and $[\mathbf{M}]$. Matrix $[\mathbf{M}]$ is factorized using library routine ldlt and the Cholesky factors are then obtained by dividing the upper triangle of the factorized $[\mathbf{M}]$ by the square root of the diagonals of $[\mathbf{D}]$ (see Section 2.4.2).

Equation (4.55) is then solved for $[\mathbf{C}]$. This is achieved by first copying the columns of $[\mathbf{K}]$ into a temporary storage vector e and forward substitution using subfor leads to the columns of $[\mathbf{C}]$. The $[\mathbf{C}]$ matrix is printed out. Then, in a very similar sequence of operations, equation (4.58) is solved for $[\mathbf{S}]$ $(=[\mathbf{S}]^T)$ by first copying the columns of $[\mathbf{C}]^T$ into e and then once again calling subfor. The final symmetrical matrix is printed out as shown in Results 4.5.

If the "largest" eigenvalue of this matrix is calculated using Program 4.1 with a tolerance of 1×10^{-5}, convergence to the eigenvalue $\lambda = 240$ will be found in 15 iterations.

If $[\mathbf{K}]$ and $[\mathbf{M}]$ are switched, the eigenvalues of the resulting symmetrical matrix found by this program would be the reciprocals of their previous values. For example, if $[\mathbf{K}]$ and $[\mathbf{M}]$ are switched and the resulting symmetrical matrix treated by Program 4.1, the "largest" eigenvalue is given as $\lambda = 0.1006$. This is the reciprocal of 9.94 which was the "smallest" eigenvalue obtained previously by Program 4.4.

4.5   Transformation methods

Returning to the standard eigenvalue equation

$$[\mathbf{A}]\{\mathbf{x}\} = \lambda\{\mathbf{x}\} \qquad (4.59)$$

it was shown in Section 4.1.2 that for any nonsingular matrix $[\mathbf{P}]$, the standard equation can be transformed into an equation with the same eigenvalues given by

$$[\mathbf{A}^*]\{\mathbf{z}\} = \lambda\{\mathbf{z}\} \qquad (4.60)$$

where

$$[\mathbf{A}^*] = [\mathbf{P}]^{-1}[\mathbf{A}][\mathbf{P}] \tag{4.61}$$

and

$$\{\mathbf{z}\} = [\mathbf{P}]^{-1}\{\mathbf{x}\} \tag{4.62}$$

The concept behind such a transformation technique is to employ it so as to make the eigenvalues of $[\mathbf{A}^*]$ easier to find than were the eigenvalues of the original $[\mathbf{A}]$.

If $[\mathbf{A}]$ is symmetrical however, it is highly unlikely that the transformation given by equation (4.61) would retain symmetry. It is easily shown though that the transformation

$$[\mathbf{A}^*] = [\mathbf{P}]^T[\mathbf{A}][\mathbf{P}] \tag{4.63}$$

would retain symmetry of $[\mathbf{A}^*]$. In order for the eigenvalues $[\mathbf{A}^*]$ given in equation (4.63) to be the same as those of $[\mathbf{A}]$, we must arrange for the additional property

$$[\mathbf{P}]^T = [\mathbf{P}]^{-1} \quad \text{or} \quad [\mathbf{P}]^T[\mathbf{P}] = [\mathbf{I}] \tag{4.64}$$

Matrices of this type are said to be "orthogonal", and a matrix which has this property is the so-called "rotation matrix"

$$[\mathbf{P}]^T = \begin{bmatrix} \cos\alpha & -\sin\alpha \\ \sin\alpha & \cos\alpha \end{bmatrix} \tag{4.65}$$

Applying this transformation to the $[\mathbf{A}]$ matrix considered in Example 4.3, we have

$$\begin{aligned} [\mathbf{A}^*] &= \begin{bmatrix} \cos\alpha & \sin\alpha \\ -\sin\alpha & \cos\alpha \end{bmatrix} \begin{bmatrix} 2 & 1 \\ 1 & 2 \end{bmatrix} \begin{bmatrix} \cos\alpha & -\sin\alpha \\ \sin\alpha & \cos\alpha \end{bmatrix} \\ &= \begin{bmatrix} 2 + 2\sin\alpha\cos\alpha & \cos^2\alpha - \sin^2\alpha \\ \cos^2\alpha - \sin^2\alpha & 2 - 2\sin\alpha\cos\alpha \end{bmatrix} \end{aligned} \tag{4.66}$$

in which $[\mathbf{A}^*]$ is clearly symmetrical for any value of α. In this case the obvious value of α to choose is the one that results in $[\mathbf{A}^*]$ being a diagonal matrix (Section 4.1.2) in which case the diagonals are the eigenvalues. Elimination of the off-diagonal terms will occur if

$$\cos^2\alpha - \sin^2\alpha = 0 \tag{4.67}$$

in which case $\tan\alpha = 1$ and $\alpha = \pi/4$, giving $\sin\alpha = \cos\alpha = 1/\sqrt{2}$.

The resulting transformed matrix is

$$[\mathbf{A}^*] = \begin{bmatrix} 3 & 0 \\ 0 & 1 \end{bmatrix} \tag{4.68}$$

indicating that the eigenvalues of $[\mathbf{A}]$ are 3 and 1.

For matrices $[\mathbf{A}]$ which are bigger than 2×2, the transformation matrix $[\mathbf{P}]$ must be "padded out" by putting ones on the leading diagonals and zeros on all off-diagonals, except those in the rows and columns corresponding to the terms to be eliminated. For example, if $[\mathbf{A}]$ is 4×4, the transformation matrix could have any of 6 forms depending on which off-diagonal terms in the initial matrix are to be eliminated, e.g.,

$$[\mathbf{P}] = \begin{bmatrix} \cos\alpha & -\sin\alpha & 0 & 0 \\ \sin\alpha & \cos\alpha & 0 & 0 \\ 0 & 0 & 1 & 0 \\ 0 & 0 & 0 & 1 \end{bmatrix} \tag{4.69}$$

$$[\mathbf{P}] = \begin{bmatrix} 1 & 0 & 0 & 0 \\ 0 & \cos\alpha & 0 & -\sin\alpha \\ 0 & 0 & 1 & 0 \\ 0 & \sin\alpha & 0 & \cos\alpha \end{bmatrix} \tag{4.70}$$

and so on.

Matrix (4.69) would eliminate terms a_{12} and a_{21} in the original matrix $[\mathbf{A}]$ after the transformation via $[\mathbf{P}]^T[\mathbf{A}][\mathbf{P}]$ while matrix (4.70) would eliminate terms a_{24} and a_{42}. The effect of the ones and zeros is to leave the other rows and columns of $[\mathbf{A}]$ unchanged. This means that off-diagonal terms which become zero during one transformation revert to nonzero values (although usually "small") on subsequent transformations and so the method is iterative as we have come to expect.

The earliest form of this type of iteration is called "Jacobi diagonalization", which proceeds by eliminating the "largest" off-diagonal terms remaining at each iteration.

Generalizing equations (4.66) for any symmetrical matrix $[\mathbf{A}]$ we have

$$[\mathbf{A}^*] = \begin{bmatrix} \cos\alpha & \sin\alpha \\ -\sin\alpha & \cos\alpha \end{bmatrix} \begin{bmatrix} a_{ii} & a_{ij} \\ a_{ji} & a_{jj} \end{bmatrix} \begin{bmatrix} \cos\alpha & -\sin\alpha \\ \sin\alpha & \cos\alpha \end{bmatrix} \tag{4.71}$$

leading to off-diagonal terms in $[\mathbf{A}^*]$ of the form

$$a_{ij}^* = a_{ji}^* = (-a_{ii} + a_{jj})\cos\alpha\sin\alpha + a_{ij}(\cos^2\alpha - \sin^2\alpha) \tag{4.72}$$

Solving for α in order to make these term equal zero we get

$$\tan 2\alpha = \frac{2a_{ij}}{a_{ii} - a_{jj}} \tag{4.73}$$

hence

$$\alpha = \frac{1}{2}\tan^{-1}\left(\frac{2a_{ij}}{a_{ii} - a_{jj}}\right) \tag{4.74}$$

To make a simple program for Jacobi diagonalization we have therefore to search for the "largest" off-diagonal term in $[\mathbf{A}]$ and find the row and column in which it lies. The "rotation angle" α can then be computed from equation (4.74) and the transformation matrix $[\mathbf{P}]$ of the type shown in equations (4.69-4.70) set up. Matrix $[\mathbf{P}]$ can then be transposed using an intrinsic library function, and the matrix products to form $[\mathbf{A}^*]$, as required by equation (4.63), carried out. This process is repeated iteratively until the leading diagonals of $[\mathbf{A}^*]$ have converged to the eigenvalues of $[\mathbf{A}]$ within acceptable tolerances.

Example 4.5

Use Jacobi diagonalization to estimate the eigenvalues of the symmetrical matrix

$$[\mathbf{A}] = \begin{bmatrix} 3.5 & -6.0 & 5.0 \\ -6.0 & 8.5 & -9.0 \\ 5.0 & -9.0 & 8.5 \end{bmatrix}$$

Solution 4.5

The following results are quoted to four decimal places but the actual calculations were performed with many more decimal places of accuracy.

First iteration,

The "largest" off-diagonal term is $a_{23} = a_{32} = -9.0$, hence from equation (4.74)

$$\alpha = \frac{1}{2}\tan^{-1}\left(\frac{2(-9)}{8.5 - 8.5}\right) = -45^o$$

The first transformation matrix will include the terms

$$p_{22} = p_{33} = \cos(-45) = 0.7071$$
$$p_{23} = -\sin(-45) = 0.7071$$
$$p_{32} = \sin(-45) = -0.7071$$

hence

$$[\mathbf{P}_1] = \begin{bmatrix} 1 & 0 & 0 \\ 0 & 0.7071 & 0.7071 \\ 0 & -0.7071 & 0.7071 \end{bmatrix}$$

The transformed matrix from equation (4.71) is given by

$$[\mathbf{A}_1] = [\mathbf{P}_1]^T[\mathbf{A}_1][\mathbf{P}_1]$$

so performing this triple matrix product in stages gives

$$[\mathbf{P}_1]^T[\mathbf{A}] = \begin{bmatrix} 1 & 0 & 0 \\ 0 & 0.7071 & -0.7071 \\ 0 & 0.7071 & 0.7071 \end{bmatrix} \begin{bmatrix} 3.5 & -6.0 & 5.0 \\ -6.0 & 8.5 & -9.0 \\ 5.0 & -9.0 & 8.5 \end{bmatrix}$$

$$= \begin{bmatrix} 3.5 & -6.0 & 5.0 \\ -7.7782 & 12.3744 & -12.3744 \\ -0.7071 & -0.3536 & -0.3536 \end{bmatrix}$$

and finally

$$[\mathbf{A}_1] = \begin{bmatrix} 3.5 & -6.0 & 5.0 \\ -7.7782 & 12.3744 & -12.3744 \\ -0.7071 & -0.3536 & -0.3536 \end{bmatrix} \begin{bmatrix} 1 & 0 & 0 \\ 0 & 0.7071 & 0.7071 \\ 0 & -0.7071 & 0.7071 \end{bmatrix}$$

$$= \begin{bmatrix} 3.5 & -7.7782 & -0.7071 \\ -7.7782 & 17.5 & 0.0 \\ -0.7071 & 0.0 & -0.5 \end{bmatrix}$$

Second iteration,

The "largest" off-diagonal term is $a_{12} = a_{21} = -7.7782$, hence from equation (4.74)

$$\alpha = \frac{1}{2}\tan^{-1}\left(\frac{2(-7.7781)}{3.5 - 17.4997}\right) = 24.0071^\circ$$

The second transformation matrix will include the terms

$$p_{11} = p_{22} = \cos(24.0075) = 0.9135$$
$$p_{12} = -\sin(24.0075) = -0.4069$$
$$p_{21} = \sin(24.0075) = 0.4069$$

hence

$$[\mathbf{P}_2] = \begin{bmatrix} 0.9135 & -0.4069 & 0 \\ 0.4069 & 0.9135 & 0 \\ 0 & 0 & 1 \end{bmatrix}$$

Similar matrix products as before lead to the transformed matrix

$$[\mathbf{A}_2] = [\mathbf{P}_2]^T[\mathbf{A}_1][\mathbf{P}_2] = \begin{bmatrix} 0.0358 & 0.0 & -0.6459 \\ 0.0 & 20.9642 & 0.2877 \\ -0.6459 & 0.2877 & -0.5000 \end{bmatrix}$$

Note that although positions $(2,3)$ *and* $(3,2)$ *are no longer zero, they are* "small" *compared with their values in the initial matrix* $[\mathbf{A}]$. *As iterations proceed, the rotation angle* $\alpha_k \longrightarrow 0$, *the transformation matrix* $[\mathbf{P}_k] \longrightarrow [\mathbf{I}]$ *and the transformed matrix* $[\mathbf{A}_k]$ *tends to a diagonal matrix with the eigenvalues on the diagonal.*

In this example, after six iterations with a convergence tolerance of 1×10^{-5}, we get

$$[\mathbf{A}_6] = \begin{bmatrix} 0.4659 & 0.0000 & 0.0000 \\ 0.0000 & 20.9681 & 0.0000 \\ 0.0000 & 0.0000 & -0.9340 \end{bmatrix}$$

thus the eigenvalues of $[\mathbf{A}]$ are $\lambda = 0.4659, 20.9681$ and -0.9340. The corresponding eigenvectors can be retrieved in the usual way by solving the singular systems of equations as discussed in Section 4.1.

Program 4.6: Jacobi diagonalization for eigenvalues of symmetrical matrices

```
PROGRAM nm46
!-Jacobi Diagonalization for Eigenvalues of Symmetrical Matrices-
USE nm_lib; USE precision; IMPLICIT NONE
INTEGER::i,iters,j,limit,n,nc,nr
REAL(iwp)::alpha,big,ct,den,d2=2.0_iwp,d4=4.0_iwp,hold,l2,      &
  one=1.0_iwp,penalty=1.E20_iwp,pi,st,small=1.E-20_iwp,tol,     &
  zero=0.0_iwp
REAL(iwp),ALLOCATABLE::a(:,:),a1(:,:),a2(:,:),enew(:),eold(:), &
  p(:,:),x(:)
OPEN(10,FILE='nm95.dat'); OPEN(11,FILE='nm95.res')
READ(10,*)n; ALLOCATE(a(n,n),a1(n,n),a2(n,n),enew(n),eold(n),  &
  p(n,n),x(n))
DO i=1,n; READ(10,*)a(i,i:n); a(i:n,i)=a(i,i:n); END DO; a2=a
READ(10,*)tol,limit; pi=d4*ATAN(one)
WRITE(11,'(A)')"-Jacobi Diagonalization for Eigenvalues of &
  &Symmetrical Matrices-"
WRITE(11,'(/,A)')"Matrix A"
DO i=1,n; WRITE(11,'(6E12.4)')a(i,:); END DO
WRITE(11,'(/,A)')"First Few Iterations"; iters=0; eold=zero
DO; iters=iters+1; big=zero
  DO i=1,n; DO j=i+1,n
    IF(ABS(a(i,j))>big)THEN
      big=ABS(a(i,j)); hold=a(i,j); nr=i; nc=j
    END IF
  END DO; END DO
  IF(ABS(big)<small)EXIT
  den=a(nr,nr)-a(nc,nc)
  IF(ABS(den)<small)THEN
    alpha=pi/d4; IF(hold<zero)alpha=-alpha
  ELSE
    alpha=ATAN(d2*hold/den)/d2
  END IF
```

```
ct=COS(alpha); st=SIN(alpha); p=zero
DO i=1,n; p(i,i)=one; END DO
p(nr,nr)=ct; p(nc,nc)=ct; p(nr,nc)=-st; p(nc,nr)=st
a=MATMUL(MATMUL(TRANSPOSE(p),a),p)
IF(iters<5)THEN
  DO i=1,n; WRITE(11,'(6E12.4)')a(i,:); END DO; WRITE(11,*)
END IF
DO i=1,n; enew(i)=a(i,i); END DO
IF(checkit(enew,eold,tol).OR.iters==limit)EXIT; eold=enew
END DO
WRITE(11,'(A,/,I5)')"Iterations to Convergence",iters
WRITE(11,'(/,A)')"Final Transformed Matrix A"
DO i=1,n; WRITE(11,'(6E12.4)')a(i,:); END DO; WRITE(11,*)
DO i=1,n; a1=a2; DO j=1,n; a1(j,j)=a1(j,j)-a(i,i); END DO
  x=zero; a1(i,i)=penalty; x(i)=penalty; x=eliminate(a1,x)
  l2=norm(x); WRITE(11,'(A,E12.4)')"Eigenvalue ",a(i,i)
  WRITE(11,'(A,6E12.4)')"Eigenvector",x/l2; WRITE(11,*)
END DO

END PROGRAM nm46


Number of equations       n
                          3

Coefficient matrix        (a(i,:),i=1,n)
                          10.    5.    6.
                                20.    4.
                                       30.


Tolerance and             tol    limit
iteration limit           1.E-5  100
```

Data 4.6: Jacobi Diagonalization for Symmetrical Matrices

```
-Jacobi Diagonalization for Eigenvalues of Symmetrical Matrices-

Matrix A
  0.1000E+02  0.5000E+01  0.6000E+01
  0.5000E+01  0.2000E+02  0.4000E+01
  0.6000E+01  0.4000E+01  0.3000E+02

First Few Iterations
  0.8338E+01  0.3751E+01 -0.7238E-15
  0.3751E+01  0.2000E+02  0.5190E+01
```

```
0.7746E-15  0.5190E+01  0.3166E+02

0.8338E+01  0.3506E+01  0.1334E+01
0.3506E+01  0.1803E+02 -0.6245E-16
0.1334E+01  0.3548E-15  0.3364E+02

0.7203E+01  0.3584E-15  0.1269E+01
-0.4033E-15  0.1916E+02  0.4111E+00
0.1269E+01  0.4111E+00  0.3364E+02

0.7142E+01 -0.1967E-01  0.4136E-16
-0.1967E-01  0.1916E+02  0.4106E+00
0.1395E-14  0.4106E+00  0.3370E+02
```

```
Iterations to Convergence
    6
```

```
Final Transformed Matrix A
 0.7142E+01 -0.4032E-18 -0.5549E-03
-0.7679E-15  0.1915E+02  0.9087E-06
-0.5549E-03  0.9087E-06  0.3371E+02
```

```
Eigenvalue    0.7142E+01
Eigenvector   0.9334E+00 -0.3032E+00 -0.1919E+00

Eigenvalue    0.1915E+02
Eigenvector   0.1967E+00  0.8796E+00 -0.4330E+00

Eigenvalue    0.3371E+02
Eigenvector   0.3002E+00  0.3664E+00  0.8807E+00
```

Results 4.6: Jacobi Diagonalization for Symmetrical Matrices

The data shown in Data 4.6 involve the number of equations followed by the upper triangle of the symmetrical array [**A**]. The convergence tolerance and iteration limit complete the data. The iteration loop takes up the rest of the program. The largest off-diagonal term is stored as hold with its position in row and column nr and nc respectively.

The rotation angle alpha is then computed from equation (4.74) and its cosine ct and sine st, followed by the explicit changes to the matrix as outlined above. The diagonals of the transformed [**A**] matrix are stored in vector enew and compared with their values at the previous iteration held in eold using subroutine checkit. Once convergence has been achieved and the diagonals are hardly changing from one iteration to the next, the final transformed diagonal matrix is written to the output file shown in Results 4.6. In this case the eigenvalues are $\lambda = 7.14$, 19.15 and 33.71.

Once the eigenvalues have been computed the program runs through a final loop computing the normalized eigenvectors that go with each of the eigenvalues. This is achieved by solving the characteristic equations of the type shown in equation (4.3). The "penalty method" (see Program 2.7) is used to fix one of the components to unity, and the function `eliminate` (see Section 2.7) completes the solution of the remaining equations. The eigenvector is then normalized so that its Euclidean norm equals unity and written to output.

List 4.6:

Scalar integers:

i	simple counter
iters	iteration counter
j	simple counter
limit	iteration limit
n	number of equations
nc	column of term to be eliminated
nr	row of term to be eliminated

Scalar reals:

alpha	rotation angle α in radians
big	term of largest absolute magnitude in array
ct	$\cos \alpha$
den	denominator of expression for α
d2	set to 2.0
d4	set to 4.0
hold	temporary store
l2	L2 norm of a vector
one	set to 1.0
penalty	set to 1×10^{20}
pi	set to π
small	set to 1×10^{-20}
st	$\sin \alpha$
tol	convergence tolerance
zero	set to 0.0

Dynamic real arrays:

a	$n \times n$ matrix of coefficients
a1	temporary $n \times n$ storage matrix
a2	temporary $n \times n$ storage matrix
enew	"new" diagonals of transformed matrix
eold	"old" diagonals of transformed matrix
p	$n \times n$ transformation matrix
x	eigenvector

4.5.1 Comments on Jacobi diagonalization

Although Program 4.6 illustrates the transformation process well for teaching purposes, it would not be used to solve large problems. One would never, in practice, store the transformation matric [**P**], but perhaps less obviously, the searching process itself becomes very time-consuming as n increases. Alternatives to the basic Jacobi method which have been proposed include serial elimination in which the off-diagonal elements are eliminated in a predetermined sequence, thus avoiding searching altogether, and a variation of this technique in which serial elimination is performed only on those elements whose modulus exceeds a certain value or "threshold". When all off-diagonal terms have been reduced to the threshold, it can be further reduced and the process continued.

Jacobi's idea can also be implemented in order to reduce [**A**] to a tridiagonal matrix [**A***] (rather than diagonal) having the same eigenvalues. This is called "Givens's method", which has the advantage of being noniterative. Of course, some method must still be found for calculating the eigenvalues of the tridiagonal [**A***]. A more popular tridiagonalisation technique is called "Householder's method" which is described in the next section.

4.5.2 Householder's transformation to tridiagonal form

Equation (4.64) gave the basic property that transformation matrices should have, namely

$$[\mathbf{P}]^T[\mathbf{P}] = [\mathbf{I}] \tag{4.75}$$

and the Householder technique involves choosing

$$[\mathbf{P}] = [\mathbf{I}] - 2\{\mathbf{w}\}\{\mathbf{w}\}^T \tag{4.76}$$

where $\{\mathbf{w}\}$ is a column vector normalized such that its Euclidean norm equals unity, thus

$$\{\mathbf{w}\}^T\{\mathbf{w}\} = 1 \tag{4.77}$$

For example, let

$$\{\mathbf{w}\} = \left\{ \begin{array}{c} \dfrac{1}{\sqrt{2}} \\[2mm] \dfrac{1}{\sqrt{2}} \end{array} \right\} \tag{4.78}$$

which has the required product. Then

$$2\{\mathbf{w}\}\{\mathbf{w}\}^T = \begin{bmatrix} 1 & 1 \\ 1 & 1 \end{bmatrix} \tag{4.79}$$

and we see that

$$[\mathbf{P}] = [\mathbf{I}] - 2\{\mathbf{w}\}\{\mathbf{w}\}^T = \begin{bmatrix} 0 & -1 \\ -1 & 0 \end{bmatrix} \tag{4.80}$$

which has the desired property that

$$[\mathbf{P}]^{-1} = [\mathbf{P}]^T = [\mathbf{P}] \tag{4.81}$$

In order to eliminate terms in the first row of $[\mathbf{A}]$ outside the tridiagonal, the vector $\{\mathbf{w}\}$ is taken as

$$\{\mathbf{w}\} = [\,0 \;\; w_2 \;\; w_3 \;\; \cdots \;\; w_n\,]^T \tag{4.82}$$

Thus the transformation matrix for row 1, assuming $[\mathbf{A}]$ is 3×3, is

$$[\mathbf{P}] = \begin{bmatrix} 1 & 0 & 0 \\ 0 & 1 - 2w_2^2 & -2w_2 w_3 \\ 0 & -2w_3 w_2 & 1 - 2w_3^2 \end{bmatrix} \tag{4.83}$$

When the product $[\mathbf{P}][\mathbf{A}][\mathbf{P}]$ is carried out, the first row of the resulting matrix contains the following three terms

$$\begin{aligned} a_{11}^* &= a_{11} \\ a_{12}^* &= a_{12} - 2w_2(a_{12}w_2 + a_{13}w_3) = r \\ a_{13}^* &= a_{13} - 2w_3(a_{12}w_2 + a_{13}w_3) = 0 \end{aligned} \tag{4.84}$$

Letting

$$h = a_{12}w_2 + a_{13}w_3 \tag{4.85}$$

the second and third of equations (4.84) can be written as

$$\begin{aligned} a_{12}^* &= a_{12} - 2w_2 h = r \\ a_{13}^* &= a_{13} - 2w_3 h = 0 \end{aligned} \tag{4.86}$$

Equation (4.77) gives us another equation in the w_i, namely

$$w_2^2 + w_3^2 = 1 \tag{4.87}$$

hence by squaring both of equations (4.86), adding them together and making the substitutions from equations (4.85) and (4.87) we get

$$r^2 = a_{12}^2 + a_{13}^2 \tag{4.88}$$

Also from equations (4.86) we can write

$$w_2 = \frac{a_{12} - r}{2h} \tag{4.89}$$

$$w_3 = \frac{a_{13}}{2h}$$

which can also be squared and added together with substitutions from equations (4.86) and (4.88) to give

$$2h^2 = r^2 - a_{12}r \qquad (4.90)$$

From equations (4.89) we can also write

$$\{\mathbf{w}\} = \frac{1}{2h}\{\mathbf{v}\} \qquad (4.91)$$

where

$$\{\mathbf{v}\} = 2h\{\mathbf{w}\} = [\, 0 \ \ (a_{12} - r) \ \ a_{13} \,]^T \qquad (4.92)$$

leading to the transformation matrix

$$[\mathbf{P}] = [\mathbf{I}] - \frac{1}{2h^2}\{\mathbf{v}\}\{\mathbf{v}\}^T \qquad (4.93)$$

In equation (4.88) for the determination of r, the sign should be chosen such that r is of opposite sign to a_{12} in this case.

For a general row i, the vector $\{\mathbf{v}\}$ will take the form

$$\{\mathbf{v}\} = [\, 0 \ \ 0 \ \ 0 \ \cdots \ (a_{i,i+1} - r) \ \ a_{i,i+2} \ \ a_{i,i+3} \ \cdots \ a_{i,n}]^T \qquad (4.94)$$

Program 4.7: Householder reduction of symmetrical matrix to tridiagonal form

```
PROGRAM nm47
!---Householder Reduction of a Symmetrical Matrix
!to Tridiagonal Form---
USE precision; IMPLICIT NONE
INTEGER::i,k,l,n; REAL(iwp)::h,one=1.0_iwp,r,zero=0.0_iwp
REAL(iwp),ALLOCATABLE::a(:,:),a1(:,:),p(:,:),v(:,:)
OPEN(10,FILE='nm95.dat'); OPEN(11,FILE='nm95.res')
READ(10,*)n; ALLOCATE(a(n,n),a1(n,n),p(n,n),v(n,1))
DO i=1,n; READ(10,*)a(i,i:n); a(i:n,i)=a(i,i:n); END DO
WRITE(11,'(A,A)')"---Householder Reduction of a Symmetrical &
   &Matrix to Tridiagonal Form---"
WRITE(11,'(/,A)')"Coefficient Matrix"
DO i=1,n; WRITE(11,'(6E12.4)')a(i,:); END DO
DO k=1,n-2
   r=zero; DO l=k,n-1; r=r+a(k,l+1)*a(k,l+1); END DO
   r=SQRT(r); IF(r*a(k,k+1)>zero)r=-r
   h=-one/(r*r-r*a(k,k+1)); v=zero
```

```
      v(k+1,1)=a(k,k+1)-r; DO l=k+2,n; v(l,1)=a(k,1); END DO
      p=MATMUL(v,TRANSPOSE(v))*h
      DO l=1,n; p(l,1)=p(l,1)+one; END DO
      a1=MATMUL(a,p); a=MATMUL(p,a1)
   END DO
   WRITE(11,'(/,A)')"Transformed Main Diagonal"
   WRITE(11,'(6E12.4)')(a(i,i),i=1,n)
   WRITE(11,'(/,A)')"Transformed Off-Diagonal"
   WRITE(11,'(6E12.4)')(a(i-1,i),i=2,n)
END PROGRAM nm47
```

List 4.7:

Scalar integers:

i	simple counter
k	simple counter
l	simple counter
n	number of equations

Scalar reals:

h	term $-1/(2h^2)$ from equation (4.93)
one	set to 1.0
r	term from equations (4.86) and (4.88)
zero	set to 0.0

Dynamic real arrays:

a	$n \times n$ matrix of coefficients
a1	temporary $n \times n$ storage matrix
p	$n \times n$ transformation matrix
v	vector $\{\mathbf{v}\}$ from equation (4.92)

```
Number of equations      n
                         4

Coefficient matrix       (a(i,:),i=1,n)
                         1.0  -3.0  -2.0   1.0
                              10.0  -3.0   6.0
                                     3.0  -2.0
                                           1.0
```

Data 4.7: Householder Reduction to Tridiagonal Form

---Householder Reduction of a Symmetrical Matrix
to Tridiagonal Form---

Coefficient Matrix
   0.1000E+01 -0.3000E+01 -0.2000E+01   0.1000E+01
  -0.3000E+01   0.1000E+02 -0.3000E+01   0.6000E+01
  -0.2000E+01 -0.3000E+01   0.3000E+01 -0.2000E+01
   0.1000E+01   0.6000E+01 -0.2000E+01   0.1000E+01

Transformed Main Diagonal
   0.1000E+01   0.2786E+01   0.1020E+02   0.1015E+01

Transformed Off-Diagonal
   0.3742E+01 -0.5246E+01 -0.4480E+01

Results 4.7: Householder Reduction to Tridiagonal Form

The input is shown in Data 4.7. The number of equations and upper triangle of symmetrical matrix [**A**] are first read in. Then $n - 2$ transformations
are made for rows designated by counter k. Values of **r**, **h** and **v** are computed
and the vector product required by equation (4.93) is carried out. Transformation matrix [**P**] can then be formed and two matrix multiplications using
MATMUL complete the transformation. Results 4.7 show the resulting tridiagonalized matrix [**A***], whose eigenvalues would then have to be computed
by some other method, perhaps by vector iteration as previously described
or by a characteristic polynomial method as shown in the following section.
Alternatively, another transformation method may be used, as shown in the
next program.

The matrix arithmetic in this algorithm has been deliberately kept simple. In practice, more involved algorithms can greatly reduce storage and
computation time in this method.

4.5.3 Lanczos transformation to tridiagonal form

In Chapter 2, we saw that some iterative techniques for solving linear e-
quations, such as the steepest descent method, could be reduced to a loop
involving a single matrix by vector multiplication followed by various simple
vector operations. The Lanczos method for reducing matrices to tridiagonal
form, while preserving their eigenvalues, involves very similar operations, and
is in fact linked to the conjugate gradient technique of Program 2.12.

The transformation matrix [**P**] is in this method constructed using mutually
orthogonal vectors. As usual we seek an eigenvalue-preserving transformation
which for symmetrical matrices was given by equation (4.63) as

$$[\mathbf{P}]^T[\mathbf{A}][\mathbf{P}] \tag{4.95}$$

A means of ensuring $[\mathbf{P}]^T[\mathbf{P}] = [\mathbf{I}]$ is to construct $[\mathbf{P}]$ from mutually orthogonal, unit length normalized vectors, say $\{\mathbf{p}\}$, $\{\mathbf{q}\}$ and $\{\mathbf{r}\}$. For a 3×3 matrix for example, we would get

$$[\mathbf{P}]^T[\mathbf{P}] = \begin{bmatrix} p_1 & p_2 & p_3 \\ q_1 & q_2 & q_3 \\ r_1 & r_2 & r_3 \end{bmatrix} \begin{bmatrix} p_1 & q_1 & r_1 \\ p_2 & q_2 & r_2 \\ p_3 & q_3 & r_3 \end{bmatrix}$$

$$= \begin{bmatrix} \{\mathbf{p}\}^T\{\mathbf{p}\} & \{\mathbf{p}\}^T\{\mathbf{q}\} & \{\mathbf{p}\}^T\{\mathbf{r}\} \\ \{\mathbf{q}\}^T\{\mathbf{p}\} & \{\mathbf{q}\}^T\{\mathbf{q}\} & \{\mathbf{q}\}^T\{\mathbf{r}\} \\ \{\mathbf{r}\}^T\{\mathbf{p}\} & \{\mathbf{r}\}^T\{\mathbf{q}\} & \{\mathbf{r}\}^T\{\mathbf{r}\} \end{bmatrix} = [\mathbf{I}] \qquad (4.96)$$

In the Lanczos method, we require $[\mathbf{P}]^T[\mathbf{A}][\mathbf{P}]$ to be a symmetrical tridiagonal matrix, say

$$[\mathbf{M}] = [\mathbf{P}]^T[\mathbf{A}][\mathbf{P}] = \begin{bmatrix} \alpha_1 & \beta_1 & 0 \\ \beta_1 & \alpha_2 & \beta_2 \\ 0 & \beta_2 & \alpha_3 \end{bmatrix} \qquad (4.97)$$

and so

$$[\mathbf{A}][\mathbf{P}] = [\mathbf{P}][\mathbf{M}] \qquad (4.98)$$

Since $[\mathbf{P}]$ is made up of the orthogonal vectors

$$[\mathbf{P}] = [\{\mathbf{p}\} \, \{\mathbf{q}\} \, \{\mathbf{r}\}] \qquad (4.99)$$

we can expand equation (4.98) to give

$$\begin{aligned} [\mathbf{A}]\{\mathbf{p}\} &= \alpha_1\{\mathbf{p}\} + \beta_1\{\mathbf{q}\} \\ [\mathbf{A}]\{\mathbf{q}\} &= \beta_1\{\mathbf{p}\} + \alpha_2\{\mathbf{q}\} + \beta_2\{\mathbf{r}\} \\ [\mathbf{A}]\{\mathbf{r}\} &= \beta_2\{\mathbf{q}\} + \alpha_3\{\mathbf{r}\} \end{aligned} \qquad (4.100)$$

Multiplying the first, second and third of equations (4.100) by $\{\mathbf{p}\}^T$, $\{\mathbf{q}\}^T$ and $\{\mathbf{r}\}^T$ respectively, and noting orthogonality of vectors, we get

$$\begin{aligned} \{\mathbf{p}\}^T[\mathbf{A}]\{\mathbf{p}\} &= \alpha_1 \\ \{\mathbf{q}\}^T[\mathbf{A}]\{\mathbf{q}\} &= \alpha_2 \\ \{\mathbf{r}\}^T[\mathbf{A}]\{\mathbf{r}\} &= \alpha_3 \end{aligned} \qquad (4.101)$$

To construct the "Lanczos vectors" $\{\mathbf{p}\}$, $\{\mathbf{q}\}$ and $\{\mathbf{r}\}$ and solve for the tridiagonal terms α_i and β_i we can follow the following algorithm:

1) Make a starting guess for vector $\{\mathbf{p}\}$ with a unit length (e.g., $[1 \ \ 0 \ \ 0]^T$)

2) From equations (4.101) compute $\alpha_1 = \{\mathbf{p}\}^T[\mathbf{A}]\{\mathbf{p}\}$

3) From equations (4.100) compute $\beta_1\{\mathbf{q}\} = [\mathbf{A}]\{\mathbf{p}\} - \alpha_1\{\mathbf{p}\}$

4) The length of $\{\mathbf{q}\}$ is unity, hence compute β_1 and $\{\mathbf{q}\}$ by normalization

5) From equations (4.101) compute $\alpha_2 = \{\mathbf{q}\}^T[\mathbf{A}]\{\mathbf{q}\}$

6) From equations (4.100) compute $\beta_2\{\mathbf{r}\} = [\mathbf{A}]\{\mathbf{q}\} - \alpha_2\{\mathbf{q}\} - \beta_1\{\mathbf{p}\}$

7) The length of $\{\mathbf{r}\}$ is unity, hence compute β_2 and $\{\mathbf{r}\}$ by normalization

8) From equations (4.101) compute $\alpha_3 = \{\mathbf{r}\}^T[\mathbf{A}]\{\mathbf{r}\}$ etc.

In general, for an $n \times n$ matrix $[\mathbf{A}]$ and denoting the orthogonal vector columns of $[\mathbf{P}]$ by $\{\mathbf{y}\}_j$, $j = 1, 2, \cdots, n$, the algorithm used by Program 4.8 is as follows (setting $\beta_0 = 0$):

$$\begin{aligned}
\{\mathbf{v}\}_j &= [\mathbf{A}]\{\mathbf{y}\}_j \\
\alpha_j &= \{\mathbf{y}\}_j^T\{\mathbf{v}\}_j \\
\{\mathbf{z}\}_j &= \{\mathbf{v}\}_j - \alpha_j\{\mathbf{y}\}_j - \beta_{j-1}\{\mathbf{y}\}_{j-1} \\
\beta_j &= \left(\{\mathbf{z}\}_j^T\{\mathbf{z}\}_j\right)^{1/2} \\
\{\mathbf{y}\}_{j+1} &= \frac{1}{\beta_j}\{\mathbf{z}\}_j
\end{aligned} \qquad (4.102)$$

Program 4.8: Lanczos reduction of symmetrical matrix to tridiagonal form

```
PROGRAM nm48
!---Lanczos Reduction of a Symmetrical Matrix
!to Tridiagonal Form---
 IMPLICIT NONE
 INTEGER,PARAMETER::iwp=SELECTED_REAL_KIND(15,300)
 INTEGER::i,j,n; REAL(iwp)::zero=0.0_iwp
 REAL(iwp),ALLOCATABLE::a(:,:),alpha(:),beta(:),v(:),y0(:),    &
   y1(:),z(:)
 OPEN(10,FILE='nm95.dat'); OPEN(11,FILE='nm95.res')
 READ(10,*)n; ALLOCATE(a(n,n),alpha(n),beta(0:n-1),v(n),y0(n),  &
   y1(n),z(n))
 DO i=1,n; READ(10,*)a(i,i:n); a(i:n,i)=a(i,i:n); END DO
 READ(10,*)y1
 WRITE(11,'(A)')"---Lanczos Reduction of a Symmetrical Matrix &
   &to Tridiagonal Form---"
```

```
WRITE(11,'(/,A)')"Coefficient Matrix"
DO i=1,n; WRITE(11,'(6E12.4)')a(i,:); END DO
WRITE(11,'(/,A)')"Guessed Starting vector"
WRITE(11,'(6E12.4)')y1; y0=zero; beta(0)=zero
DO j=1,n
  v=MATMUL(a,y1); alpha(j)=DOT_PRODUCT(y1,v); IF(j==n)EXIT
  z=v-alpha(j)*y1-beta(j-1)*y0; y0=y1
  beta(j)=SQRT(DOT_PRODUCT(z,z)); y1=z/beta(j)
END DO
WRITE(11,'(/A)')"Transformed Main Diagonal"
WRITE(11,'(6E12.4)')alpha
WRITE(11,'(/A)')"Transformed Off-Diagonal"
WRITE(11,'(6E12.4)')beta(1:)
END PROGRAM nm48
```

List 4.8:

Scalar integers:

i	simple counter
j	simple counter
n	number of equations

Scalar reals:

zero	set to 0.0

Dynamic real arrays:

a	$n \times n$ matrix of coefficients
alpha	diagonal of tridiagonal matrix from equation (4.97)
beta	off-diagonal of tridiagonal matrix from equation (4.97)
v y0 y1 z	temporary vectors, see equations (4.102)

```
Number of equations      n
                         4

Coefficient matrix       (a(i,:),i=1,n)
                          1.0  -3.0  -2.0   1.0
                               10.0  -3.0   6.0
                                      3.0  -2.0
                                            1.0
```

```
Starting vector            y1
                           1.0    0.0    0.0    0.0
```

Data 4.8: Lanczos Reduction to Tridiagonal Form

```
---Lanczos Reduction of a Symmetrical Matrix
to Tridiagonal Form---

Coefficient Matrix
   0.1000E+01 -0.3000E+01 -0.2000E+01  0.1000E+01
  -0.3000E+01  0.1000E+02 -0.3000E+01  0.6000E+01
  -0.2000E+01 -0.3000E+01  0.3000E+01 -0.2000E+01
   0.1000E+01  0.6000E+01 -0.2000E+01  0.1000E+01


Guessed Starting vector
   0.1000E+01  0.0000E+00  0.0000E+00  0.0000E+00

Transformed Main Diagonal
   0.1000E+01  0.2786E+01  0.1020E+02  0.1015E+01

Transformed Off-Diagonal
   0.3742E+01  0.5246E+01  0.4480E+01
```

Results 4.8: Lanczos Reduction to Tridiagonal Form

Note that the process is not iterative. Input and output are listed in Data and Results 4.8 respectively. The number of equations, n, is first read in, followed by the upper triangle coefficients of the symmetrical $[\mathbf{A}]$ matrix. The starting vector $\{\mathbf{y}\}_1$, which is arbitrary as long as $\{\mathbf{y}\}_1^T\{\mathbf{y}\}_1 = 1$, is then read in and β_0 is set to 0.

The main loop carries out exactly the operations of equations (4.102) to build up n values of α and the $n-1$ values of β which are printed at the end of the program. For the given starting vector $\{\mathbf{y}\}_1 = [1 \quad 0 \quad 0 \quad 0]^T$, the Lanczos method yields a slightly different tridiagonal matrix than by Householder's method (Results 4.7), but both tridiagonal matrices have the same eigenvalues.

Correct answers have been obtained in this small example but in practice for larger problems roundoff becomes a serious drawback and more elaborate algorithms which deal with this difficulty are necessary.

4.5.4   LR transformation for eigenvalues of tridiagonal matrices

A transformation method most applicable to sparsely populated (band or tridiagonalized) matrices is the so-called "**LR**" transformation. This is based on repeated $[\mathbf{L}][\mathbf{U}]$ factorization of the type described in Chapter 2.

Thus,

$$[\mathbf{A}]_k = [\mathbf{L}][\mathbf{U}] = [\mathbf{L}][\mathbf{R}] \tag{4.103}$$

for any step k of the iterative transformation. The step is completed by re-multiplying the factors in reverse order, that is

$$[\mathbf{A}]_{k+1} = [\mathbf{U}][\mathbf{L}] = [\mathbf{R}][\mathbf{L}] \tag{4.104}$$

Since from equation (4.103)

$$[\mathbf{U}] = [\mathbf{L}]^{-1}[\mathbf{A}]_k \tag{4.105}$$

the multiplication in equation (4.104) implies

$$[\mathbf{A}]_{k+1} = [\mathbf{L}]^{-1}[\mathbf{A}]_k[\mathbf{L}] \tag{4.106}$$

showing that $[\mathbf{L}]$ has the property required of a transformation matrix $[\mathbf{P}]$. As iterations proceed, the transformed matrix $[\mathbf{A}]_k$ tends to an upper triangular matrix whose eigenvalues are equal to the diagonal terms (see Section 4.1.2).

Example 4.6

*Perform "**LR**" factorization on the nonsymmetrical matrix*

$$[\mathbf{A}] = \begin{bmatrix} 4 & 3 \\ 2 & 1 \end{bmatrix}$$

Solution 4.6

$$[\mathbf{A}]_0 = \begin{bmatrix} 4 & 3 \\ 2 & 1 \end{bmatrix} = \begin{bmatrix} 1.0 & 0.0 \\ 0.5 & 1.0 \end{bmatrix} \begin{bmatrix} 4.0 & 3.0 \\ 0.0 & -0.5 \end{bmatrix}$$

$$\begin{aligned}
[\mathbf{A}]_1 &= \begin{bmatrix} 4.0 & 3.0 \\ 0.0 & -0.5 \end{bmatrix} \begin{bmatrix} 1.0 & 0.0 \\ 0.5 & 1.0 \end{bmatrix} = \begin{bmatrix} 5.5 & 3.0 \\ -0.25 & -0.5 \end{bmatrix} \\
&= \begin{bmatrix} 1.0 & 0.0 \\ -0.045 & 1.0 \end{bmatrix} \begin{bmatrix} 5.5 & 3.0 \\ 0.0 & -0.3636 \end{bmatrix}
\end{aligned}$$

$$\begin{aligned}
[\mathbf{A}]_2 &= \begin{bmatrix} 5.5 & 3.0 \\ 0.0 & -0.3636 \end{bmatrix} \begin{bmatrix} 1.0 & 0.0 \\ -0.045 & 1.0 \end{bmatrix} = \begin{bmatrix} 5.365 & 3.0 \\ 0.0164 & -0.3636 \end{bmatrix} \\
&= \begin{bmatrix} 1.0 & 0.0 \\ 0.0031 & 1.0 \end{bmatrix} \begin{bmatrix} 5.365 & 3.0 \\ 0.0 & -0.3728 \end{bmatrix}
\end{aligned}$$

$$[\mathbf{A}]_3 = \begin{bmatrix} 5.365 & 3.0 \\ 0.0 & -0.3728 \end{bmatrix} \begin{bmatrix} 1.0 & 0.0 \\ 0.0031 & 1.0 \end{bmatrix} = \begin{bmatrix} 5.3743 & 3.0 \\ -0.0012 & -0.3728 \end{bmatrix}$$

$[\mathbf{A}]_3$ is nearly upper triangular, hence its eigenvalues are approximately 5.37 and -0.37 which are exact to 2 decimal places.

Although the method would be implemented in practice using special storage strategies, it is illustrated in Program 4.9 for the simple case of a square matrix $[\mathbf{A}]$.

Program 4.9: [L][R] transformation for eigenvalues

```
PROGRAM nm49
!---LR Transformation for Eigenvalues---
USE nm_lib; USE precision; IMPLICIT NONE
INTEGER::i,iters,limit,n; REAL(iwp)::tol,zero=0.0_iwp
REAL(iwp),ALLOCATABLE::a(:,:),enew(:),eold(:),lower(:,:),      &
  upper(:,:)
OPEN(10,FILE='nm95.dat'); OPEN(11,FILE='nm95.res')
READ(10,*)n; ALLOCATE(a(n,n),upper(n,n),lower(n,n),eold(n),    &
  enew(n)); READ(10,*)a,tol,limit
WRITE(11,*)"---LR Transformation for Eigenvalues---"
WRITE(11,'(/,A)')"Coefficient Matrix"
a=TRANSPOSE(a); DO i=1,n; WRITE(11,'(6E12.4)')a(i,:); END DO
WRITE(11,'(/,A)')"First Few Iterations"; iters=0; eold=zero
DO; iters=iters+1; CALL lufac(a,lower,upper)
  a=MATMUL(upper,lower)
  IF(iters<5)THEN
    DO i=1,n; WRITE(11,'(6E12.4)')a(i,:); END DO; WRITE(11,*)
  END IF
  DO i=1,n; enew(i)=a(i,i); END DO
  IF(checkit(enew,eold,tol).OR.iters==limit)EXIT; eold=enew
END DO
WRITE(11,'(/,A,/,I5)')"Iterations to Convergence",iters
WRITE(11,'(/A)')"Final Transformed Matrix"
DO i=1,n; WRITE(11,'(6E12.4)')a(i,:); END DO
END PROGRAM nm49
```

List 4.9:

Scalar integers:

i	simple counter
iters	iteration counter
limit	iteration limit
n	number of equations

Scalar reals:

tol	convergence tolerance
zero	set to 0.0

Dynamic real arrays:

a	$n \times n$ matrix of coefficients
enew	"new" diagonals of transformed matrix
eold	"old" diagonals of transformed matrix
lower	lower triangular factor of $[\mathbf{A}]_k$
upper	upper triangular factor of $[\mathbf{A}]_k$

Number of equations	n
	4

Coefficient matrix	(a(i,:),i=1,n)

1.0000	3.7417	0.0	0.0
3.7417	2.7857	5.2465	0.0
0.0	5.2465	10.1993	4.4796
0.0	0.0	4.4796	1.0150

	tol	limit
Tolerance and iteration limit	1.E-5	100

Data 4.9: LR Transformation

---LR Transformation for Eigenvalues---

Coefficient Matrix

0.1000E+01	0.3742E+01	0.0000E+00	0.0000E+00
0.3742E+01	0.2786E+01	0.5247E+01	0.0000E+00
0.0000E+00	0.5247E+01	0.1020E+02	0.4480E+01
0.0000E+00	0.0000E+00	0.4480E+01	0.1015E+01

First Few Iterations

0.1500E+02	0.3742E+01	0.0000E+00	0.0000E+00
-0.4196E+02	-0.1367E+02	0.5247E+01	0.0000E+00
0.0000E+00	-0.5920E+01	0.1424E+02	0.4480E+01
0.0000E+00	0.0000E+00	-0.2021E+00	-0.5708E+00

```
0.4533E+01   0.3742E+01   0.0000E+00   0.0000E+00
0.8957E+01   0.6497E+01   0.5247E+01   0.0000E+00
0.0000E+00   0.8394E+01   0.4341E+01   0.4480E+01
0.0000E+00   0.0000E+00   0.1653E-01  -0.3715E+00


0.1193E+02   0.3742E+01   0.0000E+00   0.0000E+00
-0.1770E+01  -0.5004E+02   0.5247E+01   0.0000E+00
0.0000E+00  -0.5011E+03   0.5349E+02   0.4480E+01
0.0000E+00   0.0000E+00  -0.1152E-03  -0.3728E+00


0.1137E+02   0.3742E+01   0.0000E+00   0.0000E+00
0.7346E+01   0.3632E+01   0.5247E+01   0.0000E+00
-0.8438E-14   0.3741E+01   0.3681E+00   0.4480E+01
-0.2632E-17   0.1787E-16   0.1159E-03  -0.3714E+00
```

Iterations to Convergence
  48

Final Transformed Matrix
```
 0.1433E+02   0.3742E+01   0.0000E+00   0.0000E+00
 0.3551E-21   0.4457E+01   0.5247E+01   0.0000E+00
-0.1913E-40   0.7503E-04  -0.3415E+01   0.4480E+01
-0.1313E-86   0.8507E-64   0.2549E-46  -0.3714E+00
```

Results 4.9: LR Transformation

Input and output are shown in Data and Results 4.9 respectively. The program begins by reading the number of equations n, followed by the coefficients of $[\mathbf{A}]$, the convergence tolerance and iteration limit. The iteration loop is then entered, and begins with a call to lufac which completes the factorization of the current $[\mathbf{A}]_k$ into $[\mathbf{L}]$ and $[\mathbf{U}]$. These are multiplied in reverse order following equation (4.104) using MATMUL and the new estimate of the eigenvalues is found in the diagonal terms of the new $[\mathbf{A}]_{k+1}$. The data shown in Data 4.9 are the tridiagonal matrix terms produced by the Lanczos method from Program 4.8 (see Results 4.8). The output in Results 4.9 indicates that the transformed matrix following 48 iterations is essentially upper triangular, with diagonal terms indicating eigenvalues $\lambda = 14.33, 4.46, -3.42$ and -0.37. As would be expected these are also the eigenvalues of the initial matrix used in the Lanczos example, namely

$$[\mathbf{A}] = \begin{bmatrix} 1 & -3 & -2 & 1 \\ -3 & 10 & -3 & 6 \\ -2 & -3 & 3 & -2 \\ 1 & 6 & -2 & 1 \end{bmatrix}$$

4.6   Characteristic polynomial methods

At the beginning of this chapter we illustrated how the eigenvalues of a matrix form the roots of an nth order polynomial, called the "characteristic polynomial". We pointed out that the methods of Chapter 3 could, in principle, be used to evaluate these roots, but that this will rarely be an effective method of eigenvalue determination. However, there are effective methods which are based on the properties of the characteristic polynomial. These are particularly attractive when the matrix whose eigenvalues has to be found is a tridiagonal matrix, and so are especially appropriate when used in conjunction with the Householder or Lanczos transformations described in the previous sections.

4.6.1   Evaluating determinants of tridiagonal matrices

In the previous section we illustrated noniterative methods of reducing matrices to tridiagonal equivalents. The resulting eigenvalue equation for an $n \times n$ system became

$$
\begin{bmatrix}
\alpha_1 & \beta_1 & 0 & 0 & 0 & \cdots & \cdots & \cdots \\
\beta_1 & \alpha_2 & \beta_2 & 0 & 0 & \cdots & \cdots & \cdots \\
0 & \beta_2 & \alpha_3 & \beta_3 & 0 & \cdots & \cdots & \cdots \\
\vdots & \vdots & \vdots & \vdots & \vdots & \vdots & \vdots & \vdots \\
\cdots & \cdots & \cdots & \cdots & \cdots & \cdots & \alpha_{n-1} & \beta_{n-1} \\
\cdots & \cdots & \cdots & \cdots & \cdots & \cdots & \beta_{n-1} & \alpha_n
\end{bmatrix}
\begin{Bmatrix}
x_1 \\ x_2 \\ x_3 \\ \vdots \\ \vdots \\ x_n
\end{Bmatrix}
= \lambda
\begin{Bmatrix}
x_1 \\ x_2 \\ x_3 \\ \vdots \\ \vdots \\ x_n
\end{Bmatrix}
\qquad (4.107)
$$

The problem therefore becomes one of finding the roots of the determinantal equation

$$
\begin{vmatrix}
\alpha_1 - \lambda & \beta_1 & 0 & 0 & 0 & \cdots & \cdots & \cdots \\
\beta_1 & \alpha_2 - \lambda & \beta_2 & 0 & 0 & \cdots & \cdots & \cdots \\
0 & \beta_2 & \alpha_3 - \lambda & \beta_3 & 0 & \cdots & \cdots & \cdots \\
\vdots & \vdots & \vdots & \vdots & \vdots & \vdots & \vdots & \vdots \\
\cdots & \cdots & \cdots & \cdots & \cdots & \cdots & \alpha_{n-1} - \lambda & \beta_{n-1} \\
\cdots & \cdots & \cdots & \cdots & \cdots & \cdots & \beta_{n-1} & \alpha_n - \lambda
\end{vmatrix}
= 0 \qquad (4.108)
$$

Although we shall not find these roots directly, consider the calculation of the determinant on the left-hand side of equations (4.108).

For $n = 1$

$$\det_1(\lambda) = \alpha_1 - \lambda \tag{4.109}$$

For $n = 2$

$$\det_2(\lambda) = \begin{vmatrix} \alpha_1 - \lambda & \beta_1 \\ \beta_1 & \alpha_2 - \lambda \end{vmatrix} = (\alpha_1 - \lambda)(\alpha_2 - \lambda) - \beta_1^2 \tag{4.110}$$

For $n = 3$

$$\det_3(\lambda) = \begin{vmatrix} \alpha_1 - \lambda & \beta_1 & 0 \\ \beta_1 & \alpha_2 - \lambda & \beta_2 \\ 0 & \beta_2 & \alpha_3 - \lambda \end{vmatrix} = (\alpha_3 - \lambda) \begin{vmatrix} \alpha_1 - \lambda & \beta_1 \\ \beta_1 & \alpha_2 - \lambda \end{vmatrix} - \beta_2^2(\alpha_1 - \lambda) \tag{4.111}$$

and so on.

We see that a recurrence relationship builds up enabling $\det_3(\lambda)$ to be evaluated simply from a knowledge of $\det_2(\lambda)$ and $\det_1(\lambda)$. So if we let $\det_0(\lambda) = 1$, the general recurrence may be written

$$\det_n(\lambda) = (\alpha_n - \lambda) \det_{n-1}(\lambda) - \beta_{n-1}^2 \det_{n-2}(\lambda) \tag{4.112}$$

Therefore, for any value of λ, we can quickly calculate $\det_n(\lambda)$, and if we know the range within which a root $\det_n(\lambda) = 0$ must lie, its value can be computed by, for example, the Bisection method of Program 3.2. The remaining difficulty is to guide the choices of λ so as to be sure of bracketing a root. This task is made much easier due to a special property possessed by the "principal minors" of equations (4.108), which is called the "Sturm sequence" property.

4.6.2 The Sturm sequence property

A specific example of the left-hand side of equation (4.108) is shown below, for $n = 5$:

$$|\mathbf{A}| = \begin{vmatrix} 2 - \lambda & -1 & 0 & 0 & 0 \\ -1 & 2 - \lambda & -1 & 0 & 0 \\ 0 & -1 & 2 - \lambda & -1 & 0 \\ 0 & 0 & -1 & 2 - \lambda & -1 \\ 0 & 0 & 0 & -1 & 2 - \lambda \end{vmatrix} \tag{4.113}$$

The principal minors of $|\mathbf{A}|$ are the determinants of the submatrices outlined by the dotted lines, i.e., formed by eliminating the nth, $(n-1)^{th}$, etc., row and column of $[\mathbf{A}]$. The eigenvalues of $[\mathbf{A}]$ and of its principal minors will be found to be given by the following table (e.g., by using Program 4.9)

$[\mathbf{A}] =$	$[\mathbf{A}_5]$	$[\mathbf{A}_4]$	$[\mathbf{A}_3]$	$[\mathbf{A}_2]$	$[\mathbf{A}_1]$
3.732					
	3.618				
3.0		3.414			
	2.618		3.0		
2.0		2.0		2.0	
	1.382		1.0		
1.0		0.586			
	0.382				
0.268					

The characteristic polynomials of $[\mathbf{A}_i]$, $i = 0, 1, 2, \cdots, n$ from equation (4.112) are shown plotted in Figure 4.1 indicating their roots which are also their eigenvalues. From the tabular and graphical representations it can be seen that each succeeding set of eigenvalues of $[\mathbf{A}]_n$, $[\mathbf{A}]_{n-1}$, $[\mathbf{A}]_{n-2}$ etc., always "separates" the preceding set, that is, the eigenvalues of $[\mathbf{A}_{i-1}]$ always occur in the gaps between the eigenvalues of $[\mathbf{A}_i]$. This separation property is found for all symmetrical $[\mathbf{A}]$ and is called the "Sturm sequence" property.

Its most useful consequence is that for any guessed λ, the number of sign changes in $\det_i(\lambda)$ for $i = 0, 1, 2, \cdots, n$ is equal to the number of eigenvalues of $[\mathbf{A}]$ which are less than λ. When counting the sign changes it should be recalled that $\det_0(\lambda) = 1$, and noted that $\det_i(\lambda) = 0$ does *not* count as a change.

For the specific example shown in equation (4.113), let us guess a value of $\lambda = 4$ and evaluate $\det_i(4)$, $i = 0, 1, 2, \cdots, 5$ to give the table

$\det_0(4)$	1.0
$\det_1(4)$	-2.0
$\det_2(4)$	3.0
$\det_3(4)$	-4.0
$\det_4(4)$	5.0
$\det_5(4)$	-6.0

Starting at $\det_0(4) = 1.0$ and moving down the table we see 5 sign changes, hence there are 5 eigenvalues *less than* 4.

Now let's try $\lambda = 3.5$. In this case the table is

$\det_0(3.5)$	1.0
$\det_1(3.5)$	-1.5
$\det_2(3.5)$	1.25
$\det_3(3.5)$	-0.375
$\det_4(3.5)$	-0.6873
$\det_5(3.5)$	1.4060

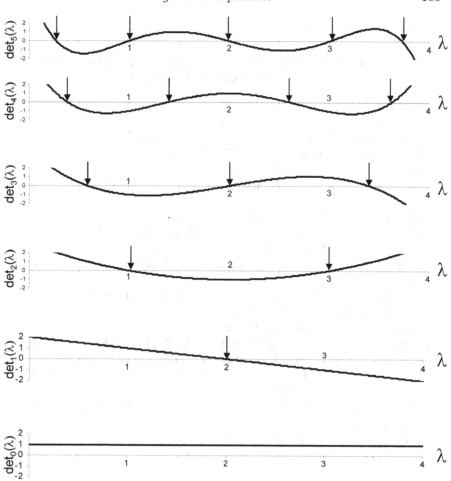

Figure 4.1: Characteristic polynomials for $[\mathbf{A}]_i$ from equations (4.112).

and we see only 4 sign changes, thus there are 4 eigenvalues less than 3.5. These two results indicate that the largest eigenvalue must lie in the range $3.5 < \lambda < 4$. The table below summarizes results for a selection of λ values.

λ	Number of sign changes = number of eigenvalues $< \lambda$
4	5
3.5	4
2.5	3
1.5	2
0.5	1

Program 4.10: Characteristic polynomial method for eigenvalues of symmetrical tridiagonal matrix

```
PROGRAM nm410
!---Characteristic Polynomial Method---
!---for Eigenvalues of Symmetrical Tridiagonal Matrix---
USE nm_lib; USE precision; IMPLICIT NONE
INTEGER::i,iters,j,limit,n,number
REAL(iwp)::al,almax,aold,half=0.5_iwp,oldl,one=1.0_iwp,sign,     &
  small=1.E-20_iwp,tol
REAL(iwp),ALLOCATABLE::alpha(:),beta(:),det(:)
OPEN(10,FILE='nm95.dat'); OPEN(11,FILE='nm95.res')
WRITE(11,'(A)')"---Characteristic Polynomial Method---"
WRITE(11,'(A)')"---for Eigenvalues of Symmetrical Tridiagonal &
  &Matrix---"
READ(10,*)n; ALLOCATE(alpha(n),beta(n-1),det(0:n))
READ(10,*)alpha; READ(10,*)beta
READ(10,*)j,al,almax,tol,limit
WRITE(11,'(/,A)')"Main Diagonal"; WRITE(11,'(6E12.4)')alpha
WRITE(11,'(/,A)')"Off-Diagonal"; WRITE(11,'(6E12.4)')beta
WRITE(11,'(/,A)')"Eigenvalue required, 1=largest,2=second &
  &largest etc."; WRITE(11,'(I8)')j
WRITE(11,'(/,A)')"Eigenvalue   Determinant   Number of roots &
  &less than"
iters=0; det(0)=one; aold=almax
DO; iters=iters+1; det(1)=alpha(1)-al
  DO i=2,n
    det(i)=(alpha(i)-al)*det(i-1)-beta(i-1)*beta(i-1)*det(i-2)
  END DO; number=0
  DO i=1,n
    IF(ABS(det(i))<small)CYCLE
    IF(ABS(det(i-1))<small)THEN
      sign=det(i)*det(i-2)
    ELSE
      sign=det(i)*det(i-1)
    END IF
    IF(sign<small)number=number+1
  END DO
  IF(number<=n-j) THEN
    oldl=al; al=half*(al+almax)
  ELSE
    almax=al; al=half*(oldl+al)
  END IF
```

```
   IF(det(n)<small)number=number-1
   IF(MOD(j,2)==0)number=number-1
   WRITE(11,'(2E12.4,I15)')al,det(n),number
   IF(check(al,aold,tol).OR.iters==limit)EXIT; aold=al
 END DO
 WRITE(11,'(/,A,/,I5)')"Iterations to Convergence",iters
END PROGRAM nm410
```

Number of equations	n
	5

Diagonals	alpha
	2.0 2.0 2.0 2.0 2.0

Off-diagonals	beta
	-1.0 -1.0 -1.0 -1.0

Eigenvalue required	j
	1

Starting value	al
	2.5

Upper estimate	almax
	5.0

Tolerance and	tol	limit
iteration limit	1.E-5	100

Data 4.10: Characteristic Polynomial Method for Tridiagonal Matrix

```
---Characteristic Polynomial Method---
---for Eigenvalues of Symmetrical Tridiagonal Matrix---

Main Diagonal
  0.2000E+01  0.2000E+01  0.2000E+01  0.2000E+01  0.2000E+01

Off-Diagonal
 -0.1000E+01 -0.1000E+01 -0.1000E+01 -0.1000E+01

Eigenvalue required, 1=largest,2=second largest etc.
     1
```

Eigenvalue	Determinant	Number of roots less than
0.3750E+01	-0.1031E+01	2
0.3125E+01	-0.2256E+00	4
0.3438E+01	0.5183E+00	4
0.3594E+01	0.1431E+01	4
0.3672E+01	0.1129E+01	4
0.3711E+01	0.6148E+00	4
0.3730E+01	0.2397E+00	4
0.3740E+01	0.1891E-01	4
0.3735E+01	-0.1003E+00	4
0.3733E+01	-0.3995E-01	4
0.3732E+01	-0.1034E-01	4
0.3732E+01	0.4332E-02	4
0.3732E+01	-0.2990E-02	4
0.3732E+01	0.6740E-03	4
0.3732E+01	-0.1157E-02	4
0.3732E+01	-0.2414E-03	4
0.3732E+01	0.2164E-03	4

```
Iterations to Convergence
   17
```

Results 4.10: Characteristic Polynomial Method for Symmetrical Tridiagonal Matrix

Input and output are shown in Data and Results 4.10 respectively. The number of equations n is first read in followed by the diagonal **alpha** and off-diagonal **beta** terms of the tridiagonal symmetrical matrix previously obtained by methods such as Householder or Lanczos. The remaining data consist of j, the required eigenvalue where **j=1** is the largest etc., a starting guess of λ (**al**), an upper limit on λ (**almax**), a convergence tolerance and iteration limit. Since symmetrical, positive definite $[\mathbf{A}]$ are implied, all of the eigenvalues will be positive.

The data relate to the example given in equations (4.113) and involve a search for the largest (**j=1**) eigenvalue. An upper limit of 5.0 is chosen as being bigger than the biggest eigenvalue, and the first guess for λ is chosen to be half this value, that is 2.5. The value of $\det_0(\lambda)$ called **det(0)** is set to 1.0 and an iteration loop for the bisection process executed. The procedure continues until the iteration limit is reached or the tolerance, **tol**, is satisfied by subroutine **check**. The value of $\det_1(\lambda)$ is set to $\alpha_1 - \lambda$ and then the other $\det_i(\lambda), i = 2, 3, \cdots, n$ are formed by recursion from equation (4.112). The number of sign changes is detected by **sign** and accumulated as **number**.

The output shown in Results 4.10 reports the trial eigenvalue λ, the value of the determinant $\det_5(\lambda)$ and the number of eigenvalues less than the current value. At convergence the largest eigenvalue is given as $\lambda = 3.732$ with

4 eigenvalues less than this value. In general more elaborate interpolation processes will be necessary, especially as numbers of equations become large.

List 4.10:

Scalar integers:

i	simple counter
iters	iteration counter
j	eigenvalue required (1=largest, 2=second largest, etc.)
limit	iteration limit
n	number of equations
number	number of equations

Scalar reals:

al	current estimate of root
almax	upper estimate of root
aold	previous estimate of root
half	set to 0.5
oldl	lower estimate of root
one	set to 1.0
sign	detects sign changes
small	set to 1×10^{-20}
tol	convergence tolerance

Dynamic real arrays:

alpha	diagonal of tridiagonal matrix from equation (4.107)
beta	off-diagonal of tridiagonal matrix from equation (4.107)
det	holds values of $\det_i(\lambda), i = 0, 1, 2, \cdots, n$

4.6.3 General symmetrical matrices, e.g., band matrices

The principles described in the previous section can be applied to general matrices, but the simple recursion formula for finding $\det(\lambda)$ no longer applies. A way of computing $\det(\lambda)$ is to factorize $[\mathbf{A}]$, using the techniques described in Program 2.3, to yield $[\mathbf{A}] = [\mathbf{L}][\mathbf{D}][\mathbf{L}]^T$. The product of the diagonal elements in $[\mathbf{D}]$ is the determinant of $[\mathbf{A}]$. Further useful information that can be derived from $[\mathbf{D}]$ is that in the factorization of $[\mathbf{A}] - \lambda[\mathbf{I}]$, the number of negative elements in $[\mathbf{D}]$ is equal to the number of eigenvalues smaller than λ.

4.7   Exercises

1. Use vector iteration to find the eigenvector corresponding to the largest eigenvalue of the matrix

$$
\begin{bmatrix}
2 & 2 & 2 \\
2 & 5 & 5 \\
2 & 5 & 11
\end{bmatrix}
$$

   Answer: $[0.2149 \ 0.4927 \ 0.8433]^T$ corresponding to eigenvalue 14.43.

2. Use vector iteration to find the largest eigenvalue of the matrix

$$
\begin{bmatrix}
3 & -1 \\
-1 & 2
\end{bmatrix}
$$

   and its associated eigenvector.

   Answer: 3.618 associated with $[0.8507 \ -0.5257]^T$

3. Use shifted vector iteration to find the smallest eigenvalue and eigenvector of the system given in Exercise 1.
   Answer: Smallest eigenvalue 0.954 corresponding to eigenvector $[0.8360 \ 0.5392 \ 0.1019]^T$.

4. The eigenvalues of the matrix

$$
\begin{bmatrix}
5 & 1 & 0 & 0 & 0 \\
1 & 5 & 1 & 0 & 0 \\
0 & 1 & 5 & 1 & 0 \\
0 & 0 & 1 & 5 & 1 \\
0 & 0 & 0 & 1 & 5
\end{bmatrix}
$$

   are $5 + 2 \cos \frac{i\pi}{6}$. Confirm these values using shifted inverse iteration with shifts of 6.7, 6.1, 5.3, 4.1 and 3.3.
   Answer: 6.732, 6.0, 5.0, 4.0, 3.268

5. Find the eigenvalues and eigenvectors of the system

$$
\begin{bmatrix}
2 & 1 \\
1 & 1
\end{bmatrix}
\begin{Bmatrix}
x_1 \\
x_2
\end{Bmatrix}
= \lambda
\begin{bmatrix}
5 & 2 \\
2 & 1
\end{bmatrix}
\begin{Bmatrix}
x_1 \\
x_2
\end{Bmatrix}
$$

   Answer: $\lambda_1 = 2.618$, $\{x_1\} = [-0.3568 \ 0.9342]^T$
   $\lambda_2 = 0.382$, $\{x_2\} = [0.9342 \ -0.3568]^T$

6. Show that the system in Exercise 5 can be reduced to the "standard form"

$$
\begin{bmatrix}
0.4 & 0.2 \\
0.2 & 2.6
\end{bmatrix}
\begin{Bmatrix}
x_1 \\
x_2
\end{Bmatrix}
= \lambda
\begin{Bmatrix}
x_1 \\
x_2
\end{Bmatrix}
$$

and hence find both of its eigenvalues. How would you recover the eigenvectors of the original system?
Answer: 0.382, 2.618. See Section 4.4.1

7. User Householder's method to tridiagonalize the matrix

$$\begin{bmatrix} 1 & 1 & 1 & 1 \\ 1 & 2 & 2 & 2 \\ 1 & 2 & 3 & 3 \\ 1 & 2 & 3 & 4 \end{bmatrix}$$

Answer:

$$\begin{bmatrix} 1.0 & -1.7321 & 0.0 & 0.0 \\ -1.7321 & 7.6667 & 1.2472 & 0.0 \\ 0.0 & 1.2472 & 0.9762 & -0.1237 \\ 0.0 & 0.0 & -0.1237 & 0.3571 \end{bmatrix}$$

8. Use Lanczos's method to tridiagonalize the matrix in Exercise 7, using the starting vector $[0.5 \ 0.5 \ 0.5 \ 0.5]^T$.
Answer:

$$\begin{bmatrix} 7.5 & 2.2913 & 0.0 & 0.0 \\ 2.2913 & 1.6429 & 0.2736 & 0.0 \\ 0.0 & 0.2736 & 0.5390 & 0.0694 \\ 0.0 & 0.0 & 0.0694 & 0.3182 \end{bmatrix}$$

9. Find the eigenvalues of the tridiagonalized matrices in Exercises 7 and 8.
Answer: 8.291, 1.00, 0.4261, 0.2832 in both cases.

10. Find all the eigenvalues of the matrix

$$\begin{bmatrix} 3 & 0 & 2 \\ 0 & 5 & 0 \\ 2 & 0 & 3 \end{bmatrix}$$

and show that the eigenvectors associated with the eigenvalues are orthogonal.
Answer: $\lambda_1 = 1$, $\lambda_2 = 5$, $\lambda_3 = 5$ associated with orthogonal eigenvectors, $[1 \ 0 \ -1]^T$, $[1 \ 0 \ 1]^T$, $[0 \ 1 \ 0]^T$ respectively.

11. Using the characteristic polynomial method for symmetrical tridiagonal matrices, calculate all the eigenvalues of the matrix below. Use a tolerance that will give solutions accurate to six decimal places.

$$\begin{bmatrix} 2 & -1 & 0 & 0 & 0 \\ -1 & 2 & -1 & 0 & 0 \\ 0 & -1 & 2 & -1 & 0 \\ 0 & 0 & -1 & 2 & -1 \\ 0 & 0 & 0 & -1 & 2 \end{bmatrix}$$

Answer: 3.732051, 3.000000, 2.000000, 1.000000, 0.267949 with a toler-ance of about 1×10^{-7}

12. Use shifted inverse iteration to estimate the eigenvalue closest to 6 of the matrix:

$$\begin{bmatrix} 5 & 2 \\ 2 & 3 \end{bmatrix}$$

Check your solution by solving the characteristic polynomial for *both* the eigenvalues and hence find also the eigenvectors normalized to a Euclidean norm of unity.
Answer: $\lambda_1 = 1.7639$, $\{x_1\} = [-0.5257 \ 0.8507]^T$; $\lambda_2 = 6.2361$, $\{x_2\} = [0.8507 \ 0.5257]^T$

13. Use an iterative method to estimate the largest eigenvalue of the matrix below using an initial guess of $[0.3 \ \ 0.4 \ \ 0.9]^T$

$$\begin{bmatrix} 10 & 5 & 6 \\ 5 & 20 & 4 \\ 6 & 4 & 30 \end{bmatrix}$$

Obtain the corresponding eigenvector and normalize it such that its length (Euclidean norm) equals unity.
Answer: $\lambda = 33.71$, $\{x\} = [0.300 \ 0.366 \ 0.881]^T$

14. An eigenvector of the matrix:

$$\begin{bmatrix} 4 & 5 & 0 \\ 5 & 4 & 5 \\ 0 & 5 & 4 \end{bmatrix}$$

is given by:

$$\begin{Bmatrix} 0.353535 \\ 0.500000 \\ 0.353535 \end{Bmatrix}$$

Find the corresponding eigenvalue and hence use the characteristic poly-nomial method to obtain the other two eigenvalues.
Answer: $\lambda_1 = 11.07$, $\lambda_1 = 4$, $\lambda_1 = -3.07$

15. Use an iterative method to find the largest eigenvalue and corresponding eigenvector of the following matrix correct to 2 decimal places.

$$\begin{bmatrix} 2 & 2 & 2 \\ 2 & 4 & 6 \\ 2 & 6 & 12 \end{bmatrix}$$

If the eigenvector corresponding to the smallest eigenvalue is given by:

$$\begin{Bmatrix} -0.54 \\ x_2 \\ -0.31 \end{Bmatrix}$$

estimate x_2.
Answer: $\lambda_1 = 15.747$, $\{x\} = [0.2254\ 0.5492\ 1]^T$, $x_2 = 0.79$

16. A vibrating system with two degrees of freedom is defined by a stiffness matrix $[K]$ and mass matrix $[M]$ as follows:

$$[K] = \begin{bmatrix} 5 & -2 \\ -2 & 2 \end{bmatrix} \quad [M] = \begin{bmatrix} 1 & 0 \\ 0 & 4 \end{bmatrix}$$

Without inverting any matrices, use an iterative method to estimate the lowest natural frequency of this system.
Answer: $\omega = 0.5365$ (Eigenvalue equals ω^2)

17. An eigenvalue of the following matrix is approximately equal to 25. Set up an iterative numerical method for obtaining a more accurate estimate of this eigenvalue, and perform the first iteration of the method by hand. Use a starting vector of $[0\ 1\ 0]^T$.

$$\begin{bmatrix} 30 & 6 & 5 \\ 6 & 40 & 7 \\ 5 & 7 & 20 \end{bmatrix}$$

Although you are asked to perform just one iteration, the operations should be performed in such a way that subsequent iterations will be facilitated. Briefly describe how you will interpret your result once a converged solution has been obtained. Do not invert any matrices.
Answer: After 1 iteration $\alpha_1 = -0.165$, $\lambda_1 = \frac{1}{-0.165} + 25 = 18.9$ and $[1\ -0.769\ -0.077]^T$.
Converged solution $\alpha = 0.418$, $\lambda = \frac{1}{0.418} + 25 = 27.4$ and $[1\ -0.558\ 0.147]^T$

18. Use a transformation method to estimate all the eigenvalues of the matrix:

$$\begin{bmatrix} 4 & 2 & 0 \\ 2 & 3 & 0 \\ 0 & 0 & 5 \end{bmatrix}$$

Find the eigenvector of the smallest eigenvalue, expressed with its largest component set equal to unity.
Answer: Eigenvalues: 5.5616, 1.4384, 5, eigenvector $[-0.7808\ 1\ 0]^T$

19. One of the eigenvalues of the matrix

$$\begin{bmatrix} 1 & 2 & 4 \\ 3 & 7 & 2 \\ 5 & 6 & 9 \end{bmatrix}$$

is known to be -0.8946. Use the characteristic polynomial to find the other two eigenvalues and hence find the eigenvector of the intermediate

eigenvalue scaled to have unit length.
Answer: Eigenvalues -0.8946, 4.1467 and 13.7480, intermediate eigen-vector $[0.2992 \; -0.7383 \quad 0.6045]^T$.

20. Set up an iterative method that would eventually lead to the smallest eigenvalue of the matrix

$$\begin{bmatrix} 2 & 2 & 2 \\ 2 & 5 & 5 \\ 2 & 5 & 11 \end{bmatrix}$$

Perform two iterations by hand with a starting vector of $[\, 1 \; -1 \quad 0\,]^T$. You can use ordinary Gaussian elimination working to four decimal places of accuracy.
Answer: After 1 iteration normalized vector is $[1.0000 \; -0.7143 \quad 0.1429]^T$. By computer solution, the smallest eigenvalue is 0.9539.

21. Working to an accuracy of four decimal places, find all the eigenvalues and the eigenvector corresponding to the smallest eigenvalue of the matrix:

$$\begin{bmatrix} 4 & 0 & 4 \\ 0 & 7 & 0 \\ 4 & 0 & 5 \end{bmatrix}$$

Normalize the eigenvector to a Euclidean norm of unity.
Answer: Eigenvalues 8.5311, 7.0, 0.4689,
eigenvector $[\, 0.7497 \;\; 0.0 \;\; -0.6618\,]^T$

Chapter 5

Interpolation and Curve Fitting

5.1 Introduction

This chapter is concerned with fitting mathematical functions to discrete data. Such data may come from measurements made during an experiment, or perhaps numerical results obtained from another computer program. The functions will usually involve polynomials, which are easy to operate on, although other types of function may also be encountered.

Two general approaches will be covered. Firstly, functions which pass exactly through every point, and secondly, functions which are a "good fit" to the points but do not necessarily pass through them. The former approach leads to "interpolating polynomials" and the latter to "curve fitting" or "approximating polynomials".

We may have several reasons for wishing to fit functions to discrete data. A common requirement is to use our derived function to interpolate between known values, or estimate derivatives and integrals. Numerical integration makes extensive use of approximating polynomials in Chapter 6, whereas estimation of derivatives from discrete data is discussed later in this chapter.

5.2 Interpolating polynomials

Firstly we consider the derivation of a function which passes exactly through a series of n_p discrete data points. While there is an infinite number of functions that could achieve this, we will focus on the simplest one which can be shown to be an nth order polynomial, where $n = n_p - 1$. We will call this function the "interpolating polynomial" of the form

$$Q_n(x) = a_0 + a_1 x + a_2 x^2 + \ldots + a_n x^n \tag{5.1}$$

Hence, if our $n + 1$ points are given as (x_i, y_i) for $i = 0, 1, 2, \ldots, n$, then

$$Q_n(x_i) = y_i \quad \text{for} \quad i = 0, 1, 2, \ldots, n \tag{5.2}$$

In the next two subsections we will describe two methods for deriving $Q_n(x)$; Both methods are quite general and work for any set of initial data points, however the Difference method will be shown to offer several advantages if the x-data is equally spaced.

5.2.1   Lagrangian polynomials

This approach works for any general set of $n+1$ data points (x_i, y_i), $i = 0, 1, 2, \ldots, n$, leading to an interpolating polynomial of the form

$$Q_n(x) = L_0(x)y_0 + L_1(x)y_1 + \ldots + L_n(x)y_n \qquad (5.3)$$

The $L_i(x)$, $i = 0, 1, 2, \ldots, n$ are themselves polynomials of degree n called "Lagrangian Polynomials" defined

$$L_i(x) = \frac{(x-x_0)(x-x_1)\ldots(x-x_{i-1})(x-x_{i+1})\ldots(x-x_{n-1})(x-x_n)}{(x_i-x_0)(x_i-x_1)\ldots(x_i-x_{i-1})(x_i-x_{i+1})\ldots(x_i-x_{n-1})(x_i-x_n)}$$
$$(5.4)$$

It can be noted from equation (5.4) that Lagrangian Polynomials have the property

$$\left. \begin{matrix} L_i(x_j) = 1 \text{ if } i = j \\ L_i(x_j) = 0 \text{ if } i \neq j \end{matrix} \right\} \ i = 0, 1, 2, \ldots, n \qquad (5.5)$$

A further property of Lagrangian Polynomials is that they sum to unity, thus

$$\sum_{i=0}^{n} L_i(x) = 1 \qquad (5.6)$$

Example 5.1

Use Lagrangian polynomials to derive a polynomial passing through the points

$$\begin{matrix} x_0 = 1 & y_0 = 1 \\ x_1 = 3 & y_1 = 5 \\ x_2 = 6 & y_2 = 10 \end{matrix}$$

and hence estimate y when $x = 4.5$.

Solution 5.1

There are three $(n_p = 3)$ data points, hence $n = 2$ and the required interpolating polynomial will be second order, thus

$$Q_2(x) = L_0(x)y_0 + L_1(x)y_1 + L_2(x)y_2$$

The three Lagrangian polynomials can be computed according to equation (5.4), hence

$$L_0(x) = \frac{(x-3)(x-6)}{(1-3)(1-6)} = \frac{1}{10}(x^2 - 9x + 18)$$

$$L_1(x) = \frac{(x-1)(x-6)}{(3-1)(3-6)} = -\frac{1}{6}(x^2 - 7x + 6)$$

$$L_2(x) = \frac{(x-1)(x-3)}{(6-1)(6-3)} = \frac{1}{15}(x^2 - 4x + 3)$$

After "weighting" each Lagrangian polynomial by the corresponding y-value from equation (5.3) we get after some simplification

$$Q_2(x) = -\frac{1}{15}(x^2 - 34x + 18)$$

As a check, the three values of x can be substituted into the interpolating polynomial to give

$$Q_2(1) = 1 \quad Q_2(3) = 5 \quad Q_2(6) = 10$$

shown plotted in Figure 5.1. The required interpolation is given by

$$Q_2(4.5) = 7.65$$

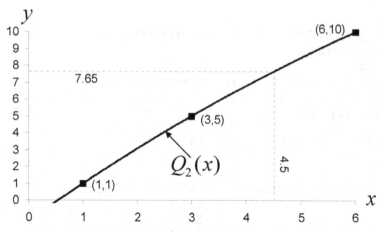

Figure 5.1: Quadratic interpolation from Example 5.1.

It may be noted that the x-values in the Lagrangian polynomial method do not need to be equidistant, nor do they need to be arranged in any particular order.

A disadvantage of the Lagrangian approach is the high number of arithmetic operations that must be carried out in order to compute an interpolate. Each Lagrangian polynomial is itself of order n, and must be evaluated at the required value of x. A further problem relating to the efficiency of the Lagrangian approach is that if new data points are added to a set that has already been operated on, no advantage can be gained from the Lagrangian polynomials already computed, and the whole process must start again from scratch.

Program 5.1: Interpolation by Lagrangian polynomials

```
PROGRAM nm51
!---Interpolation Using Lagrangian Polynomials---
 IMPLICIT NONE; INTEGER,PARAMETER::iwp=SELECTED_REAL_KIND(15,300)
 INTEGER::i,j,np; REAL(iwp)::one=1.0_iwp,term,xi,yi,zero=0.0_iwp
 REAL(iwp),ALLOCATABLE::x(:),y(:)
 OPEN(10,FILE='nm95.dat'); OPEN(11,FILE='nm95.res')
 READ(10,*)np; ALLOCATE(x(np),y(np))
 READ(10,*)(x(i),y(i),i=1,np),xi
 WRITE(11,'(A,/)')"---Interpolation Using Lagrangian &
   &Polynomials---"
 WRITE(11,'(A/A)')"       Data Points","       x               y"
 WRITE(11,'(2E12.4)')(x(i),y(i),i=1,np); yi=zero
 DO i=1,np
   term=one
   DO j=1,np; IF(j/=i)term=term*(xi-x(j))/(x(i)-x(j)); END DO
   yi=yi+term*y(i)
 END DO
 WRITE(11,'(/A/A)')"  Interpolated Point","       x               y"
 WRITE (11,'(2E12.4)')xi,yi
END PROGRAM nm51
```

Number of data points	np
	4

Data point coordinates	(x(i),y(i),i=1,np)
	1. 1.
	3. 5.
	6. 10.
	5. 9.

Interpolation point	xi
	4.5

Data 5.1: Interpolation by Lagrangian Polynomials

List 5.1:

Scalar integers:
i   simple counter
j   simple counter
np   number of input data points ($n_p = n + 1$)

Scalar reals:
one   set to 1.0
term   used to form Lagrangian polynomial terms
xi $x-$value at which interpolation is required
yi   interpolated value of y
zero   set to 0.0

Dynamic real arrays:
x   input data x-values
y   input data y-values

Program 5.1 computes the nth order interpolating polynomial $Q_n(x)$ derived from Lagrangian polynomials, and then obtains an interpolated value of y for a given value of x.

The input data and results files from Program 5.1 are given in Data and Results 5.1 respectively. Initially the number of input data points (np) is read followed by (x, y) values (np times). The last number read is the value of x at which interpolation is required (xi).

In this case there are np=4 data points, so the program obtains the interpolated value of y using a cubic $Q_3(x)$ interpolating polynomial. The output in Results 5.1 indicates the interpolated value of $y = 8.175$ corresponding to $x = 4.5$.

```
---Interpolation by Lagrangian Polynomials---

      Data Points
     x           y
  0.1000E+01  0.1000E+01
  0.3000E+01  0.5000E+01
  0.6000E+01  0.1000E+02
  0.5000E+01  0.9000E+01


   Interpolated Point
     x           y
  0.4500E+01  0.8175E+01
```

Results 5.1: Interpolation by Lagrangian Polynomials

5.2.2   Difference methods

An alternative approach to finding the interpolating polynomials that will pass exactly through n_p data points given as (x_i, y_i), $i = 0, 1, 2, \ldots, n$ (where $n = n_p - 1$) starts by writing the interpolating polynomial in the alternative form (compare equation 5.1),

$$Q_n(x) = C_0 + C_1(x - x_0) + C_2(x - x_0)(x - x_1) + \ldots$$
$$+ C_n(x - x_0)(x - x_1) \ldots (x - x_{n-2})(x - x_{n-1}) \tag{5.7}$$

where the constants C_i, $i = 0, 1, 2, \ldots, n$ can be determined from the requirement that

$$Q_n(x_i) = y_i \quad \text{for} \quad i = 0, 1, 2, \ldots, n \tag{5.8}$$

which leads after some rearrangement to

$$
\begin{aligned}
C_0 &= y_0 \\
C_1 &= \frac{y_1 - C_0}{(x_1 - x_0)} \\
C_2 &= \frac{y_2 - C_0 - C_1(x_2 - x_0)}{(x_2 - x_0)(x_2 - x_1)} \\
C_3 &= \frac{y_3 - C_0 - C_1(x_3 - x_0) - C_2(x_3 - x_0)(x_3 - x_1)}{(x_3 - x_0)(x_3 - x_1)(x_3 - x_2)}
\end{aligned}
\tag{5.9}
$$

etc.

A pattern is clearly emerging in the form of the C_i expressions, and they can readily be computed and substituted into equation (5.7). It may also be noted that unlike the Lagrangian approach, if additional data points are added to the original list, the resulting higher order polynomial is derived by the simple addition of more terms to the lower order polynomial already found.

Example 5.2

Here we repeat the example given with Program 5.1. Use the Difference method to derive a polynomial passing through the points

$$
\begin{aligned}
x_0 &= 1 \quad y_0 = 1 \\
x_1 &= 3 \quad y_1 = 5 \\
x_2 &= 6 \quad y_2 = 10
\end{aligned}
$$

and then modify the polynomial to account for the additional point

$$x_3 = 5 \quad y_3 = 9$$

Estimate the value of y when $x = 4.5$ in both cases.

Solution 5.2

The first part of the question involves three data points, hence the interpolating polynomial will be quadratic $(n = 2)$. First we compute the constants from equations (5.8)

$$C_0 = 1$$
$$C_1 = \frac{5-1}{(3-1)} = 2$$
$$C_2 = \frac{10 - 1 - 2(6-1)}{(6-1)(6-3)} = -\frac{1}{15}$$

which are then substituted into equation (5.7) to give

$$Q_2(x) = 1 + 2(x-1) - \frac{1}{15}(x-1)(x-3)$$

After simplification

$$Q_2(x) = -\frac{1}{15}(x^2 - 34x + 18)$$

which is the same result found in Example 5.1.

The fourth data point $x_3 = 5 \quad y_3 = 9$ will lead to a cubic interpolating polynomial. The additional constant is given by

$$C_3 = \frac{9 - 1 - 2(5-1) + \frac{1}{15}(5-1)(5-3)}{(5-1)(5-3)(5-6)} = -\frac{1}{15}$$

giving a cubic term from equation (5.7) which is simply added to the second order polynomial already found, thus

$$Q_3(x) = -\frac{1}{15}(x^2 - 34x + 18) - \frac{1}{15}(x-1)(x-3)(x-6)$$

or

$$Q_3(x) = -\frac{1}{15}(x^3 - 9x^2 - 7x)$$

The cubic is shown in Figure 5.2 and leads to the interpolated value of $y = 8.175$ when $x = 4.5$.

5.2.3 Difference methods with equal intervals

If the data are provided at equally spaced values of x, such that $x_i - x_{i-1} = h$, derivation of the coefficients C_i, $i = 0, 1, 2, \ldots, n$ from equation (5.7) is considerably simplified.

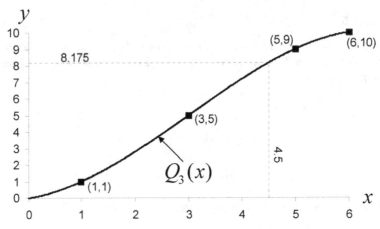

Figure 5.2: Cubic interpolation from Example 5.2.

Before proceeding, we introduce a new notation whereby, a "forward difference" is defined as

$$\Delta y_i = y_{i+1} - y_i \tag{5.10}$$

and a "backward difference" as

$$\Delta y_i = y_i - y_{i-1} \tag{5.11}$$

Furthermore, we can have "differences of differences", whereby for example, the second forward difference could be written as

$$\Delta^2 y_i = \Delta y_{i+1} - \Delta y_i \tag{5.12}$$

and more generally

$$\Delta^j y_i = \Delta^{j-1} y_{i+1} - \Delta^{j-1} y_i \tag{5.13}$$

and so on.

The following development could use either "forward" or "backward" differences, but we will use "forward difference" in the remainder of this section.

Returning to equations (5.9) we can write

$$C_0 = y_0$$

$$
\begin{aligned}
C_1 &= \frac{y_1 - C_0}{x_1 - x_0} \\
&= \frac{y_1 - y_0}{x_1 - x_0} \\
&= \frac{\Delta y_0}{h}
\end{aligned}
$$

$$
\begin{aligned}
C_2 &= \frac{y_2 - y_0 - \Delta y_0 (x_2 - x_0)/h}{(x_2 - x_0)(x_2 - x_1)} \\
&= \frac{(y_2 - y_1) - (y_1 - y_0)}{2h^2} \\
&= \frac{(\Delta y_1 - \Delta y_0)}{2h^2} \\
&= \frac{\Delta(\Delta y_0)}{2h^2} \\
&= \frac{\Delta^2 y_0}{2h^2} \quad \text{etc.}
\end{aligned}
\tag{5.14}
$$

It is easily shown that in general

$$C_j = \frac{\Delta^j y_0}{j! \, h^j} \tag{5.15}$$

TABLE 5.1: Forward Difference Table

x_0	y_0					
		Δy_0				
x_1	y_1		$\Delta^2 y_0$			
		Δy_1		$\Delta^3 y_0$		
x_2	y_2		$\Delta^2 y_1$		$\Delta^4 y_0$	
		Δy_2		$\Delta^3 y_1$		$\Delta^5 y_0$
x_3	y_3		$\Delta^2 y_2$		$\Delta^4 y_1$	
		Δy_3		$\Delta^3 y_2$		
x_4	y_4		$\Delta^2 y_3$			
		Δy_4				
x_5	y_5					

It is apparent from equations 5.11-5.13 that the $\Delta^j y_0$ terms can be evaluated recursively. For hand calculation purposes, a tabular layout is useful for this purpose as shown in Table 5.1 for an example with six data points.

The particular scheme described in Table 5.1 is known as (Newton) Forward Differences, and is characterized by subscripts remaining the same along downward sloping diagonals going from top left to bottom right. Other layouts are possible, such as Backward Differences, and Gaussian methods, but these will not be discussed here. When substituting values from Table 5.1 into equation (5.15) it should be noted that the "zeroth" difference terms are the y-values themselves, thus

$$\Delta^0 y_0 = y_0 \quad \text{etc.} \tag{5.16}$$

Example 5.3

Given the following data based on the function $y = \cos x$ where x is in degrees

x	y
20	0.93969
25	0.90631
30	0.86603
35	0.81915
40	0.76604

use a forward difference scheme to estimate $\cos 27°$.

Solution 5.3

First we arrange the data as a forward difference table, thus

x	y	Δy	$\Delta^2 y$	$\Delta^3 y$	$\Delta^4 y$
20	0.93969				
		−0.03338			
25	0.90631		−0.00690		
		−0.04028		0.00030	
30	0.86603		−0.00660		0.00007
		−0.04688		0.00037	
35	0.81915		−0.00623		
		−0.05311			
40	0.76604				

Referring to the general equation (5.7) for the interpolating polynomial, x_0 can be chosen to be any of the initial values of x. However, if we wish to include all five data points leading to a fourth order interpolating polynomial $Q_4(x)$, x_0 should be set equal to the top value in the table, i.e., $x_0 = 20$. Noting that the constant interval between x-values is given as $h = 5$, the coefficients can be evaluated in tabular form as follows

j	$\Delta^j y_0$	$C_j = \dfrac{\Delta^j y_0}{j!\, h^j}$
0	0.93969	0.93969
1	−0.03338	−0.00668
2	−0.00690	−0.00014
3	0.00030	0.00000
4	0.00007	0.00000

Working to five decimal places, the interpolating polynomial from equation (5.7) can therefore be written as

$$Q_4(x) = 0.93969 - 0.00668(x - 20) - 0.00014(x - 20)(x - 25)$$
$$+0.00000(x - 20)(x - 25)(x - 30)$$
$$+0.00000(x - 20)(x - 25)(x - 30)(x - 35)$$

and expanded to give

$$Q_4(x) = 1.00329 - 0.00038x - 0.00014x^2$$
$$+\text{negligible higher order terms}$$

which is exactly the same interpolation polynomial we would have obtained to five decimal places using Lagrangian polynomials. Substitution of the required value of $x = 27$ leads to the interpolated value

$$Q_4(27) = 0.89097$$

which is accurate to four decimal places.

It is clear from the above general expression for $Q_4(x)$ that the C_3 and C_4 terms are contributing very little to the overall solution.

The ability to truncate the interpolating polynomial if the coefficients become sufficiently small is a useful feature of the difference approach. Not only does it save computational effort, but it also gives physical insight into the theoretical origins of the data points.

In the Lagrangian approach no such saving is possible and the full n^{th} order polynomial must be derived whether it is needed or not. The Lagrangian approach does have the advantage of simplicity however, and is relatively easy to remember for hand calculation if only a few points are provided.

Program 5.2: Interpolation by forward differences

```
PROGRAM nm52
!---Interpolation by Forward Differences---
 IMPLICIT NONE
 INTEGER,PARAMETER::iwp=SELECTED_REAL_KIND(15,300)
 INTEGER::factorial,i,j,np; REAL(iwp)::h,one=1.0_iwp,term,xi,yi
 REAL(iwp),ALLOCATABLE::c(:),diff(:,:),x(:)
 OPEN(10,FILE='nm95.dat'); OPEN(11,FILE='nm95.res')
 READ(10,*)np; ALLOCATE(c(0:np-1),diff(0:np-1,0:np-1),x(0:np-1))
 READ(10,*)(x(i),diff(i,0),i=0,np-1),xi
 WRITE(11,'(A/)')"---Interpolation by Forward Differences---"
 WRITE(11,'(A/A)')"        Data Points","        x              y"
 WRITE(11,'(2E12.4)')(x(i),diff(i,0),i=0,np-1)
 h=x(1)-x(0)
 DO i=1,np-1
   DO j=0,np-1-i; diff(j,i)=diff(j+1,i-1)-diff(j,i-1); END DO
 END DO
 c(0)=diff(0,0); yi=c(0); term=one; factorial=one
 DO i=1,np-1
   factorial=factorial*i; c(i)=diff(0,i)/(factorial*h**i)
   term=term*(xi-x(i-1)); yi=yi+term*c(i)
 END DO
 WRITE(11,'(/A/A)')" Polynomial Coefficients","        C"
 WRITE(11,'(E12.4)')c
 WRITE(11,'(/A/A)')"   Interpolated Point","        x              y"
 WRITE(11,'(2E12.4)')xi,yi
END PROGRAM nm52
```

Number of data points	np	
	4	
Data point coordinates	(x(i),y(i),i=1,np)	
	0.0	1.0
	1.0	1.0
	2.0	15.0
	3.0	61.0
Interpolation point	xi	
	2.8	

Data 5.2a: Interpolation by Forward Differences (first example)

List 5.2:

Scalar integers:
i   simple counter
j   simple counter
np   number of input data points ($n_p = n + 1$)

Scalar reals:
h   data interval in x-direction
one set to 1.0
term used to form interpolating polynomial
xi $x-$value at which interpolation is required
yi   interpolated value of y

Dynamic real arrays:
c   polynomial coefficient values
diff forward difference table values
x   input data x-values

```
---Interpolation by Forward Differences---

      Data Points
    x           y
0.0000E+00  0.1000E+01
0.1000E+01  0.1000E+01
0.2000E+01  0.1500E+02
0.3000E+01  0.6100E+02

Polynomial Coefficients
       C
0.1000E+01
0.0000E+00
0.7000E+01
0.3000E+01

   Interpolated Point
    x           y
0.2800E+01  0.4838E+02
```

Results 5.2a: Interpolation by Forward Differences (first example)

Program 5.2 computes an interpolated value of y for a given value of x based on n_p input data points that are equally spaced along the x-axis. The interpolate is computed using the nth order (where $n = n_p - 1$) interpolating polynomial $Q_n(x)$ derived from a forward difference table using equation (5.7).

```
Number of data points        np
                             5

Data point coordinates       (x(i),y(i),i=1,np)
                             0.0    1.0
                             1.0    1.0
                             2.0    15.0
                             3.0    61.0
                             4.0    100.0

Interpolation point          xi
                             2.8
```

Data 5.2b: Interpolation by Forward Differences (second example)

```
---Interpolation by Forward Differences---

      Data Points
     x            y
 0.0000E+00  0.1000E+01
 0.1000E+01  0.1000E+01
 0.2000E+01  0.1500E+02
 0.3000E+01  0.6100E+02
 0.4000E+01  0.1000E+03

Polynomial Coefficients
       C
 0.1000E+01
 0.0000E+00
 0.7000E+01
 0.3000E+01
-0.2375E+01

   Interpolated Point
     x            y
 0.2800E+01  0.5029E+02
```

Results 5.2b: Interpolation by Forward Differences (second example)

The data for Program 5.2 have a format that is exactly the same as for Program 5.1. Initially the number of input data points (np) is read followed by (x, y) values (np times). The last number read is the value of x at which interpolation is required (xi).

The example data shown in Data 5.4a have np=4 data points, so the program obtains the interpolated value of y using a cubic $Q_3(x)$ interpolating polynomial. The output in Results 5.4a indicates the interpolated value of $y = 48.38$ corresponding to $x = 2.8$. It may be noted that the polynomial coefficient $C_1 = 0$ in this case.

A second example shown in Data 5.5b adds an additional data point $x = 4$ and $y = 100$, so np=5. As shown in Results 5.5b, the resulting fourth order $Q_4(x)$ polynomial involves an additional coefficient $C_4 = -2.375$ but the other coefficients are unaffected. The interpolated value in this case is $y = 50.29$.

5.3 Interpolation using cubic spline functions

The interpolating methods described so far in this chapter lead to high order polynomials, in general of order equal to one less than the number of data points. In addition to being laborious computationally, these high order polynomials can lead to undesirable maxima and minima between the given data points. An alternative approach to interpolation involves fitting low order polynomials from point to point across the range in a "piece-wise" manner. The order of the polynomials is up to the user. For example if the polynomials are chosen to be linear, the data points are simply connected by straight lines with "corners" at each point where the straight lines join. It should be noted that the discontinuous nature of linear functions may not be a problem in engineering applications if there are numerous data points close together.

The most popular piecewise polynomial method involves cubic functions fitted between neighboring points, which are able to preserve second derivative continuity at the joins. Such cubic polynomials are often referred to as "Spline Functions" because the interpolating function can be visualized as a flexible elastic beam (or spline), initially straight, deformed in such a way that it passes through the required points $(x_i, y_i), i = 0, 1, \ldots, n$.

Figure 5.3 shows how interpolation through four points might be achieved by the use of three cubic functions $f_1(x)$, $f_2(x)$ and $f_3(x)$. In general, if we have n_p points, n cubic spline functions will be required (where $n = n_p - 1$), which can be written in the form

$$f_i(x) = A_{0i} + A_{1i}x + A_{2i}x^2 + A_{3i}x^3, \quad i = 1, 2, \ldots, n \qquad (5.17)$$

The $4n$ unknown coefficients A_{ji} can be determined from the following $4n$ conditions:

1. The cubics must meet at all internal points leading to the 2n equations.

$$f_i(x_i) = y_i, \quad i = 1, 2, \ldots, n$$
$$f_{i+1}(x_i) = y_i, \quad i = 0, 1, \ldots, n-1 \tag{5.18}$$

2. The first derivative must be continuous at all internal points leading to the $n-1$ equations.

$$f_i'(x_i) = f_{i+1}'(x_i), \quad i = 1, 2, \ldots, n-1 \tag{5.19}$$

3. The second derivative must also be continuous at all internal points leading to further $n-1$ equations.

$$f_i''(x_i) = f_{i+1}''(x_i), \quad i = 1, 2, \ldots, n-1 \tag{5.20}$$

4. The final two conditions refer to the two ends of the spline, where the second derivative is set to zero, thus

$$f_1''(x_0) = 0$$
$$f_n''(x_n) = 0 \tag{5.21}$$

These final boundary conditions preserve the "structural" analogy, and imply a zero bending moment at the ends of the "beam".

Although $4n$ equations in $4n$ unknowns could be solved to obtain the required coefficients, conditions 1. and 3. above can be combined to give a rearranged version of the original cubic function from equation (5.17) as

$$f_i(x) = \frac{f''(x_{i-1})(x_i - x)^3 + f''(x_i)(x - x_{i-1})^3}{6\Delta x_{i-1}}$$
$$+ \left(\frac{y_{i-1}}{\Delta x_{i-1}} - \frac{f''(x_{i-1})\Delta x_{i-1}}{6} \right)(x_i - x) \tag{5.22}$$
$$+ \left(\frac{y_i}{\Delta x_{i-1}} - \frac{f''(x_i)\Delta x_{i-1}}{6} \right)(x - x_{i-1})$$

where $i = 1, 2, \ldots, n$, $x_{i-1} \le x \le x_i$ and $\Delta x_{i-1} = x_i - x_{i-1}$ which is the same "forward difference" notation used previously from equation (5.10).

Differentiation of equation (5.22) and imposition of condition 2. above for continuity of first derivatives leads to

$$\Delta x_{i-1}f''(x_{i-1}) + 2(\Delta x_{i-1} + \Delta x_i)f''(x_i) + \Delta x_i f''(x_{i+1})$$
$$= 6\left(\frac{\Delta y_i}{\Delta x_i} - \frac{\Delta y_{i-1}}{\Delta x_{i-1}} \right) \tag{5.23}$$

where $i = 1, 2, \ldots, n-1$ and $\Delta y_{i-1} = y_i - y_{i-1}$.

This is now equivalent to a system of $n-1$ linear equations in the unknown second derivatives at the internal points as follows

$$
\begin{bmatrix}
2(\Delta x_0 + \Delta x_1) & \Delta x_1 & 0 & \cdots & 0 \\
\Delta x_1 & 2(\Delta x_1 + \Delta x_2) & \Delta x_2 & \cdots & 0 \\
0 & \Delta x_2 & 2(\Delta x_2 + \Delta x_3) & \cdots & 0 \\
\vdots & \vdots & \vdots & \vdots & \vdots \\
0 & 0 & 0 & \cdots & 2(\Delta x_{n-2} + \Delta x_{n-1})
\end{bmatrix}
$$

$$
\begin{Bmatrix}
f''(x_1) \\
f''(x_2) \\
f''(x_3) \\
\vdots \\
f''(x_{n-1})
\end{Bmatrix}
= 6
\begin{Bmatrix}
\Delta y_1/\Delta x_1 - \Delta y_0/\Delta x_0 \\
\Delta y_2/\Delta x_2 - \Delta y_1/\Delta x_1 \\
\vdots \\
\Delta y_{n-1}/\Delta x_{n-1} - \Delta y_{n-2}/\Delta x_{n-2}
\end{Bmatrix}
\tag{5.24}
$$

It may be noted that the coefficient matrix is symmetric and tridiagonal, which influences the chosen solution method (see Chapter 2). Once the second derivatives $f''(x_i)$, $i = 1, 2, \ldots, n-1$ have been obtained, they can be combined with the boundary conditions $f''(x_0) = f''(x_n) = 0$, enabling all the cubic functions from equation (5.22) to be evaluated if required.

To obtain an estimate of y corresponding to a particular value of x, it must first be determined which cubic function to use by observing the location of x relative to the original data points, x_i, $i = 0, 1, \ldots, n$. Once this is done, the appropriate values of $f''(x_i)$ and $f''(x_{i-1})$ can be substituted into equation (5.22) and the function evaluated at the required point.

Example 5.4

Given the four data points

i	x_i	y_i
0	0.0	2.0
1	1.0	1.3
2	1.5	0.4
3	2.3	0.9

use cubic spline functions to estimate y when $x = 1.3$.

Solution 5.4

In this example there will be three cubic splines spanning the four coordinates. We will assume that the second derivatives at $x = 0.0$ and $x = 2.3$ equal zero, hence two second derivatives remain to be found at $x = 1.0$ and $x = 1.5$.

From equation (5.24) we can write for this case

$$
\begin{bmatrix}
2(\Delta x_0 + \Delta x_1) & \Delta x_1 \\
\Delta x_1 & 2(\Delta x_1 + \Delta x_2)
\end{bmatrix}
\begin{Bmatrix}
f''(x_1) \\
f''(x_2)
\end{Bmatrix}
= 6
\begin{Bmatrix}
\Delta y_1/\Delta x_1 - \Delta y_0/\Delta x_0 \\
\Delta y_2/\Delta x_2 - \Delta y_1/\Delta x_1
\end{Bmatrix}
$$

The required forward difference terms can be obtained in tabular form as follows (compare this with Table 5.1 and Solution 5.3)

x	Δx	y	Δy
0.0		2.0	
	1.0		−0.7
1.0		1.3	
	0.5		−0.9
1.5		0.4	
	0.8		0.5
2.3		0.9	

which after substitution lead to the pair of equations

$$\begin{bmatrix} 3.0 & 0.5 \\ 0.5 & 2.6 \end{bmatrix} \begin{Bmatrix} f''(x_1) \\ f''(x_2) \end{Bmatrix} = \begin{Bmatrix} -6.60 \\ 14.55 \end{Bmatrix}$$

which are easily solved to give

$$\begin{Bmatrix} f''(x_1) \\ f''(x_2) \end{Bmatrix} = \begin{Bmatrix} -3.2364 \\ 6.2185 \end{Bmatrix}$$

All second derivatives are now known as $f''(0) = 0$, $f''(1) = -3.2364$, $f''(1.5) = 6.2185$ and $f''(2.3) = 0$, hence the three cubic functions can be retrieved from equation (5.22). They are shown plotted in Figure 5.3

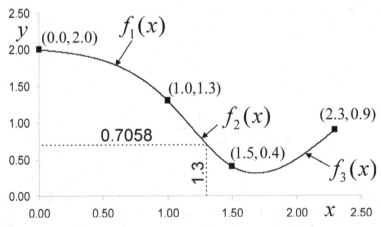

Figure 5.3: Cubic splines from Example 5.3.

Interpolation is required at $x = 1.3$ which lies in the range $1.0 < x < 1.5$ covered by cubic spline $f_2(x)$. By setting $i = 2$ this function from equation

(5.22) is given by

$$f_2(x) = \frac{f''(x_1)(x_2 - x)^3 + f''(x_2)(x - x_1)^3}{6\Delta x_1}$$
$$+ \left(\frac{y_1}{\Delta x_1} - \frac{f''(x_1)\Delta x_1}{6}\right)(x_2 - x)$$
$$+ \left(\frac{y_2}{\Delta x_1} - \frac{f''(x_2)\Delta x_1}{6}\right)(x - x_1)$$

hence

$$f_2(x) = \frac{-3.2364(1.5 - x)^3 + 6.2185(x - 1.0)^3}{6(0.5)}$$
$$+ \left(\frac{1.3}{0.5} - \frac{-3.2364(0.5)}{6}\right)(1.5 - x)$$
$$+ \left(\frac{0.4}{0.5} - \frac{6.2185(0.5)}{6}\right)(x - 1.0)$$
$$= 3.1517x^3 - 11.0732x^2 + 10.9126x - 1.6911$$

and

$$f_2(1.3) = 0.7058$$

Program 5.3: Interpolation by cubic spline functions

```
PROGRAM nm53
!---Interpolation by Cubic Spline Functions---
USE nm_lib; USE precision; IMPLICIT NONE
INTEGER::i,np; REAL(iwp)::d6=6.0_iwp,xi,yi,two=2.0_iwp,        &
   zero=0.0_iwp
REAL(iwp),ALLOCATABLE::diffx(:),diffy(:),kv(:),rhs(:),x(:),y(:)
INTEGER,ALLOCATABLE::kdiag(:)
OPEN(10,FILE='nm95.dat'); OPEN(11,FILE='nm95.res')
READ(10,*)np
ALLOCATE(diffx(0:np-2),diffy(0:np-2),kdiag(np-2),             &
   kv(2*(np-2)-1),rhs(0:np-1),x(0:np-1),y(0:np-1))
READ(10,*)(x(i),y(i),i=0,np-1),xi
WRITE(11,'(A/)')"---Interpolation by Cubic Spline Functions---"
WRITE(11,'(A/A)')"        Data Points","        x              y"
WRITE(11,'(2E12.4)')(x(i),y(i),i=0,np-1)
DO i=0,np-2
  diffx(i)=x(i+1)-x(i); diffy(i)=y(i+1)-y(i)
END DO
rhs=zero
```

```
DO i=1,np-2
  kdiag(i)=2*i-1; kv(kdiag(i))=two*(diffx(i-1)+diffx(i))
  rhs(i)=d6*(diffy(i)/diffx(i)-diffy(i-1)/diffx(i-1))
END DO
DO i=1,np-3; kv(2*i)=diffx(i); END DO
CALL sparin(kv,kdiag); CALL spabac(kv,rhs,kdiag)
WRITE(11,'(/A/A)')"   Interpolated Point","       x             y"
DO i=1,np-1
  IF(xi<x(i))THEN; yi=(rhs(i-1)*(x(i)-xi)**3+                        &
    rhs(i)*(xi-x(i-1))**3)/(d6*diffx(i-1))+                         &
    (y(i-1)/diffx(i-1)-rhs(i-1)*diffx(i-1)/d6)*(x(i)-xi)+           &
    (y(i)/diffx(i-1)-rhs(i)*diffx(i-1)/d6)*(xi-x(i-1))
    EXIT
  END IF
END DO
WRITE(11,'(2E12.4)')xi,yi
END PROGRAM nm53
```

List 5.3:

Scalar integers:

i	simple counter
np	number of input data points $(n_p = n + 1)$

Scalar reals:

d6	set to 6.0
xi	$x-$value at which interpolation is required
yi	interpolated value of y
two	set to 2.0
zero	set to 0.0

Dynamic real arrays:

diffx	forward difference table values in x
diffy	forward difference table values in y
kv	holds equation coefficients in skyline form
rhs	holds right hand side and solution vector
x	input data x-values
y	input data y-values

Dynamic integer arrays:

kdiag	diagonal term location vector

```
Number of data points          np
                               5

Data point coordinates         (x(i),y(i),i=1,np
                               0.0    0.0000E+00
                               2.5   -0.4538E-02
                               5.0   -0.5000E-02
                               8.0    0.0000E+00
                              10.0    0.4352E-02

Interpolation point            xi
                               6.5
```

Data 5.3: Interpolation by Cubic Spline Functions


```
---Interpolation by Cubic Spline Functions---

      Data Points
     x            y
 0.0000E+00   0.0000E+00
 0.2500E+01  -0.4538E-02
 0.5000E+01  -0.5000E-02
 0.8000E+01   0.0000E+00
 0.1000E+02   0.4352E-02

   Interpolated Point
     x            y
 0.6500E+01  -0.2994E-02
```

Results 5.3: Interpolation by Cubic Spline Functions

Program 5.3 computes an interpolated value of y for a given value of x based on cubic spline interpolation between the input data points. The input data do not have to be equally spaced along the x-axis, but must be presented in ascending values of x. The interpolate is computed using the particular spline function that spans the range within which the required interpolated value lies. The program includes two SUBROUTINES from the nm_lib library called sparin and spabac, which perform factorization and forward/back-substitution on symmetric coefficient matrices stored in skyline form (see Chapter 2). The data for Program 5.3 has a format that is exactly the same as for Programs 5.1 and 5.2. Initially the number of input data points (np) is read followed by (x, y) values (np times). The last number read is the value of x at which interpolation is required (xi).

The example data shown in Data 5.7 have np=5 data points, so the program will compute three internal second derivatives, enabling four cubic splines to be derived between each pair of adjacent points. The output in Results 5.7 indicates an interpolated value of $y = -2.994 \times 10^{-3}$ corresponding to $x = 6.5$. It may be noted that the fourth order interpolating polynomial given by Program 5.1 for this input data leads to the slightly different interpolated value of $y = -3.078 \times 10^{-3}$.

Users wishing to generate a plot such as the one in Figure 5.3 showing the whole interpolating function can do so by adding an additional DO-loop to the main program in which x is increased incrementally across the full range of input data (see Chapter 1 for advice on creating $x - y$ plots).

5.4   Numerical differentiation

Numerical differentiation involves estimating derivatives from a set of discrete data points that might be produced from laboratory measurements, or from some other numerical computations.

It should be realized that numerical differentiation can be an unreliable process which is highly sensitive to small fluctuations in data. This is in contrast to numerical integration covered in the next chapter, which is an inherently accurate process.

As an example, consider the case of a particle moving with uniform acceleration. Under experimental conditions, the velocity is measured as a function of time at discrete intervals. The results of these readings are plotted in Figure 5.4a. The response is essentially a straight line, but due to inevitable small fluctuations in the readings the line is ragged. If these data are differentiated to give accelerations, it is clear that the small ripples will have a dramatic effect on the calculated derivative as shown in Figure 5.4b. The ripples have virtually no influence on the displacement plot however, which is obtained by integration of the velocity data as shown in Figure 5.4c.

Although numerical differentiation of discrete data is vulnerable to fluctuations, especially from measured data, with "well behaved" data the process can lead to reasonable estimates of derivatives.

Numerical differentiation of discrete data essentially involves finding the interpolating polynomial $Q_n(x)$, as discussed in Section 5.2 and differentiating it as required. The number of points (n_d) used in the differentiation formula and hence the order of the interpolating polynomial n (where $n = n_d - 1$) is up to the user. Bearing in mind that we are generally interested in first and second derivatives in engineering analysis, low order polynomials (e.g., $n \leq 4$) are usually preferred.

The notation used throughout this section is that x_0 will be the location at which the derivative y_0', y_0'', etc. is to be estimated. It will also be assumed that the x-values of the discrete data are equally spaced as in $x_{i+1} - x_i = h$.

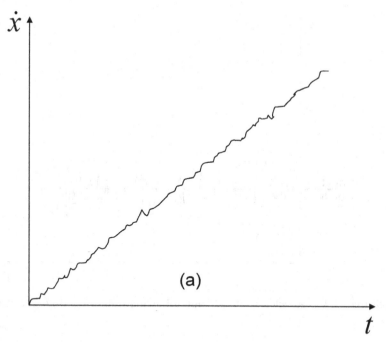

Figure 5.4a: Measured data of velocity vs. time.

5.4.1 Interpolating polynomial method

Consider three data points (x_0, y_0), (x_1, y_1) and (x_2, y_2), which may be a subset of a much longer list, from which we wish to estimate y_0' and y_0''. First we use either the Lagrangian or the Difference method to derive the $Q_2(x)$ interpolating polynomial. Let us use the Difference Method from Section 5.2.3 hence

$$Q_2(x) = C_0 + C_1(x - x_0) + C_2(x - x_0)(x - x_1) \qquad (5.25)$$

where

$$\begin{aligned} C_0 &= y_0 \\ C_1 &= \frac{y_1 - y_0}{h} \\ C_2 &= \frac{y_2 - 2y_1 + y_0}{2h^2} \end{aligned} \qquad (5.26)$$

Differentiation of (5.25) gives

$$Q_2'(x) = C_1 + C_2(2x - (x_0 + x_1)) \qquad (5.27)$$

and

$$Q_2''(x) = 2C_2 \qquad (5.28)$$

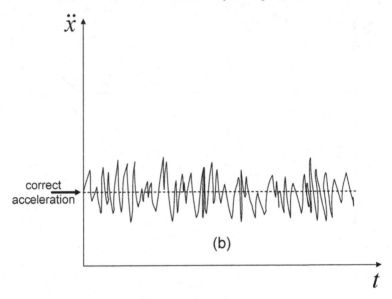

Figure 5.4b: Acceleration vs. time following numerical differentiation.

Substitution of the coefficients from equation (5.26) and letting $x = x_0$ leads to

$$Q_2'(x_0) = y_0' = \frac{1}{2h}(-3y_0 + 4y_1 - y_2) \tag{5.29}$$

and

$$Q_2''(x_0) = y_0'' = \frac{1}{h^2}(y_0 - 2y_1 + y_2) \tag{5.30}$$

Equations (5.29) and (5.30) are our first examples of "Numerical Differentiation Formulas". Equation (5.29) is a "Three-point, forward difference, first derivative formula", and Equation (5.30) is a "Three-point, forward difference, second derivative formula".

The term "forward" in this context means that the y-values in the formula are all to the right (in the positive x-direction) of the point at which the derivative is required. Two other types of formula are available. Central difference formulas take equal account of points both to the right and left of the point at which the derivative is required, and backward difference formulas, which are the mirror image of forward difference formulas, look only to the left (in the negative x-direction). Figure 5.5 summarizes the different types of formulas.

There are many different types of numerical differentiation formulas depending on the following three descriptors:

- The number of points included in the formula (n_d).

- The "direction" of the formula (forward, central or backward).

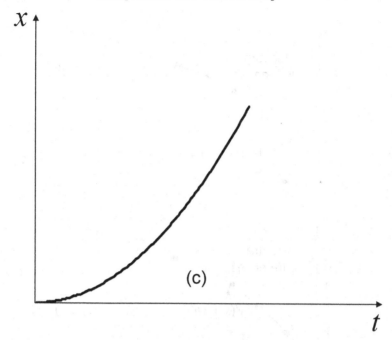

Figure 5.4c: Displacement vs. time following numerical integration.

- The required derivative (first, second, etc.).

If we wanted to estimate derivatives in the middle of the three points given in our original example, we would redefine our subscripts as (x_{-1}, y_{-1}), (x_0, y_0) and (x_1, y_1) to ensure that a zero subscript corresponded to the place where derivatives are required. In this case, equations (5.25-5.30) become

$$Q_2(x) = C_0 + C_1(x - x_{-1}) + C_2(x - x_{-1})(x - x_0) \tag{5.31}$$

where

$$\begin{aligned} C_0 &= y_{-1} \\ C_1 &= \frac{y_0 - y_{-1}}{h} \\ C_2 &= \frac{y_1 - 2y_0 + y_{-1}}{2h^2} \end{aligned} \tag{5.32}$$

Differentiation of (5.31) gives

$$Q_2'(x) = C_1 + C_2(2x - (x_{-1} + x_0)) \tag{5.33}$$

and

$$Q_2''(x) = 2C_2 \tag{5.34}$$

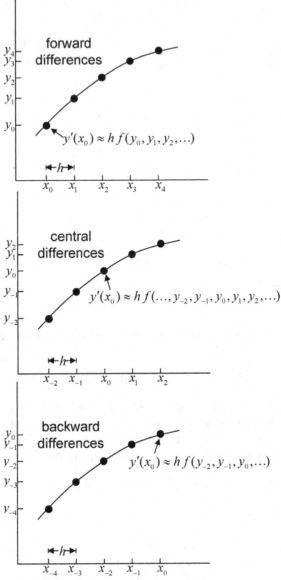

Figure 5.5: Forward, central and backward difference formulas.

Substitution of the coefficients from equation (5.32) and letting $x = x_0$ leads to

$$Q_2'(x_0) = y_0' = \frac{1}{2h}(-y_{-1} + y_1) \tag{5.35}$$

and

$$Q_2''(x_0) = y_0'' = \frac{1}{h^2}(y_{-1} - 2y_0 + y_1) \tag{5.36}$$

Equations (5.35) and (5.36) are examples of "Three-point, central difference, first and second derivative formulas" respectively.

It should be noted that equations (5.30) and (5.36) are identical, since the second derivative derived from three points is constant. Further differentiation of $Q_2(x)$ from equations (5.25) and (5.31) would be pointless, because the third and higher derivatives equal zero. In general, if we include n_d data points, the interpolating polynomial will be of order $n = n_d - 1$, and the nth derivative will be the highest that can be deduced. If for example we wanted to derive a third derivative formula, at least four points would be required.

5.4.2 Taylor series method

The previous section showed how the interpolating polynomial could be used to derive differentiation formulas. While this method was quite straightforward, it gave us no indication of the accuracy of the resulting formula. This can be overcome by the use of the Taylor series which offers a general approach for deriving numerical differentiation formulas and their dominant error terms.

Given a general function $f(x)$ passing through the point (x_0, y_0), the Taylor series can be used to estimate the value of $f(x)$ at a neighboring point $x_0 + ih$ (i is a positive or negative integer) as follows

$$f(x_0 + ih) \approx f(x_0) + (ih)\, f'(x_0) + \frac{(ih)^2}{2!} f''(x_0) + \frac{(ih)^3}{3!} f'''(x_0) + \dots \tag{5.37}$$

Using the notation that $y_0 = f(x_0)$, $y_i = f(x_0 + ih)$, $y_0' = f'(x_0)$, and so on, the Taylor expansion from equation (5.37) can be written in the shortened form as

$$y_i \approx y_0 + (ih)\, y_0' + \frac{(ih)^2}{2!} y_0'' + \frac{(ih)^3}{3!} y_0''' + \dots \tag{5.38}$$

The derivation of a particular numerical differentiation formula now proceeds by using the Taylor series from equation (5.38) to generate estimates of the y_i-values required by the formula. Following this, derivative terms that are not needed are eliminated while the required derivative is made the subject of the formula. A summary of forward, central and backward differentiation formulas, together with the dominant error terms derived by this method, is given in Tables 5.5, 5.6 and 5.7 respectively.

Example 5.5

Use the Taylor series method to derive the three point, backward difference, first and second derivative formulas, and their dominant error terms.

TABLE 5.2: Forward Difference Derivative Formulas

	n_d	C	w_0	w_1	w_2	w_3	w_4	E
y_0'	2	$\dfrac{1}{h}$	-1	1				$-\dfrac{1}{2}hy''$
	3	$\dfrac{1}{2h}$	-3	4	-1			$\dfrac{1}{3}h^2y'''$
	4	$\dfrac{1}{6h}$	-11	18	-9	2		$-\dfrac{1}{4}h^3y^{iv}$
	5	$\dfrac{1}{12h}$	-25	48	-36	16	-3	$\dfrac{1}{5}h^4y^v$
y_0''	3	$\dfrac{1}{h^2}$	1	-2	1			$-hy'''$
	4	$\dfrac{1}{h^2}$	2	-5	4	-1		$\dfrac{11}{12}h^2y^{iv}$
	5	$\dfrac{1}{12h^2}$	35	-104	114	-56	11	$-\dfrac{5}{6}h^3y^v$
y_0'''	4	$\dfrac{1}{h^3}$	-1	3	-3	1		$-\dfrac{3}{2}hy^{iv}$
	5	$\dfrac{1}{2h^3}$	-5	18	-24	14	-3	$\dfrac{7}{4}h^2y^v$
y_0^{iv}	5	$\dfrac{1}{h^4}$	1	-4	6	-4	1	$-2hy^v$

$$y^{(k)} = C\sum_{i=0}^{n_d-1} w_i y_i + E$$

Solution 5.5

We will assume the data have a uniform spacing of h in the x-direction, so in this case we are looking for formulas of the form:

$$y_0' \approx f_1(h, y_{-2}, y_{-1}, y_0)$$

and

$$y_0'' \approx f_2(h, y_{-2}, y_{-1}, y_0)$$

TABLE 5.3: Central Difference Derivative Formulas

	n_d	C	w_{-3}	w_{-2}	w_{-1}	w_0	w_1	w_2	w_3	E
y_0'	3	$\dfrac{1}{2h}$			-1	0	1			$-\dfrac{1}{6}h^2 y'''$
	5	$\dfrac{1}{12h}$		1	-8	0	8	-1		$\dfrac{1}{30}h^4 y^v$
	7	$\dfrac{1}{60h}$	-1	9	-45	0	45	-9	1	$-\dfrac{1}{140}h^6 y^{vii}$
y_0''	3	$\dfrac{1}{h^2}$			1	-2	1			$-\dfrac{1}{12}h^2 y^{iv}$
	5	$\dfrac{1}{12h^2}$		-1	16	-30	16	-1		$\dfrac{1}{90}h^4 y^{vi}$
	7	$\dfrac{1}{180h^2}$	2	-27	270	-490	270	-27	2	$-\dfrac{1}{560}h^6 y^{viii}$
y_0'''	5	$\dfrac{1}{2h^3}$		-1	2	0	-2	1		$-\dfrac{1}{4}h^2 y^v$
	7	$\dfrac{1}{8h^3}$	1	-8	13	0	-13	8	-1	$\dfrac{7}{120}h^4 y^{vii}$
y_0^{iv}	5	$\dfrac{1}{h^4}$		1	-4	6	-4	1		$-\dfrac{1}{6}h^2 y^{vi}$
	7	$\dfrac{1}{6h^4}$	-1	12	-39	56	-39	12	-1	$\dfrac{7}{240}h^4 y^{viii}$

$$y^{(k)} = C\sum_{i=-m}^{m} w_i y_i + E \quad \text{where} \quad m = \frac{n_d - 1}{2}$$

First we expand the Taylor series about y_0 to estimate y_{-1} and y_{-2}, noting that in a backward difference formula the shift is negative (e.g., $-h$ and $-2h$), thus

$$y_{-1} = y_0 - h\,y_0' + \frac{h^2}{2!}\,y_0'' - \frac{h^3}{3!}\,y_0''' \ldots \qquad (i)$$

$$y_{-2} = y_0 - 2h\,y_0' + \frac{4h^2}{2!}\,y_0'' - \frac{8h^3}{3!}\,y_0''' \ldots \qquad (ii)$$

TABLE 5.4: Backward Difference Derivative Formulas

	n_d	C	w_{-4}	w_{-3}	w_{-2}	w_{-1}	w_0	E
y_0'	2	$\dfrac{1}{h}$				-1	1	$\dfrac{1}{2}hy''$
	3	$\dfrac{1}{2h}$			1	-4	3	$\dfrac{1}{3}h^2y'''$
	4	$\dfrac{1}{6h}$		-2	9	-18	11	$\dfrac{1}{4}h^3y^{iv}$
	5	$\dfrac{1}{12h}$	3	-16	36	-48	25	$\dfrac{1}{5}h^4y^v$
y_0''	3	$\dfrac{1}{h^2}$			1	-2	1	hy'''
	4	$\dfrac{1}{h^2}$		-1	4	-5	2	$\dfrac{11}{12}h^2y^{iv}$
	5	$\dfrac{1}{12h^2}$	11	-56	114	-104	35	$\dfrac{5}{6}h^3y^v$
y_0'''	4	$\dfrac{1}{h^3}$		-1	3	-3	1	$\dfrac{3}{2}hy^{iv}$
	5	$\dfrac{1}{2h^3}$	3	-14	24	-18	5	$\dfrac{7}{4}h^2y^v$
y_0^{iv}	5	$\dfrac{1}{h^4}$	1	-4	6	-4	1	$2hy^v$

$$y^{(k)} = C\sum_{i=-(n_d-1)}^{0} w_iy_i + E$$

To obtain the three point, backward difference, first derivative formula we will eliminate y_0'' from the above two equations by multiplying equation (i) by four and subtracting equation (ii) from equation (i), thus

$$4y_{-1} - y_{-2} = 3y_0 - 2hy_0' + \frac{2h^3}{3}y_0'''$$

which after rearrangement becomes

$$y_0' = \frac{1}{2h}(y_{-2} - 4y_{-1} + 3y_0) + \frac{1}{3}h^2 y_0'''$$

Similarly, to obtain the three point, backward difference, second derivative formula we will eliminate y_0' from the above two equations by multiplying equation (i) by two and subtracting equation (ii) from equation (i), thus

$$2y_{-1} - y_{-2} = y_0 - h^2 y_0'' + h^3 y_0'''$$

which after rearrangement becomes

$$y_0'' = \frac{1}{h^2}(y_{-2} - 2y_{-1} + y_0) + h y_0'''$$

The dominant error terms do not give the absolute error, but are useful for comparing the errors generated by different formulas. The dominant error terms for these first and second derivative formulas are $\frac{1}{3}h^2 y_0'''$ and $h y_0'''$ respectively. The presence of a third derivative in both these error terms indicates that the methods are of similar order of accuracy, thus both formulas would be exact if the third derivative disappeared. This would happen, for example, if the data were derived from a second order (or lower) polynomial. The coefficients of $\frac{1}{3}h^2$ and h from the two formulas indicate that for "small" h, the first derivative formula is the more accurate.

Example 5.6

Use the Taylor series method to derive the five-point, central difference, first derivative formula and its dominant error term.

Solution 5.6

We will assume the data have a uniform spacing of h in the x-direction, so in this case we are looking for a formula of the form,

$$y_0' \approx f(h, y_{-2}, y_{-1}, y_0, y_1, y_2)$$

In this case we expand the Taylor series about y_0 to estimate y_{-2}, y_{-1}, y_1 and y_2, thus

$$y_{-2} = y_0 - 2h\,y_0' + \frac{4h^2}{2!}\,y_0'' - \frac{8h^3}{3!}\,y_0''' + \frac{16h^4}{4!}\,y_0^{iv} - \frac{32h^5}{5!}\,y_0^{v} \cdots \quad (i)$$

$$y_{-1} = y_0 - h\,y_0' + \frac{h^2}{2!}\,y_0'' - \frac{h^3}{3!}\,y_0^{iii} + \frac{h^4}{4!}\,y_0^{iv} - \frac{h^5}{5!}\,y_0^{v} \cdots \quad (ii)$$

$$y_1 = y_0 + h\,y_0' + \frac{h^2}{2!}\,y_0'' + \frac{h^3}{3!}\,y_0^{iii} + \frac{h^4}{4!}\,y_0^{iv} + \frac{h^5}{5!}\,y_0^{v} \cdots \quad (iii)$$

$$y_2 = y_0 + 2h\,y_0' + \frac{4h^2}{2!}\,y_0'' + \frac{8h^3}{3!}\,y_0''' + \frac{16h^4}{4!}\,y_0^{iv} + \frac{32h^5}{5!}\,y_0^{v} \cdots \quad (iv)$$

To obtain the five point, central difference, first derivative formula we need to eliminate y_0'', y_0''' and y_0^{iv} from the above four equations and make y_0' the subject. This can be achieved by multiplying equations by constants and subtracting them from each other, in a similar way to that used in the previous example. It is left to the reader to show that the elimination of terms in this case leads to

$$y_0' = \frac{1}{12h}(y_{-2} - 8y_{-1} + 8y_1 - y_2) + \frac{1}{30}h^4 y^v$$

Note that this is still a "five-point" formula although the coefficient of the y_0 term in this case happens to equal zero.

Example 5.7

The following data rounded to four decimal places were taken from the function $y = \sin x$ with x expressed in radians.

x	y
0.4	0.3894
0.6	0.5646
0.8	0.7174
1.0	0.8415
1.2	0.9320

Use forward differences to estimate $y'(0.4)$ and $y''(0.4)$, central differences to estimate $y'(0.8)$ and backward differences to estimate $y'(1.2)$.

Solution 5.7

The data are equally spaced in x with the constant interval given by $h = 0.2$. In order to estimate $y'(0.4)$ we have a choice of formulas, depending on how many points we wish to include. As an example, from Table 5.2, the four-point $(n_d = 4)$ forward difference formula for $y'(0.4)$ gives the following expression

$$y'(0.4) \approx \frac{1}{6(0.2)}(-11(0.3894) + 18(0.5646) - 9(0.7174) + 2(0.8415)) = 0.9215$$

The exact solution is given by $\cos(0.4) = 0.9211$, hence the relative error in this case is

$$E_{rel} = \frac{0.9215 - 0.9211}{0.9211} \times 100 = 0.04\%$$

Similarly, the four-point $(n_d = 4)$ forward difference formula for $y''(0.4)$ gives

$$y''(0.4) \approx \frac{1}{0.2^2}(2(0.3894) - 5(0.5646) + 4(0.7174) - 1(0.8415)) = -0.4025$$

To illustrate the influence of the number of points included in the formulas of Tables 5.2–5.4, results are presented in tabular form for the cases considered, together with the relative errors.

Forward differences for $y'(0.4)$ *(Exact: 0.9211)*

n_d	$y'(0.4)$	$E_{rel}\%$
2	0.8760	−4.90
3	0.9320	1.18
4	0.9215	0.04
5	0.9198	−0.15

Forward differences for $y''(0.4)$ *(Exact: −0.3894)*

n_d	$y''(0.4)$	$E_{rel}\%$
3	−0.5600	43.80
4	−0.4025	3.36
5	−0.3704	−4.88

Central differences for $y'(0.8)$ *(Exact: 0.6967)*

n_d	$y'(0.8)$	$E_{rel}\%$
3	0.6923	−0.64
5	0.6969	0.03

Backward differences for $y'(1.2)$ *(Exact: 0.3624)*

n_d	$y''(1.2)$	$E_{rel}\%$
2	0.4525	24.88
3	0.3685	1.70
4	0.3603	−0.56
5	0.3621	−0.08

In the forward difference calculation of $y'(0.4)$, the greatest accuracy was apparently obtained with the four-point formula. The larger error recorded in the "more accurate" five-point formula occurred as a result of the tabulated data having been rounded after four decimal places.

The three-point forward difference result of −0.5600 for $y''(0.4)$ was very poor with a relative error of nearly 49%. This is because three-point formulas predict a constant value for the second derivative over the range under consideration (in this case $0.4 < x < 0.8$). With the second derivative varying quite rapidly over this range ($-0.3894 > y''(x) > -0.7174$) the three-point formula would only be able to give a reasonable result at $x = 0.6$ where $y''(0.6) = -0.5646$.

The error terms in Tables 5.2–5.4 show that for a given number of data points, the central difference formulas are always the most accurate. This is confirmed by the small relative errors computed in this example.

The relative error recorded in the two-point backward difference calculation of $y'(1.2)$ was very significant at nearly 25 per cent. This was due to a relatively large value of the second derivative in the vicinity of the required solution. As shown in Table 5.4, the dominant error term for this formula is given by

$$E = \frac{1}{2}hy'' = -\frac{1}{2}(0.2)\sin(1.2) = -0.093$$

which approximately equals the difference between the exact (0.3624) and computed (0.4525) values.

5.5 Curve fitting

If we are seeking a function to follow a large number of data points that might be measured in an experiment or generated by a computer program, it is often more practical to seek a function which represents a "best fit" to the data rather than one which passes through all points exactly. Various strategies are possible for minimizing the error between the individual data points and the approximating function. One of the best known is the Method of Least Squares which gives the user the ability to choose the functions to be used in the curve fitting exercise. The method is also known as "linear regression" or "multiple linear regression" in the case of two or more independent variables.

5.5.1 Least squares

Consider n_p data points $(\tilde{\mathbf{x}}_1, y_1), (\tilde{\mathbf{x}}_2, y_2), \ldots, (\tilde{\mathbf{x}}_{n_p}, y_{n_p})$, where $\tilde{\mathbf{x}}$ represents the independent variables $[x_1, x_2, \ldots, x_{n_v}]^T$ and y represents the dependent variable. The number of independent variables, n_v, can take any desired value, although in conventional linear regression it would be set to 1.

The required "best fit" function can be written in the form

$$F(\tilde{\mathbf{x}}) = C_1 f_1(\tilde{\mathbf{x}}) + C_2 f_2(\tilde{\mathbf{x}}) + \ldots + C_k f_k(\tilde{\mathbf{x}}) \tag{5.39}$$

where $f_j(\tilde{\mathbf{x}})$, $j = 1, 2, \ldots, k$ are chosen functions of $\tilde{\mathbf{x}}$ and the $C_j, j = 1, 2, \ldots, k$ are constants that will be optimized by the least squares process. The term "linear" in linear regression refers only on the model's dependence in equation (5.39) on the C_j constants. The $f_j(\tilde{\mathbf{x}})$ functions, which are under the user's control, can be nonlinear if required.

The goal is to make $F(\tilde{\mathbf{x}})$ as close as possible to y, so consider the sum of the squares of the differences between these quantities at each of the n_p data points as follows

$$E = \sum_{i=1}^{n_p} \left\{ [F(\tilde{\mathbf{x}}_i) - y_i]^2 \right\} = \sum_{i=1}^{n_p} \left\{ [C_1 f_1(\tilde{\mathbf{x}}_i) + C_2 f_2(\tilde{\mathbf{x}}_i) + \ldots + C_k f_k(\tilde{\mathbf{x}}_i) - y_i]^2 \right\}$$

$$(5.40)$$

The error term given by equation (5.40) can be minimized by partial differentiation of E with respect to each of the constants C_j in turn and equating the result to zero, thus

$$\frac{\partial E}{\partial C_1} = 2 \sum_{i=1}^{n_p} \left\{ [C_1 f_1(\tilde{\mathbf{x}}_i) + C_2 f_2(\tilde{\mathbf{x}}_i) + \ldots + C_k f_k(\tilde{\mathbf{x}}_i) - y_i] f_1(\tilde{\mathbf{x}}_i) \right\} = 0$$

$$\frac{\partial E}{\partial C_2} = 2 \sum_{i=1}^{n_p} \left\{ [C_1 f_1(\tilde{\mathbf{x}}_i) + C_2 f_2(\tilde{\mathbf{x}}_i) + \ldots + C_k f_k(\tilde{\mathbf{x}}_i) - y_i] f_2(\tilde{\mathbf{x}}_i) \right\} = 0$$

$$\vdots \qquad (5.41)$$

$$\frac{\partial E}{\partial C_k} = 2 \sum_{i=1}^{n_p} \left\{ [C_1 f_1(\tilde{\mathbf{x}}_i) + C_2 f_2(\tilde{\mathbf{x}}_i) + \ldots + C_k f_k(\tilde{\mathbf{x}}_i) - y_i] f_k(\tilde{\mathbf{x}}_i) \right\} = 0$$

This symmetric system of k linear simultaneous equations can be written in matrix form as

$$\begin{bmatrix} \sum_{i=1}^{n_p} \{f_1(\tilde{\mathbf{x}}_i) f_1(\tilde{\mathbf{x}}_i)\} & \sum_{i=1}^{n_p} \{f_1(\tilde{\mathbf{x}}_i) f_2(\tilde{\mathbf{x}}_i)\} & \cdots & \sum_{i=1}^{n_p} \{f_1(\tilde{\mathbf{x}}_i) f_k(\tilde{\mathbf{x}}_i)\} \\ \sum_{i=1}^{n_p} \{f_2(\tilde{\mathbf{x}}_i) f_1(\tilde{\mathbf{x}}_i)\} & \sum_{i=1}^{n_p} \{f_2(\tilde{\mathbf{x}}_i) f_2(\tilde{\mathbf{x}}_i)\} & \cdots & \sum_{i=1}^{n_p} \{f_2(\tilde{\mathbf{x}}_i) f_k(\tilde{\mathbf{x}}_i)\} \\ \vdots & \vdots & \vdots & \vdots \\ \sum_{i=1}^{n_p} \{f_k(\tilde{\mathbf{x}}_i) f_1(\tilde{\mathbf{x}}_i)\} & \sum_{i=1}^{n_p} \{f_k(\tilde{\mathbf{x}}_i) f_2(\tilde{\mathbf{x}}_i)\} & \cdots & \sum_{i=1}^{n_p} \{f_k(\tilde{\mathbf{x}}_i) f_k(\tilde{\mathbf{x}}_i)\} \end{bmatrix}$$

$$\begin{Bmatrix} C_1 \\ C_2 \\ \vdots \\ C_k \end{Bmatrix} = \begin{Bmatrix} \sum_{i=1}^{n_p} \{f_1(\tilde{\mathbf{x}}_i) y_i\} \\ \sum_{i=1}^{n_p} \{f_2(\tilde{\mathbf{x}}_i) y_i\} \\ \vdots \\ \sum_{i=1}^{n_p} \{f_k(\tilde{\mathbf{x}}_i) y_i\} \end{Bmatrix} \qquad (5.42)$$

and solved for $C_1, C_2, \ldots, C_k$ using an appropriate method from Chapter 2. Finally, the optimized C_j constants are substituted back into the curve fitting equation (5.39) which can then be used as required for further interpolation.

Example 5.8

Use the Least Squares method to derive the best fit straight line to the n_p data points given by $(x_1, y_1), (x_2, y_2), \ldots, (x_{n_p}, y_{n_p})$.

Solution 5.8

This example has just one independent variable x, thus $n_v = 1$. Furthermore, if a linear equation is to be fitted to the data, the following function involving two unknown constants $(k = 2)$ could be used where

$$F(x) = C_1 f_1(x) + C_2 f_2(x)$$
$$= C_1 + C_2 x$$

where $f_1(x) = 1$ and $f_2(x) = x$.

From the matrix equation (5.41) we have

$$
\begin{bmatrix}
n_p & \sum_{i=1}^{n_p} x_i \\[2ex]
\sum_{i=1}^{n_p} x_i & \sum_{i=1}^{n_p} \{x_i^2\}
\end{bmatrix}
\begin{Bmatrix}
C_1 \\[2ex]
C_2
\end{Bmatrix}
=
\begin{Bmatrix}
\sum_{i=1}^{n_p} y_i \\[2ex]
\sum_{i=1}^{n_p} \{x_i y_i\}
\end{Bmatrix}
$$

which can be solved to give

$$
C_1 = \frac{\sum\limits_{i=1}^{n} y_i - C_2 \sum\limits_{i=1}^{n} x_i}{n}
$$

$$
C_2 = \frac{n \sum\limits_{i=1}^{n} \{x_i y_i\} - \sum\limits_{i=1}^{n} x_i \sum\limits_{i=1}^{n} y_i}{n \sum\limits_{i=1}^{n} \{x_i^2\} - \left(\sum\limits_{i=1}^{n} x_i\right)^2}
$$

After substitution of C_1 and C_2 into the above expression for $F(x)$, the classical linear regression equation is obtained.

It may be noted that the correlation coefficient for the data is given as

$$r = \frac{n \sum\limits_{i=1}^{n} \{x_i y_i\} - \sum\limits_{i=1}^{n} x_i \sum\limits_{i=1}^{n} y_i}{\left\{ \left[n \sum\limits_{i=1}^{n} \{x_i^2\} - \left(\sum\limits_{i=1}^{n} x_i \right)^2 \right] \left[n \sum\limits_{i=1}^{n} \{y_i^2\} - \left(\sum\limits_{i=1}^{n} y_i \right)^2 \right] \right\}^{\frac{1}{2}}}$$

The Coefficient of Determination is given by r^2 and is a measure of the degree of linear dependence of the data.

5.5.2 Linearization of data

In order to use the least squares method, the proposed curve fitting function must be in the "linear" standard form of equation (5.39). Some quite commonly encountered curve fitting functions useful in engineering analysis are not initially in this form, but can be transformed into the standard form by a simple change of variable.

Consider for example attempting to fit a "power law" function of the form

$$y = Ax^B \tag{5.43}$$

to some data. The function given by equation (5.43) is not in the standard form of equation (5.39) because the constant B appears as an exponent, and is not a simple multiplying coefficient. By taking logs of both sides of the equation we get

$$\ln y = \ln(Ax^B)$$
$$= \ln A + B \ln x \tag{5.44}$$

Now if we make the substitutions $X = \ln x$ and $Y = \ln y$ we have

$$Y = \ln A + BX \tag{5.45}$$

or in standard form as

$$F(X) = C_1 + C_2 X \tag{5.46}$$

where $C_1 = \ln A$ and $C_2 = B$. The change of variables has delivered a linear relationship between $\ln x$ and $\ln y$ in the standard form. The least squares method will now lead to optimized coefficients C_1 and C_2 with the original constants retrieved from $A = e^{C_1}$ and $B = C_2$.

The process just described is called "linearization", and can be applied to a number of functions that do not initially fit the standard form. In some cases linearization can be achieved by transforming just one of the variables. A summary of some transformations that lead to linearization is given in Table 5.5. In some cases it may be noted that more than one transformation is possible.

TABLE 5.5: Examples of transformations leading to linearization

Function $y = f(x)$	Linearized form $Y = C_1 + C_2 X$	Variables (X, Y)	Constants (C_1, C_2)
$y = \dfrac{A}{x + B}$	$y = \dfrac{A}{B} - \dfrac{1}{B}xy$	$X = xy$ $Y = y$	$C_1 = \dfrac{A}{B}$ $C_2 = -\dfrac{1}{B}$
$y = \dfrac{A}{x + B}$	$\dfrac{1}{y} = \dfrac{B}{A} + \dfrac{1}{A}x$	$X = x$ $Y = \dfrac{1}{y}$	$C_1 = \dfrac{B}{A}$ $C_2 = \dfrac{1}{A}$
$y = \dfrac{1}{Ax + B}$	$\dfrac{1}{y} = B + Ax$	$X = x$ $Y = \dfrac{1}{y}$	$C_1 = B$ $C_2 = A$
$y = \dfrac{x}{Ax + B}$	$\dfrac{1}{y} = A + B\dfrac{1}{x}$	$X = \dfrac{1}{x}$ $Y = \dfrac{1}{y}$	$C_1 = A$ $C_2 = B$
$y = Ae^{Bx}$	$\ln y = \ln A + Bx$	$X = x$ $Y = \ln y$	$C_1 = \ln A$ $C_2 = B$
$y = Ax^B$	$\ln y = \ln A + B \ln x$	$X = \ln x$ $Y = \ln y$	$C_1 = \ln A$ $C_2 = B$
$y = \dfrac{1}{(Ax + B)^2}$	$\dfrac{1}{\sqrt{y}} = B + Ax$	$X = x$ $Y = \dfrac{1}{\sqrt{y}}$	$C_1 = B$ $C_2 = A$
$y = Axe^{-Bx}$	$\ln \dfrac{y}{x} = \ln A - Bx$	$X = x$ $Y = \dfrac{y}{x}$	$C_1 = \ln A$ $C_2 = -B$
$y = \dfrac{L}{1 + Ae^{Bx}}$ $L = \text{constant}$	$\ln\left(\dfrac{L}{y} - 1\right) =$ $\ln A + Bx$	$X = x$ $Y = \ln\left(\dfrac{L}{y} - 1\right)$	$C_1 = \ln A$ $C_2 = B$

Example 5.9

A damped oscillator has a natural frequency of $\omega = 91.7 \text{ s}^{-1}$. During free vibration, measurements indicate that the amplitude (y) decays with time (t) as follows:

t (s)	y (m)
0.00	0.05
0.07	0.0139
0.14	0.0038
0.21	0.0011

A curve fit of the form:

$$f(t) = y_o e^{-\zeta \omega t}$$

has been suggested. Use a Least Squares approach to find this function, and hence estimate the fraction of critical damping ζ.

Solution 5.9

This exponentially decaying curve fitting function is not in standard form so it needs to be transformed. The initial equation is given by

$$y = y_o e^{-91.7 \zeta t}$$

which with reference to Table 5.5 can be written in the linearized form

$$\ln y = \ln y_0 - 91.7 \zeta t$$

and hence

$$Y = C_1 + C_2 X$$

where $Y = \ln y$, $X = t$, $C_1 = \ln y_0$ and $C_2 = -91.7\zeta$. With reference to equation (5.39), the functions that will be used to develop the simultaneous equations are $f_1(X) = 1$ and $f_2(X) = X$.

For hand calculation a tabular approach is recommended, thus

$X = t$	$Y = \ln y$	f_1	f_2	$f_1 f_1$	$f_1 f_2$	$f_2 f_2$	$f_1 Y$	$f_2 Y$
0.00	−2.9957	1	0.00	1	0.00	0.0000	−2.9957	0.0000
0.07	−4.2759	1	0.07	1	0.07	0.0049	−4.2759	−0.2993
0.14	−5.5728	1	0.14	1	0.14	0.0196	−5.5728	−0.7802
0.21	−6.8124	1	0.21	1	0.21	0.0441	−6.8124	−1.4306
			Σ	4	0.42	0.0686	−19.6568	−2.5101

From equation (5.42),

$$\begin{bmatrix} 4 & 0.42 \\ 0.42 & 0.0686 \end{bmatrix} \left\{ \begin{array}{c} C_1 \\ C_2 \end{array} \right\} = \left\{ \begin{array}{c} -19.6568 \\ -2.5101 \end{array} \right\}$$

leading to solutions

$$\left\{ \begin{matrix} C_1 \\ C_2 \end{matrix} \right\} = \left\{ \begin{matrix} -3.002 \\ -18.210 \end{matrix} \right\}$$

Since $C_2 = -18.210$ *it follows that* $\zeta = \dfrac{-18.210}{-91.7} = 0.20.$

Program 5.4: Curve fitting by least squares

```
PROGRAM nm54
!---Curve Fitting by Least Squares---
 USE nm_lib; USE precision; IMPLICIT NONE
 INTEGER::i,ic,j,k,nc,np,nv; INTEGER,ALLOCATABLE::kdiag(:)
 REAL(iwp)::es,my,r2,sd,yi,zero=0.0_iwp
 REAL(iwp),ALLOCATABLE::c(:),f(:),kv(:),x(:,:),y(:)
 OPEN(10,FILE='nm95.dat'); OPEN(11,FILE='nm95.res')
 READ(10,*)np,nv,nc
 ALLOCATE(kdiag(nc),kv(nc*(nc+1)/2),f(nc),c(nc),x(np,nv),y(np))
 READ(10,*)(x(i,:),y(i),i=1,np)
 WRITE(11,'(A)')"---Curve Fitting by Least Squares---"
 c=zero; kv=zero; my=SUM(y)/np
 DO i=1,np
 CALL f54(x(i,:),f); ic=0
   DO j=1,nc
     c(j)=c(j)+f(j)*y(i)
     DO k=1,j; ic=ic+1; kv(ic)=kv(ic)+f(j)*f(k); END DO
   END DO
 END DO
 DO i=1,nc; kdiag(i)=i*(i+1)/2; END DO
 CALL sparin(kv,kdiag); CALL spabac(kv,c,kdiag)
 WRITE(11,'(/A,/,5E12.4)')" Optimized Function Coefficients",c
 WRITE(11,'(/A)')" Data points and fitted point"
 WRITE(11,'(A,I1,A)')"  (x(i),i=1,",nv,"), y, yi"
 sd=zero; es=zero
 DO i=1,np; CALL f54(x(i,:),f); yi=DOT_PRODUCT(c,f)
  sd=sd+(y(i)-my)**2; es=es+(y(i)-yi)**2
  WRITE(11,'(5E12.4)')x(i,:),y(i),yi
 END DO
 r2=(sd-es)/sd
 WRITE(11,'(/A,/,E12.4)')"  r-squared",r2
 CONTAINS

 SUBROUTINE f54(x,f)
  IMPLICIT NONE
```

```
 REAL(iwp),INTENT(IN)::x(:)
 REAL(iwp),INTENT(OUT)::f(:)
 f(1)=1.0_iwp
 f(2)=LOG(x(1))
!  f(2)=x(1)
!  f(3)=x(2)
 RETURN
END SUBROUTINE f54

END PROGRAM nm54
```

Number of data points,	np	nv	nc
independent variables	5	1	2
and functions (constants)			

Data points

```
        (x(i,:),y(i),i=1,np)
        29.   1.6
        50.   23.5
        74.   38.0
        103.  46.4
        118.  48.9
```

Data 5.4a: Curve Fitting by Least Squares (first example)

```
---Curve Fitting by Least Squares---

 Optimized Function Coefficients
-0.1111E+03  0.3402E+02

 Data Points and Fitted Point
 (x(i),i=1,1), y, yi
 0.2900E+02  0.1600E+01  0.3428E+01
 0.5000E+02  0.2350E+02  0.2196E+02
 0.7400E+02  0.3800E+02  0.3530E+02
 0.1030E+03  0.4640E+02  0.4655E+02
 0.1180E+03  0.4890E+02  0.5117E+02

 r-squared
 0.9881E+00
```

Results 5.4a: Curve Fitting by Least Squares (first example)

Program 5.4 performs multiple linear regression based on input involving several independent variable. The program includes two SUBROUTINES from the nm_lib library called sparin and spabac, which perform factorization

---

List 5.4:

Scalar integers:

i	simple counter
ic	simple counter
j	simple counter
k	simple counter
nc	number of constants in curve fitting function
np	number of input data points
nv	number of independent variables

Scalar reals:

es	running total of the squared error
my	mean of the initial y-values
r2	the coefficient of determination, r^2
sd	running total of the squared deviation from the mean
yi	interpolated value of y
zero	set to 0.0

Dynamic integer arrays:

kdiag	diagonal term location vector

Dynamic real arrays:

c	coefficients to be found by least squares
f	function values
kv	holds equation coefficients in skyline form
x	input data x-values
y	input data y-values

---

and forward/back-substitution on the symmetric system of equations created by equation (5.42). There is a SUBROUTINE called f54 at the end of the main program which evaluates the values of the user supplied functions $f_j(\tilde{\mathbf{x}}), j = 1, 2, \ldots, k$ at each of the input data points. The functions required for the curve fitting must be input by the user into f54 and will vary from one problem to the next.

The data start with np the number of data points, nv the number of independent variables and nc the number of constants to be evaluated. This is followed by the np data values, which each consist of nv values of x and one value of y.

In the first example, a function of the form

$$F(x) = C_1 + C_2 \ln x$$

is to be fitted to data given by

x	y
29.0	1.6
50.0	23.5
74.0	38.0
103.0	46.4
118.0	48.9

The data shown in Data 5.4a indicate that five (np=5) data points are read, involving just one (nv=1) independent variable. The curve fitting function $F(x)$ consists of two coefficients so nc=2 with the functions programmed into SUBROUTINE f54 as

```
f(1)=1.0_iwp
f(2)=LOG(x(1))
```

noting that x(1) represents the only independent variable in this case.

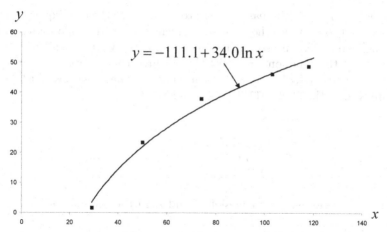

$$y = -111.1 + 34.0 \ln x$$

Figure 5.6: Curve fit from first Program 5.4 example.

The output shown in Results 5.4b gives the optimized coefficients, C_1 and C_2 together with a table of the original data points $(\tilde{x}, y)$ together with the fitted values of $F(x)$ in the right column.

In this case $C_1 = -111.1$ and $C_2 = 34.02$, hence the "best fit" function is

$$f(x) = -111.1 + 34.02 \ln x$$

which is shown plotted, together with the original points in Figure 5.6. The output also reports the *coefficient of determination* r^2 reflecting the degree of linear dependence of the data

The data and results from a second example involving two independent variables are given in Data and Results 5.4. The data come from a stability analysis in geotechnical engineering, in which the Factor of Safety (FS) of a particular embankment has been calculated for a range of soil shear strength values given by a cohesion (c') and friction angles (ϕ').

c'	ϕ'	FS
1.0	20.0	1.37
2.0	25.0	2.07
1.5	23.0	1.75
2.2	22.0	1.99
0.8	21.0	1.32
0.5	25.0	1.41
0.1	22.0	1.02

A function of the form

$$FS = C_1 + C_2 c' + C_3 \phi'$$

is thought to represent a reasonable fit to the data, so the least squares method will be used to compute the optimal coefficients. The data file shown in Data 5.4b indicates there are seven (np=7) data points, two independent variables (nv=2) and the function to be fitted has three coefficients (nc=3). This is followed by the (c', ϕ', FS) values (np times). The functions must be programmed into the SUBROUTINE f54 as follows

```
f(1)=1.0_iwp
f(2)=x(1)
f(3)=x(2)
```

where x(1) represents the cohesion c' and x(2) the friction angle ϕ'.

```
Number of data points,        np   nv   nc
independent variables          7    2    3
and functions (constants)

Data points                  (x(i,:),y(i),i=1,np)
                              1.0   20.   1.37
                              2.0   25.   2.07
                              1.5   23.   1.75
                              2.2   22.   1.99
                              0.8   21.   1.32
                              0.5   25.   1.41
                              0.1   22.   1.02
```

Data 5.4b: Curve Fitting by Least Squares (second example)

```
---Curve Fitting by Least Squares---

Optimized Function Coefficients
-0.1808E+00  0.4544E+00  0.5389E-01

Data Points and Fitted Point
(x(i),i=1,2), y, yi
0.1000E+01  0.2000E+02  0.1370E+01   0.1351E+01
0.2000E+01  0.2500E+02  0.2070E+01   0.2075E+01
0.1500E+01  0.2300E+02  0.1750E+01   0.1740E+01
0.2200E+01  0.2200E+02  0.1990E+01   0.2005E+01
0.8000E+00  0.2100E+02  0.1320E+01   0.1314E+01
0.5000E+00  0.2500E+02  0.1410E+01   0.1394E+01
0.1000E+00  0.2200E+02  0.1020E+01   0.1050E+01


r-squared
0.9979E+00
```

Results 5.4b: Curve Fitting by Least Squares (second example)

The fitted values shown in the output given in Results 5.4b clearly represent a reasonable fit to the original data. The coefficient of determination equals 0.998 in this case. Users interested in more rigorous measures of "goodness-of-fit" are invited to make the necessary modifications to the main program.

5.6 Exercises

1. Given the data

θ	$F(\theta)$
0	1.00
$\pi/4$	2.12
$\pi/2$	0.00

   express the Lagrangian polynomials as function of θ, and hence estimate the value of $F(\pi/3)$.
   Answer: $F(\pi/3) = 1.773$

2. Use Lagrangian polynomials to obtain an interpolating polynomial for the data

x	y
0.0	0.1
0.1	0.1005
0.2	0.1020
0.3	0.1046

and use it to estimate the value of y when $x = 0.4$. Note: This question involves extrapolation which is less reliable than interpolation.
Answer: $y(0.4) = 0.1084$

3. Given the cubic
$$f(x) = x^3 - 2x^2 + 3x + 1$$
derive an interpolating polynomial $g(x)$ that coincides with $f(x)$ at $x = 1, 2$ and 3, and hence show that

$$\int_1^3 f(x)\,dx = \int_1^3 g(x)\,dx$$

Answer: $g(x) = 4x^2 - 8x + 7$, Simpson's Rule is able to integrate cubics exactly.

4. Derive a polynomial which passes through the points

x	y
0	1
1	1
2	15
3	61

using (a) Lagrangian polynomials and (b) the Difference Method.
Answer: $f(x) = 3x^3 - 2x^2 - x + 1$

5. Rework Exercise 4 using the Difference Method to obtain the fourth order interpolating polynomial, if the additional point $(4, 100)$ is included.
Answer: $f(x) = -2.375x^4 + 17.25x^3 - 28.125x^2 + 13.25x + 1$

6. Given the "sampling points"

$$x_1 = -\sqrt{\frac{3}{5}} \ , x_2 = 0 \ \ x_3 = \sqrt{\frac{3}{5}}$$

form the Lagrangian polynomials $L_1(x)$, $L_2(x)$ and $L_3(x)$ and show that

$$\int_{-1}^{-1} L_1\,dx = \int_{-1}^{-1} L_3\,dx = \frac{5}{9}, \quad \int_{-1}^{-1} L_2\,dx = \frac{8}{9}$$

7. The following data are based on the function $y = \sin x$ where x is in degrees

x	y
10	0.17365
15	0.25882
20	0.34202
25	0.42262
30	0.50000

Set up a table of forward differences, and by including more terms each time, estimate $\sin 28°$ using first, second, third and fourth order interpolating polynomials.
Answer: First order 0.4690, second order 0.4694, third order 0.4695, fourth order 0.4695.

8. Rework Exercise 1 using a cubic spline function and hence estimate $F(\pi/3)$.
Answer: $F(\pi/3) = 1.713$

9. Rework Exercise 2 using a cubic spline function and hence estimate the value of y when $x = 0.15$.
Answer: $y(0.15) = 0.1011$

10. Rework Exercise 3 to derive a cubic spline function which coincides with $f(x)$ at $x = 1$, 2 and 3. Compute the value of the cubic spline and $f(x)$ at $x = 1.5$ and $x = 2.5$.
Answer: **spline**(1.5) $= 4.25$, $f(1.5) = 4.375$; **spline**(2.5) $= 12.25$, $f(2.5) = 11.625$

11. Rework Exercise 4 using cubic spline functions to estimate the value of y when $x = 2.5$.
Answer: $y(2.5) = 35.15$

12. Rework Exercise 6 using a cubic spline function to estimate the value of y when $x = 17$.
Answer: $y(17) = 0.2924$

13. Given the data of Exercise 1 estimate $F'(0)$, $F'(\pi/4)$ and $F'(\pi/2)$ making use of as much of the data as possible.
Answer: $F'(0) \approx 3.49$, $F'(\pi/4) \approx -0.64$, $F'(\pi/2) \approx -4.76$

14. Given the data from Exercise 7 estimate $y'(30)$ using 2-, 3-, 4- and 5-point backward difference formulas. Compare your results with the exact solution of 0.86603.
Answer: 0.887, 0.868, 0.866, 0.866

15. Given the data from Exercise 7 estimate $y'(20)$ using 3- and 5-point central difference formulas. Compare your results with the exact solution of 0.93969.
    Answer: 0.939, 0.940

16. Given the data from Exercise 7 estimate $y''(10)$ using 3-, 4- and 5-point forward difference formulas. Compare your results with the exact solution of -0.17365.
    Answer: -0.259, -0.176, -0.175

17. Given the data from Exercise 7 estimate $y'''(10)$ and $y^{iv}(10)$ using 5-points in each case. Compare your results with the exact solutions of -0.98481 and 0.17365 respectively.
    Answer: 0.971, 0.172

18. In geotechnical engineering, the relationship between Prandtl's bearing capacity factor N_c and the friction angle ϕ is summarized in the table below

ϕ	N_c
0	5.14
5	6.49
10	8.35
15	10.98
20	14.83
25	20.72
30	30.14

    Use least squares to fit first, second and third order polynomials to this data, and use these to estimate N_c when $\phi = 28$.
    Answer: First order 24.02, second order 26.08, third order 26.01

19. Use least squares to fit a straight line to the data

x	1.3	2.1	3.4	4.3	4.7	6.0	7.0
y	2.6	3.5	3.2	3.0	4.1	4.6	5.2

    and hence estimate the value of y when $x = 2.5$ and $x = 5.5$.
    Answer: $f(x) = 2.083 + 0.4035x$, hence $y(2.5) \approx 3.092$, $y(5.5) \approx 4.302$.

20. Use least squares to fit a power equation of the form, $y = ax^b$ to the data

x	2.5	3.5	5.0	6.0	7.5	10.0	12.5
y	4.8	3.5	2.1	1.7	1.2	0.9	0.7

    Answer: Following linearization, $f(x) = 15.34e^{-1.23x}$

21. Use least squares to fit an exponential equation of the form

x	0.6	1.0	2.3	3.1	4.4	5.8	7.2
y	3.6	2.5	2.2	1.5	1.1	1.3	0.9

Answer: Following linearization, $f(x) = 3.24e^{-0.188x}$

22. Given the following data, use the difference method to estimate the value of y when $x = -1.5$ using two, three and all four points from the table (3 answers in total).

x	y
-1	-2
-2	-18
-3	-52
-4	-110

Answer: $-10, -7.75, -8.125$

23. Derive the three-point backward difference formula for a first derivative y'_o and the dominant error term.

Answer: $y'_o = \dfrac{1}{2h}(3y_o - 4y_{-1} + y_{-2}) + \dfrac{1}{3}h^2 y'''_o$

24. Measured readings from a 1-d consolidation test on a saturated clay gave the following values of the Time Factor (T) and the Average Degree of Consolidation (U):

T	U
0.196	0.5
0.286	0.6
0.403	0.7
0.567	0.8
0.848	0.9

It has been suggested that an equation of the form

$$T = a\log_{10}(1 - U) + b$$

would represent a suitable fit to these results. Estimate the values of a and b.

Answer: $a = -0.931, b = -0.084$

25. Given the following data:

x	y
1.0	-2.000
0.8	-0.728
0.6	0.016
0.4	0.424
0.2	0.688

use a forward difference table to derive the interpolating polynomial. What do the coefficients of the interpolating polynomial indicate about the nature of the tabulated data?

Answer: $Q_3(x) = 0.688 - 1.320(x - 0.2) - 1.8009(x - 0.2)(x - 0.4) - 4(x - 0.2)(x - 0.4)(x - 0.6)$, 4th order terms do not participate in the solution.

26. Find the least-squares parabola, $f(x) = Ax^2 + Bx + C$ for the set of data:

x	y
-3	15
-1	5
1	1
3	5

Answer: $f(x) = 0.875x^2 - 1.7x + 2.125$

27. The function $y = L/(1 + Ce^{Ax})$ has been suggested as a suitable fit to the data given below. Linearize the function, and hence estimate A and C. The constant $L = 1000$.

x	y
0	200
1	400
2	650
4	950

Answer: $A = -1.086$, $C = 1.459$, $y(3) = 858.0$

28. Use least squares to fit a power equation of the form $f(x) = ax^b$ to the data:

x	y
2.5	4.8
3.5	3.5
5.0	2.1
6.0	1.7
7.5	1.2

Answer: $y = 16.14x^{-1.27}$

29. Form a difference table with the data below to generate a polynomial function that passes through all the points:

x	y
2.5	4.8
3.0	3.5
3.5	2.1
4.0	1.7

Modify your polynomial if the additional point $(4.5, 1.2)$ is added to the table.

Answer: $Q_3(x) = 4.8 - 2.6(x - 2.5) - 0.2(x - 2.5)(x - 3) + 1.4667(x - 2.5)(x-3)(x-3.5)$, $Q_4(x) = Q_3(x) - 1.4667(x-2.5)(x-3)(x-3.5)(x-4)$

Chapter 6

Numerical Integration

6.1 Introduction

Numerical integration or "quadrature" is used whenever analytical integration approaches are either inconvenient or impossible to perform. Numerical integration techniques can be applied to mathematical functions, or discrete data that might be measured in an engineering experiment. Initially in this chapter, we will consider methods for numerical integration of a function of one variable, i.e.,

$$\int_a^b f(x)\ dx \tag{6.1}$$

although area and volume integrals of functions of more than one variable will also be considered subsequently.

This chapter will cover several numerical integration formulas (also known as "rules") for numerical integration. The methods will usually not depend on the type of function being integrated, although some "customized" approaches will also be discussed.

Although numerical integration sometimes leads to an exact solution, especially if the function under consideration is a simple polynomial, our solutions will often be approximate. This is especially true when integrating combinations of transcendental functions (e.g., sine, cosine, logarithm, etc.) which may in any case have no exact analytical solution. Once approximate solutions have been obtained, it will then be important to get a sense of the magnitude of the errors. In some cases the numerical integration rule can have an adaptive feature that seeks a target level of accuracy.

The first step in numerical integration methods is to replace the function to be integrated, $f(x)$, by a simple polynomial $Q_{n-1}(x)$ of degree $n-1$ which coincides with $f(x)$ at n points x_i where $i = 1, 2, \ldots, n$.

Thus,

$$\int_a^b f(x)\ dx \approx \int_a^b Q_{n-1}(x)\ dx \tag{6.2}$$

where

$$Q_{n-1}(x) = a_0 + a_1 x + a_2 x^2 + \ldots + a_{n-1} x^{n-1} \qquad (6.3)$$

and the a_i are constant coefficients.

Polynomials such as $Q_{n-1}(x)$ are easily integrated analytically, and are also exactly integrated numerically. Hence, provided the approximation given by equation (6.2) is reasonable, we should be able to obtain reasonable estimates of the required integral.

Numerical integration "rules" are usually expressed as a summation of the form

$$\int_a^b Q_{n-1}(x)\, dx = \sum_{i=1}^n w_i Q_{n-1}(x_i) \qquad (6.4)$$

hence

$$\int_a^b f(x)\, dx \approx \sum_{i=1}^n w_i f(x_i) \qquad (6.5)$$

The x_i are called "sampling points", being those places where the function $f(x)$ is evaluated, and the w_i are constant "weighting coefficients". Equation (6.5) forms the basis for all our numerical integration methods in a single variable.

Although $Q_{n-1}(x)$ is the function that is actually being integrated from equation (6.4), we do not usually need to know its exact form since the sampling points will be substituted into $f(x)$.

All that distinguishes one numerical integration rule from another is the number and location of the sampling points and the corresponding weighting coefficients.

Methods in which the sampling points are equally spaced within the range of integration are called "Newton-Cotes" rules, and these are considered first. It is later shown that "Gaussian" rules, in which the sampling points are optimally spaced, lead to considerable improvements in accuracy and efficiency.

Other aspects of numerical integration dealt with in this chapter include adaptive methods with error control, special integrals in which the integrand contains an exponentially decaying function or a certain type of singularity, and multiple integrals in which functions of more that one variable are integrated numerically over areas or volumes.

With reference to equation (6.5), sampling points usually lie within the range of integration, thus $a \le x_i \le b$; however rules

will also be described that sample outside the range of integration. Such rules have applications to the solution of ordinary differential equations that will be covered in Chapter 7.

6.2   Newton-Cotes rules

6.2.1   Introduction

Newton-Cotes rules are characterized by sampling points that are equally spaced within the range of integration and include the limits of integration themselves. In the following subsections we will describe the first few members of the Newton-Cotes family, starting with the simplest method that uses just one sampling point.

6.2.2   Rectangle rule, $(n = 1)$

In this simple method, which is also known as the "Rectangle rule", the function to be integrated from equation (6.2) is approximated by a "zero order" polynomial $Q_0(x)$ which coincides with $f(x)$ at one point only, namely the lower limit of integration as shown in Figure 6.1.

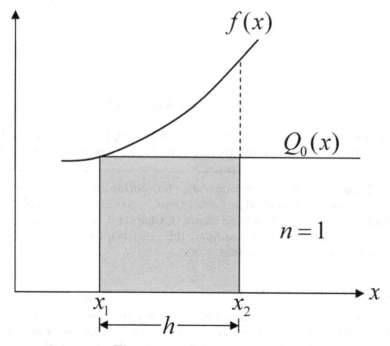

Figure 6.1: Rectangle rule.

Since the function being integrated is constant, the solution is the area of the shaded rectangle, thus

$$\text{Rectangle rule:} \qquad \int_{x_1}^{x_2} f(x) \, dx \approx h f(x_1) \qquad (6.6)$$

Clearly the method is not particularly accurate, and will only give exact solutions if $f(x)$ is itself a zeroth order polynomial, i.e., a line parallel to the x-axis.

Example 6.1

Estimate the value of

$$I = \int_{\pi/4}^{\pi/2} \sin x \, dx$$

using the Rectangle rule.

Solution 6.1

$$h = \frac{\pi}{4}$$

hence

$$I \approx \frac{\pi}{4} \sin \frac{\pi}{4}$$

$$= 0.5554 \ (cf. \ exact \ solution \ 0.7071)$$

The solution is very poor in this case due to sampling at a location that is so unrepresentative of the function over the range to be integrated.

While the method is not recommended for routine integration of functions, the simplicity of the method has attractions, especially when combined with a small step length h. As will be shown in Chapter 7, an important group of numerical methods for solving ordinary differential equations called "explicit" algorithms are based on this rule.

6.2.3   Trapezoid rule, $(n = 2)$

This is a popular approach in which $f(x)$ is approximated by a first order polynomial $Q_1(x)$ which coincides with $f(x)$ at both limits of integration. As shown in Figure 6.2, the integral is approximated by the area of a trapezoid.

$$\text{Trapezoid rule:} \qquad \int_{x_1}^{x_2} f(x) \, dx \approx \frac{1}{2} h (f(x_1) + f(x_2)) \qquad (6.7)$$

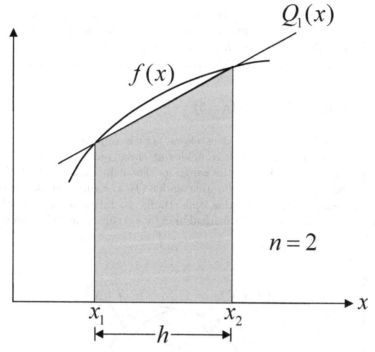

Figure 6.2: Trapezoid rule.

The formula will be exact for $f(x)$ of degree 1 or less.

Example 6.2

Estimate the value of

$$I = \int_{\pi/4}^{\pi/2} \sin x \; dx$$

using the Trapezoid rule.

Solution 6.2

$$h = \frac{\pi}{4}$$

hence

$$I \approx \frac{1}{2}\frac{\pi}{4}\left[\sin\left(\frac{\pi}{4}\right) + \sin\left(\frac{\pi}{2}\right)\right]$$

$$= 0.6704 \; (\text{cf. exact solution } 0.7071)$$

While better than the result produced by the Rectangle rule, the numerical solution is still poor. More accurate estimates can be found by using "repeated" rules as will be described in a later section.

6.2.4 Simpson's rule, $(n = 3)$

Another well known method is where $f(x)$ is approximated by a second order polynomial $Q_2(x)$ which coincides at three points, namely the limits of integration and middle of the range as shown in Figure 6.3. Unlike the previous examples however, the area under $Q_2(x)$ cannot be deduced from simple geometry, so we use this opportunity to introduce a powerful and general method for deriving numerical integration rules called the "Polynomial Substitution Method."

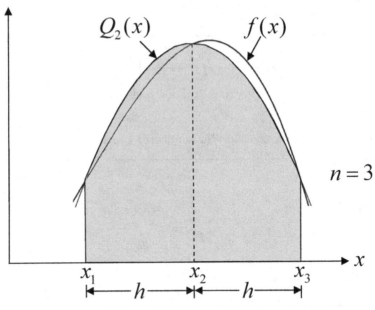

Figure 6.3: Simpson's rule.

6.2.4.1 Polynomial Substitution Method

In this example we will use the Polynomial Substitution Method to derive the weighting coefficients of a three-point $(n = 3)$ Newton-Cotes rule of the general form

$$\int_{x_1}^{x_3} f(x)\ dx \approx w_1 f(x_1) + w_2 f(x_2) + w_3 f(x_3) \qquad (6.8)$$

where $x_2 = (x_1 + x_3)/2$ and $x_3 - x_2 = x_2 - x_1 = h$.

From the pattern established by the first two Newton-Cotes rules, it can be assumed that Simpson's rule with three sampling points from equation (6.8) will be exact if $f(x)$ is a polynomial of order 0,1 or 2. This will enable us to generate 3 equations in the unknown w_1, w_2 and w_3.

Before starting the derivation however, it is convenient and algebraically simpler to make a temporary transformation of the problem such that the mid-point of the range is at the origin, and the limits of integration[1] lie at $\pm h$. In order to do this, we shift all x-values by x_2 leading to a transformed function $F(x)$, where

$$\int_{x_1}^{x_3} f(x)\ dx \equiv \int_{-h}^{h} F(x)\ dx \qquad (6.9)$$

and

$$F(x) = f(x + x_2) \qquad (6.10)$$

Let $F(x) = 1$

$$\int_{-h}^{h} 1\ dx = 2h = w_1 + w_2 + w_3 \qquad (6.11)$$

Let $F(x) = x$

$$\int_{-h}^{h} x\ dx = 0 = w_1(-h) + w_2(0) + w_3(h) = -w_1 h + w_3 h \qquad (6.12)$$

Let $F(x) = x^2$

$$\int_{-h}^{h} x^2\ dx = \frac{2}{3}h^3 = w_1(-h)^2 + w_2(0)^2 + w_3(h)^2 = w_1 h^2 + w_3 h^2 \qquad (6.13)$$

Solution of equations (6.11), (6.12) and (6.13) leads to

$$w_1 = w_3 = \frac{1}{3}h \qquad \text{and} \qquad w_2 = \frac{4}{3}h \qquad (6.14)$$

thus

---

[1] As discussed later in this chapter, a special case used by some generalized integration rules is obtained by setting $h = 1$.

$$\int_{-h}^{h} F(x)\, dx \approx \frac{1}{3}F(-h) + \frac{4}{3}F(0) + \frac{1}{3}F(h) \qquad (6.15)$$

which after transformation back to the original function and limits, gives

$$\int_{x_1}^{x_3} f(x)\, dx \approx \frac{1}{3}f(x_1) + \frac{4}{3}f(x_2) + \frac{1}{3}f(x_3) \qquad (6.16)$$

Example 6.3

Estimate the value of

$$I = \int_{\pi/4}^{\pi/2} \sin x\, dx$$

using Simpson's rule.

Solution 6.3

$$h = \frac{\pi}{8}$$

hence

$$I \approx \frac{1}{3}\frac{\pi}{8}\left[\sin\left(\frac{\pi}{4}\right) + 4\sin\left(\frac{3\pi}{8}\right) + \sin\left(\frac{\pi}{2}\right)\right]$$

$$= 0.7072 \ (cf. \ exact \ solution \ 0.7071)$$

The reader is encouraged to re-derive the one and two-point rules given by equations (6.6) and (6.7) to gain confidence in the generality of the Polynomial Substitution Method.

It turns out that Simpson's rule given by equation (6.16) is exact for $f(x)$ up to degree 3, although we did not need to assume this in the derivation. This anomaly, and other issues of accuracy, will be discussed in Section (6.2.6).

6.2.5   Higher order Newton-Cotes rules $(n > 3)$

There is no limit to the number of sampling points that could be incorporated in a Newton-Cotes rule. For example, following the pattern already established in previous sections, a four-point rule would fit a cubic to $f(x)$ and is exact for $f(x)$ of degree 3 or less. A five-point rule would fit a quartic to $f(x)$ and is exact for $f(x)$ of degree 5 or less and so on.[2]

---

[2] As will be shown, all odd-numbered Newton-Cotes rules with $n \geq 3$ give an extra order of accuracy.

These higher order rules are rarely used in practice however, as "repeated" lower order methods (see Section 6.2.8) are often preferred.

6.2.6 Accuracy of Newton-Cotes rules

Rules with an even number of sampling points (i.e., $n = 2, 4, \ldots$) will exactly integrate polynomials of degree up to one less than the number of sampling points (i.e., $n - 1$). Rules with an odd number of sampling points (i.e., $n = 3, 5, \ldots$) are more efficient, in that they will exactly integrate polynomials of degree up to the number of sampling points (i.e., n). The reason for this difference is demonstrated by Figure 6.3 for Simpson's rule ($n = 3$), where it is easily shown that if $f(x)$ is cubic, it will be integrated exactly because the errors introduced above and below the approximating polynomial $Q_2(x)$ cancel out.

In order to assess the accuracy of Newton-Cotes rules more formally, we need to compare the result generated by the rule, with the exact solution written as a Taylor series. The largest term from the series ignored by the approximate solution is known as the "dominant error term".

6.2.6.1 Error in the Rectangle rule ($n = 1$)

Consider the Taylor series expansion of $f(x)$ about the lower limit of integration x_1, i.e.,

$$f(x) = f(x_1) + (x - x_1)f^{'}(x_1) + \frac{(x - x_1)^2}{2!}f^{''}(x_1) + \ldots \qquad (6.17)$$

Integration of both sides of this equation gives

$$\int_{x_1}^{x_2} f(x)\, dx = \left[xf(x_1 + \frac{1}{2}(x - x_1)^2 f^{'}(x_1) + \frac{1}{6}(x - x_1)^3 f^{''}(x_1) + \ldots \right]_{x_1}^{x_2} \qquad (6.18)$$

which after substitution of the limits of integration and noting that $h = x_2 - x_1$ becomes

$$\int_{x_1}^{x_2} f(x)\, dx = hf(x_1) + \frac{1}{2}h^2 f^{'}(x_1) + \frac{1}{6}h^3 f^{''}(x_1) + \ldots \qquad (6.19)$$

Comparing equations (6.19) and (6.6) shows that the Rectangle rule truncates the series after the first term, hence the Dominant Error Term is $\frac{1}{2}h^2 f^{'}(x_1)$.

6.2.6.2 Error in the Trapezoid rule ($n = 2$)

Noting again that $h = x_2 - x_1$, consider the Taylor series expansion of $f(x)$ about the lower limit of integration x_1 to obtain $f(x_2)$, i.e.,

$$f(x_2) = f(x_1) + hf^{'}(x_1) + \frac{h^2}{2!}f^{''}(x_1) + \ldots \qquad (6.20)$$

After multiplying through by $\frac{1}{2}h$ and rearranging, we can write

$$\frac{1}{2}h^2 f'(x_1) = \frac{1}{2}hf(x_2) - \frac{1}{2}hf(x_1) - \frac{1}{4}h^3 f''(x_1) - \dots \tag{6.21}$$

Substitution of equation (6.21) into (6.19) gives

$$\int_{x_1}^{x_2} f(x)\,dx = \frac{1}{2}h\left(f(x_1) + f(x_2)\right) - \frac{1}{12}h^3 f''(x_1) - \dots \tag{6.22}$$

which after comparison with (6.7) indicates a Dominant Error Term for this rule of $-\frac{1}{12}h^3 f''(x_1)$.

6.2.6.3 Error in Simpson's rule ($n = 3$)

An alternative approach to finding the dominant error term can be used, if the highest order of polynomial for which the rule is exact is known in advance.

Simpson's rule exactly integrates cubics, but will only approximately integrate quartics, so it follows that the dominant error term must contain a fourth derivative, thus

$$\int_{-h}^{h} f(x)\,dx = \frac{1}{3}f(-h) + \frac{4}{3}f(0) + \frac{1}{3}f(h) + \alpha f^{iv}(x) \tag{6.23}$$

Letting $f(x) = x^4$ we get

$$\int_{-h}^{h} x^4\,dx = \frac{2}{5}h^5 = \frac{1}{3}h^5 + 0 + \frac{1}{3}h^5 + 24\alpha \tag{6.24}$$

hence

$$\alpha = \frac{h^5}{24}\left(\frac{2}{5} - \frac{2}{3}\right) = -\frac{1}{90}h^5 \tag{6.25}$$

so Simpson's rule has a Dominant Error Term $-\frac{1}{90}h^5 f^{iv}(x)$.

6.2.7 Summary of Newton-Cotes rules

All Newton-Cotes rules can be written in the form

$$\int_a^b f(x)\,dx = h\sum_{i=1}^{n} W_i f(x_i) + Ch^{k+1} f^k(x) \tag{6.26}$$

where h represents the distance between sampling points $x_i = a + (i-1)h$ with coefficients W_i for $i = 1, 2, \dots, n$ and C is a constant. $f^k(x)$ represents the kth derivative of $f(x)$. A summary of values in equation (6.26) is given in Table 6.1.

TABLE 6.1: Summary of Newton-Cotes rules

n	Name	h	W_1	W_2	W_3	W_4	W_5	C	k
1	Rectangle	$b-a$	1					$\frac{1}{2}$	1
2	Trapezoid	$b-a$	$\frac{1}{2}$	$\frac{1}{2}$				$-\frac{1}{12}$	2
3	Simpson's	$\frac{b-a}{2}$	$\frac{1}{3}$	$\frac{4}{3}$	$\frac{1}{3}$			$-\frac{1}{90}$	4
4	4-point	$\frac{b-a}{3}$	$\frac{3}{8}$	$\frac{9}{8}$	$\frac{9}{8}$	$\frac{3}{8}$		$-\frac{3}{80}$	4
5	5-point	$\frac{b-a}{4}$	$\frac{14}{45}$	$\frac{64}{45}$	$\frac{24}{45}$	$\frac{64}{45}$	$\frac{14}{45}$	$-\frac{8}{945}$	6

An alternative way of portraying the Newton-Cotes rules that facilitates comparison with the Gaussian rules to be described later in this chapter is to consider a normalized problem with limits of integration of ± 1, thus

$$\int_{-1}^{1} f(x)\, dx = \sum_{i=1}^{n} w_i f(x_i) + C f^k(x) \qquad (6.27)$$

In this case the interval h is always known, so it is included directly in the weighting coefficients w_i. A summary of values in equation (6.27) is given in Table 6.2 and Table 6.3 gives a summary of sampling points and weights in decimal form for the first five normalized Newton-Cotes rules.

It should be noted that the weighting coefficients are always symmetrical about the midpoint of the range of integration and their sum must always equal the range of integration.

Although the coefficient C of the dominant error term is a guide to the accuracy of these rules, the actual dominant error term includes both h raised to some power, and a derivative of the function being integrated. Clearly as h gets smaller (for $h < 1$), the error term gets smaller still; however care must be taken with certain functions to ensure that the higher derivatives $f^k(x)$ are not becoming excessively large within the range of integration.

6.2.8 Repeated Newton-Cotes rules

If the range to be integrated is large, an alternative to trying to fit a higher order polynomial over the full range is to break up the range into strips and

TABLE 6.2: Normalized Newton-Cotes rules

n	Name	h	w_1	w_2	w_3	w_4	w_5	C	k
1	Rectangle	2	2					2	1
2	Trapezoid	2	1	1				$-\frac{2}{3}$	2
3	Simpson's	1	$\frac{1}{3}$	$\frac{4}{3}$	$\frac{1}{3}$			$-\frac{1}{90}$	4
4	4-point	$\frac{2}{3}$	$\frac{1}{4}$	$\frac{3}{4}$	$\frac{3}{4}$	$\frac{1}{4}$		$-\frac{2}{405}$	4
5	5-point	$\frac{1}{2}$	$\frac{7}{45}$	$\frac{32}{45}$	$\frac{12}{45}$	$\frac{32}{45}$	$\frac{7}{45}$	$-\frac{1}{30240}$	6

to use lower order polynomials over each strip. All of the methods described so far can be used in this "repeated" mode.

6.2.8.1 Repeated Rectangle rule

As shown in Figure 6.4, the range of integration $[A, B]$ is split into k strips with widths of $h_i, i = 1, 2, \ldots, k$. Although it is simpler to make all the strips the same width, the strips can be made narrower in regions where the function is changing rapidly. In "adaptive" quadrature methods the width of the strips can be modified automatically as will be described later in this chapter.

In the Repeated Rectangle rule the area of each strip is approximated by a rectangle with a height given by the value of the function at the lower limit of the strip. These are then added together to give the overall solution, thus

$$\int_A^B f(x)\, dx \approx h_1 f(A) + h_2 f(A + h_1) + h_3 f(A + h_1 + h_2) + \ldots + h_k f(B - h_k)$$

(6.28)

It can be seen from Figure 6.4 that the method amounts to replacing a smooth continuous function $f(x)$ by a sequence of horizontal lines. The more strips taken, the more closely the actual shape of the function is reproduced. If all the strips are the same width h, the formula becomes

$$\int_A^B f(x)\, dx \approx h\left(f(A) + f(A + h) + f(A + 2h) + \ldots + f(B - h)\right) \quad (6.29)$$

TABLE 6.3: Sampling Points and Weights for Normalized Newton-Cotes rules

n	Name	x_i	w_i
1	Rectangle	-1.000000000000000	2.000000000000000
2	Trapezoid	-1.000000000000000	1.000000000000000
		1.000000000000000	1.000000000000000
3	Simpson's	-1.000000000000000	0.333333333333333
		0.000000000000000	1.333333333333333
		1.000000000000000	0.333333333333333
4	4-point	-1.000000000000000	0.250000000000000
		-0.333333333333333	0.750000000000000
		0.333333333333333	0.750000000000000
		1.000000000000000	0.250000000000000
5	5-point	-1.000000000000000	0.155555555555556
		-0.500000000000000	0.711111111111111
		0.000000000000000	0.266666666666667
		0.500000000000000	0.711111111111111
		1.000000000000000	0.155555555555556

Example 6.4

Estimate the value of

$$I = \int_{\pi/4}^{\pi/2} \sin x \, dx$$

using the Repeated Rectangle rule with three $(k = 3)$ strips of equal width.

Solution 6.4

From equation (6.29)

$$h = \frac{\pi}{12}$$

hence

$$I \approx \frac{\pi}{12} \left(\sin(\pi/4) + \sin(\pi/3) + \sin(5\pi/12) \right)$$
$$= 0.6647 \ (cf. \ exact \ solution \ 0.7071)$$

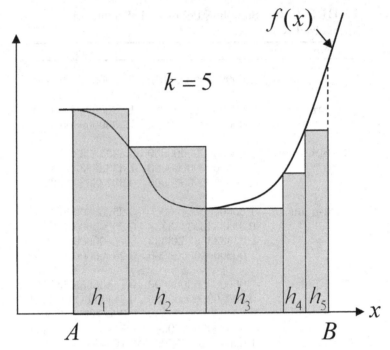

Figure 6.4: Repeated Rectangle rule.

It can be seen that while the numerical solution is poor, it represents a significant improvement over the result of 0.5554 obtained with a single application of this rule. Better accuracy still could be obtained by taking more strips.

6.2.8.2   Repeated Trapezoid rule

In the Repeated Trapezoid rule the area of each strip is approximated by a trapezoid as shown in Figure 6.5.

Each trapezoid coincides with the function at two locations. These are then added together to give the overall solution, thus

$$\int_A^B f(x)\,dx \approx \frac{1}{2}h_1\left(f(A) + f(A + h_1)\right) + \frac{1}{2}h_2\left(f(A + h_1) + \right.$$

$$\left. f(A + h_1 + h_2)\right) + \ldots + \frac{1}{2}h_k\left(f(B - h_k) + f(B)\right) \quad (6.30)$$

It can be seen from Figure 6.5 that the method amounts to replacing a smooth continuous function $f(x)$ by a sequence of linear line segments. If all the strips are the same width h, the formula becomes

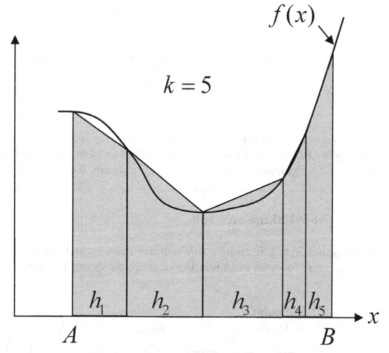

Figure 6.5: Repeated Trapezoid.

$$\int_A^B f(x)\,dx \approx \frac{1}{2}h(f(A) + 2f(A + h) + 2f(A + 2h) + \dots$$
$$\dots + 2f(B - h) + f(B)) \qquad (6.31)$$

Example 6.5

Estimate the value of

$$I = \int_{\pi/4}^{\pi/2} \sin x\,dx$$

using the Repeated Trapezoid rule with three $(k = 3)$ strips of equal width.

Solution 6.5

From equation (6.31)

$$h = \frac{\pi}{12}$$

hence

$$I \approx \frac{1}{2}\frac{\pi}{12}\left(\sin(\pi/4) + 2\sin(\pi/3) + 2\sin(5\pi/12) + \sin(\pi/2)\right)$$
$$= 0.7031 \ (cf. \ exact \ solution \ 0.7071)$$

It can be seen that in this case the numerical solution is much improved by the linear approximations. It may be recalled that a single application of the Trapezoid rule in Example 6.2 ($k = 1$) gave the solution 0.6704.

6.2.8.3   Repeated Simpson's rule

A single application of Simpson's rule requires three sampling points, so the repeated rule must have an even number of strips as shown in Figure 6.6.

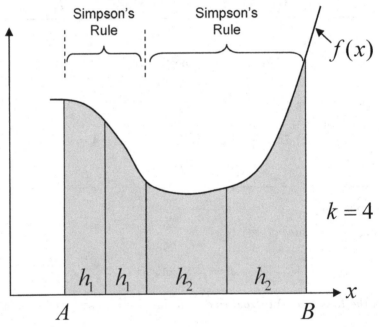

Figure 6.6: Repeated Simpson's rule.

Each pair of strips must have the same width, but widths may differ from pair to pair. The repeated Simpson's rule fits a parabola over each set of three sampling points, leading to the following expression assuming k (even) strips

$$\int_A^B f(x)\, dx \approx \frac{1}{3}h_1(f(A) + 4f(A+h_1) + f(A+2h_1)) +$$

$$\frac{1}{3}h_2(f(A+2h_1) + 4f(A+2h_1+h_2) + f(A+2h_1+2h_2)) + \ldots$$

$$\ldots + \frac{1}{3}h_k(f(B-2h_k) + 4f(B-h_k) + f(B)) \qquad (6.32)$$

If all strips are of equal width h, the rule simplifies to

$$\int_A^B f(x)\, dx \approx \frac{1}{3}h(f(A) + 4f(A+h) + 2f(A+2h) + 4f(A+3h) + \ldots$$

$$\ldots + 2f(B-2h) + 4f(B-h) + f(B)) \qquad (6.33)$$

Example 6.6

Estimate the value of

$$I = \int_{\pi/4}^{\pi/2} \sin x\, dx$$

using the Repeated Simpson's rule with four $(k=4)$ strips of equal width.

Solution 6.6

From equation (6.33)

$$h = \frac{\pi}{16}$$

hence

$$I \approx \frac{1}{3}\frac{\pi}{16}\left(\sin(\pi/4) + 4\sin(5\pi/16) + 2\sin(3\pi/8) + 4\sin(7\pi/16) + \sin(\pi/2)\right)$$

$$= 0.7071 \ (\text{cf. exact solution } 0.7071)$$

In this case the result is accurate to 4DP. It may be recalled that a single application of Simpson's rule in Example 6.3 $(k=1)$ was almost as accurate, giving 0.7072.

Program 6.1: Repeated Newton-Cotes rules

```
PROGRAM p61
!---Repeated Newton-Cotes Rules---
 USE nm_lib; USE precision; IMPLICIT NONE
 INTEGER::i,j,nr,nsp
 REAL(iwp)::a,area,b,cr,hr,pt5=0.5_iwp,wr,zero=0.0_iwp
 REAL(iwp),ALLOCATABLE::samp(:,:),wt(:)
 OPEN(10,FILE='nm95.dat'); OPEN(11,FILE='nm95.res')
 READ(10,*)a,b,nsp,nr; ALLOCATE(samp(nsp,1),wt(nsp))
 CALL newton_cotes(samp,wt)
 wr=(b-a)/nr; hr=pt5*wr; area=zero
 DO i=1,nr
   cr=a+(i-1)*wr+hr
   DO j=1,nsp; area=area+wt(j)*hr*f61(cr+samp(j,1)*hr); END DO
 END DO
 WRITE(11,'(A)')"---Repeated Newton-Cotes Rules---"
 WRITE(11,'(/,A,2F12.4)')"Limits of integration",a,b
 WRITE(11,'(A,I7)')"Newton Cotes Rule     ",nsp
 WRITE(11,'(A,I7)')"Number of repetitions",nr
 WRITE(11,'(A,F12.4)' )"Computed result       ",area
 CONTAINS

 FUNCTION f61(x)
  IMPLICIT NONE
  REAL(iwp),INTENT(IN)::x; REAL(iwp)::f61
  f61=SIN(x)*SIN(x)+x
  RETURN
 END FUNCTION f61

END PROGRAM p61
```

Program 6.1 estimates the value of the integral

$$I = \int_a^b f(x)\,dx$$

using a Repeated Newton-Cotes rule of the user's choice. Input data consist of the limits of integration a and b, the number of sampling points nsp in each application of the rule and the number of repetitions nr of the rule across the range. Each application of the Newton-Cotes rule acts over strips of width h which are assumed constant.

The sampling points and weighting coefficients for each repetition are held,

---

List 6.1:

Scalar integers:

i	simple counter
j	simple counter
nr	number of rule repetitions across the range
nsp	number of sampling points (=2 for Trapezoid Rule, etc.)

Scalar reals:

a	lower limit of integration
area	holds running total of area from each repetition
b	upper limit of integration
cr	central coordinate of each repetition
hr	half the range of each repetition
pt5	set to 0.5
wr	width of each repetition
zero	set to 0.0

Dynamic real arrays:

samp	holds sampling points
wt	holds weighting coefficients

---

respectively, in the first columns of array samp and in wt, provided by subroutine newton_cotes held in library nm_lib.

All subroutines used in this book are described and listed in Appendices A and B.

To illustrate use of the program, the following problem is solved using the Trapezoid rule repeated five times.

$$I = \int_{0.25}^{0.75} (\sin^2 x + x)\, dx$$

The input and output files from Program 6.1 are given in Data 6.1 and Results 6.1 respectively. The function to be integrated, $f(x) = \sin^2 x + x$, has been programmed into f61 at the end of the main program.

```
Limits of integration          a       b
                              0.25    0.75

Number of sampling points     nsp
per repetition                 2

Number of repetitions          nr
                               5
```

Data 6.1: Repeated Newton-Cotes rules

```
---Repeated Newton-Cotes Rules---
```

Limits of integration	0.2500	0.7500
Newton Cotes Rule	2	
Number of repetitions	5	
Computed result	0.3709	

Results 6.1: Repeated Newton-Cotes rules

As shown in Results 6.1 with nsp=2 and nr=5, the method gives a value for the above integral of 0.3709, which compares quite well with the analytical solution of 0.3705. In problems of this type, especially when no analytical solution is available, it is good practice to repeat the calculation with more strips to ensure that the solution is converging with sufficient accuracy. For example, if the above problem is recalculated with nr=10, the computed result improves to 0.3706.

6.2.9   Remarks on Newton-Cotes rules

The choice of a suitable method for numerical integration is never completely straightforward. If the analyst is free to choose the position of the sampling points, then Newton-Cotes methods should probably not be used at all, as the Gaussian rules covered in the next section are more efficient. The Newton-Cotes have the advantage of simplicity however, especially the lower order members of the family such as the Trapezoid and Simpson's rules.

Generally speaking, frequently repeated low order methods are preferred to high order methods, and the repeated Trapezoid rule presented in Program 6.1 will give acceptable solutions to many problems.

The choice of the interval h in a repeated rule presents a further difficulty. Ideally, h should be "small" but not so small that excessive computer time is required, or that the accuracy is affected by computer word-length limitations. When using Program 6.1 to integrate a function, it is recommended that two or three solutions are obtained using an increasing number of repetitions. This will indicate whether the result is converging. Lack of convergence suggests a "badly behaved" function which could contain a singularity requiring special treatment.

The repeated Trapezoid rule is also suitable for integrating tabulated data from an experiment where, for example, the interval at which measurements are made is not constant. In such cases, a formula similar to equation (6.30) would be used.

6.3 Gauss-Legendre rules

6.3.1 Introduction

Newton-Cotes rules were easy to use, because their sampling points were evenly spaced within the range of integration, and the weighting coefficients easy to remember (at least up to Simpson's rule).

Gaussian rules allow the sampling points to be optimally spaced within the range of integration and achieve greater accuracy for a given number of sampling points. These optimal sampling point locations are still symmetrical about the middle of the range of integration, but the positions are not intuitive and certainly harder to remember. This latter point may be a slight hindrance to "hand" calculation using Gaussian methods, but not when the information can be stored in a subroutine library.

In Gaussian rules the summation notation adopted previously is still applicable, thus

$$\int_a^b f(x)\,dx \approx \sum_{i=1}^n w_i f(x_i) \tag{6.34}$$

however when using polynomial substitution to develop Gaussian rules, not only the w_i but also the x_i are initially treated as unknowns. In order to solve for the resulting $2n$ equations it will be necessary to force the formula to be exact for integration of $f(x)$ up to degree $2n - 1$. This represents a considerable improvement over the equivalent Newton-Cotes rule, which, for the same number of sampling points, will only be exact for $f(x)$ up to degree $n - 1$ (or n if n is odd and greater than 1).

There are various types of Gaussian rules, but the most important and widely used are the Gauss-Legendre rules, which will be considered first.

Instead of using limits of a and b as indicated in equation (6.34) the development of these rules is greatly simplified and generalized by considering normalized limits of ± 1. A similar approach was considered for Newton-Cotes rules leading to Table 6.3. This in no way limits the generality of these methods because solutions corresponding to the actual limits can be easily retrieved by a simple transformation.

In the following subsections we will describe the first few members of the Gauss-Legendre family, starting with the simplest method that is commonly known as the Midpoint rule.

6.3.2 Midpoint rule, $(n = 1)$

If we can sample the function to be integrated at one point only, it is intuitively obvious that the point should be in the middle of the range of integration as shown in Figure 6.7.

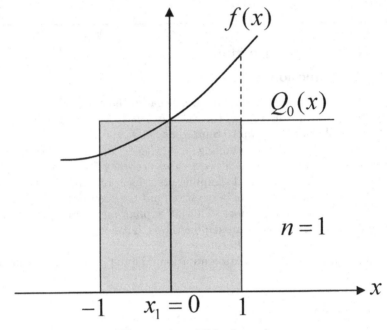

Figure 6.7: Midpoint rule.

The function $f(x)$ is approximated by $Q_0(x)$ which runs parallel to the x-axis and coincides with the function to be integrated at the mid-point of the range. The formula given by equation (6.35) will be exact for $f(x)$ up to degree 1 and takes the form

$$\int_{-1}^{1} f(x)\,dx \approx 2f(0) \tag{6.35}$$

Example 6.7

Estimate the value of

$$I = \int_{\pi/4}^{\pi/2} \sin x \, dx$$

using the Midpoint rule.

Solution 6.7

Although the limits of integration are not ±1, we can solve the problem by a simple transformation of the x-coordinate followed by a scaling of weighting coefficients. The sampling point will be at the mid-point of the range of *integration, and the weighting coefficient must equal the range of integration thus*

$$x_1 = \frac{3\pi}{8} \text{ and } w_1 = \frac{\pi}{4}$$

. hence

$$I \approx \frac{\pi}{4} \sin \frac{3\pi}{8}$$

$$= 0.7125 \ (cf. \ exact \ solution \ 0.7071)$$

The Midpoint rule with its optimally located sampling point clearly gives a significantly improved result compared with the Rectangle rule (which gave 0.5554), in spite of both methods involving the same amount of computational effort.

6.3.3 Two-point Gauss-Legendre rule, $(n = 2)$

In this case we have no idea where the optimal locations of the sampling points will be, so we will use the Polynomial Substitution Method to find them.

The formula will be of the form

$$\int_{-1}^{1} f(x) \, dx \approx w_1 f(x_1) + w_2 f(x_2) \tag{6.36}$$

which contains four ($2n$) unknowns, w_1, x_1, w_2 and x_2. In order to generate four equations to solve for these values, we must force the formula given by equation (6.35) to be exact for $f(x) = 1$, $f(x) = x$, $f(x) = x^2$ and $f(x) = x^3$.

Let $f(x) = 1$, hence

$$\int_{-1}^{1} 1 \, dx = 2 = w_1 + w_2 \tag{6.37}$$

Let $f(x) = x$, hence

$$\int_{-1}^{1} x \, dx = 0 = w_1 x_1 + w_2 x_2 \tag{6.38}$$

Let $f(x) = x^2$, hence

$$\int_{-1}^{1} x^2 \, dx = \frac{2}{3} = w_1 x_1^2 + w_2 x_2^2 \tag{6.39}$$

Let $f(x) = x^3$, hence

$$\int_{-1}^{1} x^3 \, dx = 0 = w_1 x_1^3 + w_2 x_2^3 \tag{6.40}$$

From equations (6.37) and (6.39) we can write

$$\frac{w_1}{w_2} = -\frac{x_2}{x_1} = -\frac{x_2^3}{x_1^3} \tag{6.41}$$

$$\text{hence } x_2 = \pm x_1 \tag{6.42}$$

Assuming the sampling points do not coincide, we must have $x_2 = -x_1$ so from equation (6.37)

$$w_1 = w_2 \text{ and from equation (6.36) } w_1 = w_2 = 1 \tag{6.43}$$

Finally from equation (6.38)

$$x_1^2 = x_2^2 = \frac{1}{3} \tag{6.44}$$

$$\text{thus } x_1 = -\frac{1}{\sqrt{3}} \text{ and } x_2 = \frac{1}{\sqrt{3}} \tag{6.45}$$

The two-point rule with normalized limits of integration can therefore be written as

$$\int_{-1}^{1} f(x) \, dx \approx f(-\frac{1}{\sqrt{3}}) + f(\frac{1}{\sqrt{3}}) \tag{6.46}$$

and will be exact for polynomial $f(x)$ up to degree 3. The method is actually finding the area under a linear function $Q_1(x)$ that coincides with $f(x)$ at the strategic locations $x = \pm\frac{1}{\sqrt{3}}$ as shown in Figure 6.8.

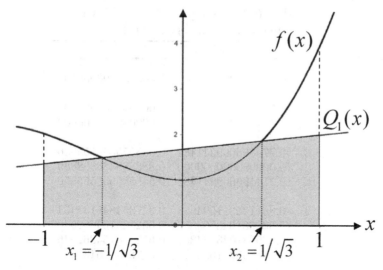

Figure 6.8: Two-point Gauss-Legendre.

Example 6.8

Estimate the value of

$$I = \int_{-1}^{1} (x^3 + 2x^2 + 1)\, dx$$

using the two-point Gauss-Legendre rule.

Solution 6.8

This question is posed with the normalized limits of integration, so the weights and sampling points can be applied directly, thus

$$I \approx \left[\left(-\frac{1}{\sqrt{3}} \right)^3 + 2\left(-\frac{1}{\sqrt{3}} \right)^2 + 1 \right] + \left[\left(\frac{1}{\sqrt{3}} \right)^3 + 2\left(\frac{1}{\sqrt{3}} \right)^2 + 1 \right]$$

$$= \frac{10}{3} \quad \text{(which is exact)} \tag{6.47}$$

As expected, the two-point Gauss-Legendre rule is able to integrate cubics exactly. It should be noted that in order to integrate a cubic exactly with a Newton-Cotes rule, we needed Simpson's rule with three sampling points.

TABLE 6.4: Sampling Points and Weights for Normalized Gauss-Legendre rules

n	x_i	w_i
1	0.000000000000000	2.000000000000000
2	-0.577350269189626	1.000000000000000
	0.577350269189626	1.000000000000000
3	-0.774596669241484	0.555555555555556
	0.000000000000000	0.888888888888889
	0.774596669241484	0.555555555555556
4	-0.861136311594053	0.347854845137454
	-0.339981043584856	0.652145154862546
	0.339981043584856	0.652145154862546
	0.861136311594053	0.347854845137454
5	-0.906179845938664	0.236926885056189
	-0.538469310105683	0.478628670499366
	0.000000000000000	0.568888888888889
	0.538469310105683	0.478628670499366
	0.906179845938664	0.236926885056189

6.3.4   Three-point Gauss-Legendre rule, $(n = 3)$

This rule will take the form

$$\int_{-1}^{1} f(x)\, dx \approx w_1 f(x_1) + w_2 f(x_2) + w_3 f(x_3) \qquad (6.48)$$

and will be exact for $f(x)$ up to degree $(2n - 1) = 5$. Although there are six unknowns in equation (6.47) which would require polynomial substitution of $f(x)$ up to degree 5, advantage can be taken of symmetry to reduce the number of equations. As with Newton-Cotes rules, the weighting coefficients and sampling points are always symmetric about the midpoint of the range of integration, hence it can be stated from inspection that

$$w_1 = w_3$$
$$x_1 = -x_3 \qquad (6.49)$$
$$x_2 = 0 \qquad (6.50)$$

The six equations from Polynomial Substitution must still be generated; however there are only three independent unknowns given by x_1, w_1 and w_2. It is left as an exercise for the reader to show that the final form of the three-point Gauss-Legendre rule is as follows

$$\int_{-1}^{1} f(x)\,dx \approx \frac{5}{9}f(-\sqrt{\frac{3}{5}}) + \frac{8}{9}f(0) + \frac{5}{9}f(\sqrt{\frac{3}{5}}) \qquad (6.51)$$

Figure 6.9 shows the locations of the three sampling points for this rule. Following the same pattern as before, the actual integration is performed under a second order polynomial $Q_2(x)$ (not shown in the figure) that coincides with $f(x)$ at the three sampling points.

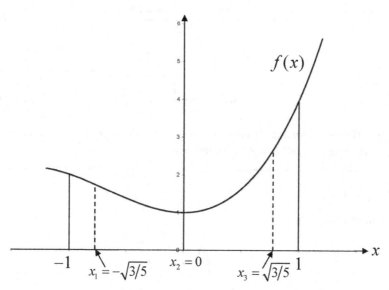

Figure 6.9: Three-point Gauss-Legendre.

We have described the first three members of the Gauss-Legendre family, which are summarized together with the next two members of the family (up to $n = 5$) in Table 6.4. The method is so called because the sampling points are the roots of a family of polynomials called Legendre polynomials which take the form

$$P_n(x) = \frac{1}{2^n n!} \frac{d^n}{dx^n} (x^2 - 1)^n = 0 \qquad (6.52)$$

6.3.5 Changing the limits of integration

All the Gauss-Legendre rules described in this section used limits of integration of ± 1. Naturally we usually wish to integrate functions between other limits, so in order to use the normalized weights and sampling points we need

to transform our problem. By a change of variable we can arrange for the normalized problem to
yield the same solution as the original problem, thus

$$\int_a^b f(x)\, dx \equiv \int_{-1}^1 f(\xi)\, d\xi \tag{6.53}$$

In order to achieve this transformation we make the substitutions

$$x = \frac{(b-a)\xi + (b+a)}{2} \tag{6.54}$$

and

$$dx = \frac{(b-a)}{2}\, d\xi \tag{6.55}$$

Once the problem has been transformed in this way, the familiar Gauss-Legendre weights and sampling points can be applied.

Example 6.9

Estimate the value of

$$I = \int_1^4 x \cos x\, dx$$

using the three-point ($n = 3$) Gauss-Legendre rule with a change of variable.

Solution 6.9

Since the limits are not ± 1 we must first transform the problem by making the substitutions given in equations (6.52) and (6.53), thus

$$x = \frac{3\xi + 5}{2} \quad \text{and} \quad dx = \frac{3}{2}\, d\xi$$

After substitution, the problem becomes

$$I = \int_1^4 x \cos x\, dx \equiv \frac{3}{4}\int_{-1}^1 (3\xi + 5)\cos\left(\frac{3\xi + 5}{2}\right) d\xi$$

which can be solved directly using the values in Table 6.4. In this example we will use the version given in equation (6.50) which leads to

$$I \approx \frac{3}{4} \left\{ \frac{5}{9} \left(-3\sqrt{\frac{3}{5}} + 5 \right) \cos \left(\frac{1}{2} \left(-3\sqrt{\frac{3}{5}} + 5 \right) \right) \right.$$

$$+ \frac{8}{9} 5 \cos \frac{5}{2}$$

$$\left. + \frac{5}{9} \left(3\sqrt{\frac{3}{5}} + 5 \right) \cos \left(\frac{1}{2} \left(3\sqrt{\frac{3}{5}} + 5 \right) \right) \right\}$$

$$\approx -5.0611 \ (cf. \ exact \ solution \ -5.0626)$$

The result is quite accurate; however since the function being integrated is transcendental, it can never be exactly integrated by conventional rules.

An alternative approach which uses the normalized values in Table 6.4 is to proportion the sampling points symmetrically about the midpoint of the range. This method involves no change of variable and the original limits of integration are maintained. The weights must also be scaled such that they add up to the actual range of integration. The normalized weights for any given rule always add up to 2, so for general limits these normalized values must be multiplied by half the range of integration. Example 6.9 is now repeated using this approach.

Example 6.10

Estimate the value of

$$I = \int_1^4 x \cos x \, dx$$

using the three-point ($n = 3$) Gauss-Legendre rule with proportioning of sampling points and weights.

Solution 6.10

As shown in Figure 6.10, the sampling points are located by proportioning the normalized values from Table 6.4 about the center of the range of integration.

For $n = 3$ and using the version given in equation (6.50) the normalized sampling points are given as

$$x_1 = -\sqrt{\frac{3}{5}} \quad x_2 = 0 \quad x_3 = \sqrt{\frac{3}{5}}$$

and the normalized weights as

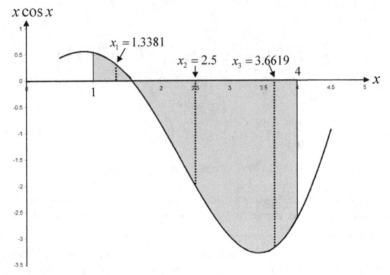

Figure 6.10: Sampling point locations in three-point Gauss-Legendre.

$$w_1 = \frac{5}{9} \qquad w_2 = \frac{8}{9} \qquad w_3 = \frac{5}{9}$$

To find the actual sampling points, the normalized sampling points are multi-plied by half the range of integration, i.e., 3/2, and positioned relative to the midpoint of the range, i.e., 5/2. The standard weights are also multiplied by half the range of integration so that they will add up to the range, i.e., 3.

Hence the actual sampling points are given by

$$x_1 = \frac{5}{2} - \frac{3}{2}\sqrt{\frac{3}{5}} = 1.33810, \qquad x_2 = \frac{5}{2} = 2.5, \qquad x_3 = \frac{5}{2} + \frac{3}{2}\sqrt{\frac{3}{5}} = 3.66190$$

and the actual weights by

$$w_1 = \frac{3}{2}\left(\frac{5}{9}\right) = 0.83333 \qquad w_2 = \frac{3}{2}\left(\frac{8}{9}\right) = 1.33333 \qquad w_3 = \frac{3}{2}\left(\frac{5}{9}\right) = 0.83333$$

The required integral is given by

$$I \approx 0.8333(1.3381)\cos 1.3381 + 1.3333(2.5)\cos 2.5 + 0.8333(3.6619)\cos 3.6619$$
$$= -5.0611 \ (cf. \ exact \ solution \ -5.0626)$$

Program 6.2: Repeated Gauss-Legendre rules

```
PROGRAM p62
!---Repeated Gauss-Legendre Rules---
 USE nm_lib; USE precision; IMPLICIT NONE
 INTEGER::i,j,nr,nsp
 REAL(iwp)::a,area,b,cr,hr,pt5=0.5_iwp,wr,zero=0.0_iwp
 REAL(iwp),ALLOCATABLE::samp(:,:),wt(:)
 OPEN(10,FILE='nm95.dat'); OPEN(11,FILE='nm95.res')
 READ(10,*)a,b,nsp,nr; ALLOCATE(samp(nsp,1),wt(nsp))
 CALL gauss_legendre(samp,wt); wr=(b-a)/nr; hr=pt5*wr; area=zero
 DO i=1,nr
   cr=a+(i-1)*wr+hr
   DO j=1,nsp; area=area+wt(j)*hr*f62(cr+samp(j,1)*hr); END DO
 END DO
 WRITE(11,'(A)')"---Repeated Gauss-Legendre Rules---"
 WRITE(11,'(/,A,2F12.4)')"Limits of Integration",a,b
 WRITE(11,'(A,I7)')"Gauss-Legendre Rule   ",nsp
 WRITE(11,'(A,I7)')"Number of Repetitions",nr
 WRITE(11,'(A,F12.4)' )"Computed Result       ",area
 CONTAINS

FUNCTION f62(x)
  IMPLICIT NONE
  REAL(iwp),INTENT(IN)::x; REAL(iwp)::f62
  f62=x*COS(x)
  RETURN
END FUNCTION f62

END PROGRAM p62
```

For larger ranges of integration, Gauss-Legendre rules can be repeated as described earlier for Newton-Cotes rules. The second program in this chapter estimates the value of the integral

$$I = \int_a^b f(x)\,dx$$

using a Repeated Gauss-Legendre rule of the user's choice. Input data consist of the limits of integration a and b, the number of sampling points nsp in each application of the rule and the number of repetitions nr of the rule across the range. Each application of the Gauss-Legendre rule acts over a strip of width h which is assumed constant.

List 6.2:

Scalar integers:
i   simple counter
j   simple counter
nr   number of rule repetitions across the range
nsp   number of Gauss-Legendre points

Scalar reals:
a   lower limit of integration
area   holds running total of area from each repetition
b   upper limit of integration
cr   central coordinate of each repetition
hr   half the range of each repetition
pt5   set to 0.5
wr   width of each repetition
zero   set to 0.0

Dynamic real arrays:
samp   holds sampling points
wt   holds weighting coefficients

Program 6.2 is essentially the same as Program 6.1 except for subroutine `gauss_legendre` that provides the sampling points and weighting coefficients. The function $f(x)$ to be integrated has been inserted directly into function `f62(x)` at the end of the main program. To illustrate use of the program, the same problem considered in Examples 6.9 and 6.10 is considered again but this time with two repetitions (`nr=2`) of the 3-point Gauss-Legendre rule (`nsp=3`). The problem to be solved is

$$I = \int_1^4 (x \cos x) \, dx$$

The input and output for Program 6.2 are given in Data 6.2 and Results 6.2 respectively. The function $x \cos(x)$ has been programmed into `f62` at the end of the main program.

```
Limits of integration          a       b
                              1.00    2.00

Number of sampling points    nsp
per repetition                2

Number of repetitions         nr
                              5
```

Data 6.2: Repeated Gauss-Legendre rules

```
---Repeated Gauss-Legendre Rules---

Limits of Integration      1.0000      4.0000
Gauss-Legendre Rule        3
Number of Repetitions      2
Computed Result            -5.0626
```

Results 6.2: Repeated Gauss-Legendre rules

As shown in Results 6.2, the repeated rule gives a value for the above integral of -5.0626 which agrees with the exact solution to four decimal places.

6.3.6   Accuracy of Gauss-Legendre rules

Gauss-Legendre rules with n sampling points will exactly integrate polynomials up to degree $2n - 1$. Since we know the highest degree for which the rules will be exact, we can find the dominant error term using the method described in Section 6.2.6 for Simpson's rule.

The two-point Gauss-Legendre rule will integrate cubic functions exactly, but not quartics, so the dominant error term must include a fourth derivative, hence

$$\int_{-1}^{1} f(x)\, dx = f(-\frac{1}{\sqrt{3}}) + f(\frac{1}{\sqrt{3}}) + C f^{iv}(x) \tag{6.56}$$

Substitution of $f(x) = x^4$ in which $f^{iv}(x) = 4! = 24$ gives

$$\int_{-1}^{1} x^2\, dx = \frac{2}{5} = \frac{1}{9} + \frac{1}{9} + 24C \tag{6.57}$$

therefore $C = \frac{1}{135}$.

In order to compare Newton-Cotes and Gauss-Legendre rules we can return to the general form first expressed in equation (6.27) as

$$\int_{-1}^{1} f(x)\, dx = \sum_{i=1}^{n} w_i f(x_i) + C f^k(x) \tag{6.58}$$

For Gauss-Legendre rules, $k = 2n$ and it is easily shown that the coefficient of the dominant error term can be written as

$$C = \frac{2^{2n+1}(n!)^4}{(2n+1)[(2n)!]^3} \tag{6.59}$$

Table 6.5 summarizes the dominant error from equation (6.58) for the two families of rules up to $n = 5$. The greater accuracy of Gauss-Legendre in terms of both the coefficient C and the order k is striking.

TABLE 6.5: Comparison of Dominant Error Terms in Normalized Newton-Cotes and Gauss-Legendre rules

	Newton-Cotes		Gauss-Legendre	
n	C	k	C	k
1	2	1	$\frac{1}{3}$	2
2	$-\frac{2}{3}$	2	$\frac{1}{135}$	4
3	$-\frac{1}{90}$	4	$\frac{1}{15750}$	6
4	$-\frac{2}{405}$	4	$\frac{1}{3472875}$	8
5	$-\frac{1}{30240}$	6	$\frac{1}{1237732650}$	10

6.4 Adaptive integration rules

In the discussion of repeated rules in Section 6.2.8, the possibility was raised of adapting the widths of the strips to reflect the behavior of the function being integrated. For example if the function was changing rapidly over some region, the strips might be made narrower, whereas if the function was varying gradually, wider strips might achieve the same level of accuracy. It would clearly be tedious to implement these alterations manually, so in this section we introduce an adaptive algorithm to adjust the widths of the strips automatically.

The basis of this adaptive approach is an estimate of the error over each strip obtained by comparing the results obtained by two rules that have significantly different levels of accuracy. The more accurate rule is assumed to give the "exact" result, so the error is estimated as the difference between the two results. If this error estimate exceeds user-specified error tolerances, the strip under consideration is divided into two and the process repeated.

The algorithm starts with a single strip across the full range, which is bisected into smaller strips iteratively until the error criteria are satisfied for all strips.

A typical sequence is shown in Figure 6.11. Figure 6.11(a) shows the first iteration which involves integration over the full range. The letter F (for "false") indicates that the error exceeded the tolerance, so the range is split into two strips of equal width. Figure 6.11(b) shows the second iteration which involves integration over both strips. The letter T (for "true") indicates that the right strip has satisfied the error criteria; however the F means the left strip has not, so that range is again split into two smaller strips of equal width. Figure 6.11(c) shows the third iteration which involves integration over both strips in the left half of the range. Once more, the right half has satisfied the error bound but the left half has not, requiring a further subdivision of the strip marked F. The fourth and final iteration in this demonstration shown in Figure 6.11(d) indicates that all strips are now marked T, thus all strips satisfy the error bounds. The solutions over each strip using the more accurate rule are then summed and printed. It should be noted that once a strip has satisfied the error tolerance and is marked T, it is not revisited.

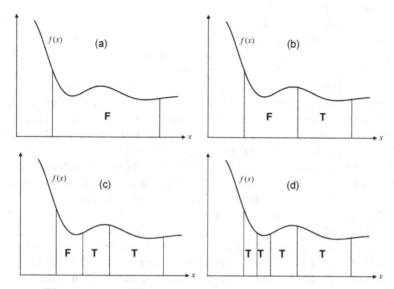

Figure 6.11: Adaptive integration with error control.

The algorithm demonstrates the need for both absolute and relative error criteria (see Section 1.8). The relative error is the most important from a engineering viewpoint since it limits the errors to some user-specified proportion of the area being estimated. The absolute error is also needed however as a safeguard, in the event of the area of any strip becoming very close to zero, in which case the relative error tolerance might be unreasonably severe. In this case the absolute error would take precedence.

Program 6.3: Adaptive Gauss-Legendre rules

```
PROGRAM p63
!---Adaptive Gauss-Legendre Rules---
 USE nm_lib; USE precision; IMPLICIT NONE
 INTEGER::cf,ct,i,j,inew,nr1=1,nr2,nsp1,nsp2
 REAL(iwp)::a,abserr,ans,area,area1,area2,b,errest,hr,relerr,    &
   one=1.0_iwp,pt5=0.5_iwp,tol,tot_err,wr,zero=0.0_iwp
 LOGICAL::verdict
 REAL(iwp),ALLOCATABLE::                                         &
   answer1(:),err1(:),limits1(:,:),samp1(:,:),wt1(:),            &
   answer2(:),err2(:),limits2(:,:),samp2(:,:),wt2(:)
 LOGICAL,ALLOCATABLE::conv1(:),conv2(:)
 OPEN(10,FILE='nm95.dat'); OPEN(11,FILE='nm95.res')
 ALLOCATE(answer1(nr1),err1(nr1),conv1(nr1),limits1(nr1,2))
 ALLOCATE(answer2(nr1),err2(nr1),conv2(nr1),limits2(nr1,2))
!---Rule 2 Should be More Accurate than Rule 1 so nsp2 >> nsp1
 READ(10,*)limits1,abserr,relerr,nsp1,nsp2
 ALLOCATE(samp1(nsp1,1),samp2(nsp2,1),wt1(nsp1),wt2(nsp2))
 CALL gauss_legendre(samp1,wt1); CALL gauss_legendre(samp2,wt2)
 WRITE(11,'(A)')"---Adaptive Gauss-Legendre Rules---"
 WRITE(11,'(/A,E12.4)')'Absolute Error Tolerance       ',abserr
 WRITE(11,'(A,E12.4)') 'Relative Error Tolerance       ',relerr
 WRITE(11,'(/A,2F12.4)')'Limits of Integration      ',limits1
 WRITE(11,'(A,I5)')'Low  Order Gauss-Legendre Rule',nsp1
 WRITE(11,'(A,I5)')'High Order Gauss-Legendre Rule',nsp2
 conv1=.FALSE.
 DO
   area=zero; tot_err=zero; ct=0; cf=0
   DO i=1,nr1
     IF(.NOT.conv1(i))THEN; a=limits1(i,1); b=limits1(i,2)
       nsp1=UBOUND(samp1,1); nsp2=UBOUND(samp2,1)
       wr=b-a; hr=pt5*wr; area1=zero; area2=zero
       DO j=1,nsp1
         area1=area1+wt1(j)*hr*f63(a+hr*(one-samp1(j,1)))
       END DO
       DO j=1,nsp2
         area2=area2+wt2(j)*hr*f63(a+hr*(one-samp2(j,1)))
       END DO
       errest=area1-area2; tol=MAX(abserr,relerr*ABS(area2))
       ans=area2; verdict=.FALSE.
       IF(ABS(errest)<tol)verdict=.TRUE.
       answer1(i)=ans; conv1(i)=verdict; err1(i)=errest
```

```
      END IF
      IF(conv1(i))THEN; ct=ct+1; ELSE; cf=cf+1; END IF
      area=area+answer1(i); tot_err=tot_err+err1(i)
    END DO
    IF(cf==0)THEN
      WRITE(11,'(A,I5)')"Number of Repetitions        ",nr1
      WRITE(11,'(/A)' )" *****Strip Limits*****    Strip Area&
        &         Error"
      DO i=1,nr1
        WRITE(11,'(2E12.4,2E16.8)')limits1(i,:),answer1(i),err1(i)
      END DO
      WRITE(11,'(/A,2E16.8)')"Solution and Total Error",        &
        area,tot_err; EXIT
    END IF
    limits2=limits1; answer2=answer1; conv2=conv1; err2=err1
    nr2=nr1; nr1=ct+2*cf
    DEALLOCATE(answer1,conv1,err1,limits1)
    ALLOCATE(answer1(nr1),conv1(nr1),err1(nr1),limits1(nr1,2))
    conv1=.FALSE.; inew=0
    DO i=1,nr2
      IF(conv2(i))THEN; inew=inew+1
        limits1(inew,:)=limits2(i,:); answer1(inew)=answer2(i)
        err1(inew)=err2(i); conv1(inew)=.TRUE.
      ELSE
        inew=inew+1; limits1(inew,1)=limits2(i,1)
        limits1(inew,2)=(limits2(i,1)+limits2(i,2))*pt5
        inew=inew+1
        limits1(inew,1)=(limits2(i,1)+limits2(i,2))*pt5
        limits1(inew,2)=limits2(i,2)
      END IF
    END DO
    DEALLOCATE(answer2,conv2,err2,limits2)
    ALLOCATE(answer2(nr1),conv2(nr1),err2(nr1),limits2(nr1,2))
  END DO
  CONTAINS

  FUNCTION f63(x)
   IMPLICIT NONE
   REAL(iwp),INTENT(IN)::x; REAL(iwp)::f63
   f63=x**(1.0_iwp/7.0_iwp)/(x**2.0_iwp+1.0_iwp)
   RETURN
  END FUNCTION f63

END PROGRAM p63
```

List 6.3:

Scalar integers:

cf	counts .FALSE. strips
ct	counts .TRUE. strips
i	simple counter
inew	simple counter
nr1	number of strips in current iteration
nr2	number of strips in next iteration
nsp1	number of sampling points in rule number 1
nsp2	number of sampling points in rule number 2

Scalar reals:

a	lower limit of integration
abserr	absolute error bound
ans	area of an individual strip
area	total area
b	upper limit of integration
errest	estimated error of an individual strip
one	set to 1.0
pt5	set to 0.5
relerr	relative error bound
tot_err	estimated total error
zero	set to 0.0

Scalar logicals:

verdict	.TRUE. if error bounds of strip satisfied

Dynamic real arrays:

answer1	holds strip areas in current iteration
answer2	holds strip areas in next iteration
err1	holds errors of strips in current iteration
err2	holds errors of strips in next iteration
limits1	holds limits of integration in current iteration
limits2	holds limits of integration in next iteration
samp1	holds sampling points for rule number 1
samp2	holds sampling points for rule number 2
wt1	holds weighting coefficients for rule number 1
wt2	holds weighting coefficients for rule number 2

Dynamic logical arrays:

conv1	holds .TRUE. or .FALSE. for all strips in current iteration
conv2	holds .TRUE. or .FALSE. for all strips in next iteration

The program makes use of two library subroutines from `nm_lib`, namely `gauss_legendre` as used in Program 6.2, and `adapt`. Subroutine `adapt` estimates the area of each strip and returns the `LOGICAL` variable `verdict` that is `.TRUE.` if the accuracy criteria are satisfied and `.FALSE.` if not.

The program estimates the integral

$$I = \int_a^b f(x)\, dx$$

to an accuracy defined by the user in the form of the maximum permissible absolute and relative errors over each strip. The program uses two Gauss-Legendre rules of different accuracy levels.

The example problem used here to illustrate the adaptive approach is the integral

$$I = \int_0^1 \frac{x^{1/7}}{(x^2+1)}\, dx$$

where the function to be integrated has been programmed into the `f63` at the end of the main program.

Input for Program 6.3 is given in Data 6.3 and consists of the limits of integration 0.0 and 1.0 which are read into the array `limits1`, the absolute and relative error bounds which are read into `abserr=1.e-3` and `relerr=1.e-3` respectively, and finally the number of sampling points for each of the Gauss-Legendre rules to be applied over each strip read into `nsp1` and `nsp2`.

Limits of integration	a	b
	0.0	1.0

Error bounds	abserr	relerr
	1.e-3	1.e-3

Number of sampling points in the two rules	nsp1	nsp2
	1	4

Data 6.3: Adaptive Gauss-Legendre rules

```
---Adaptive Gauss-Legendre Rules---

Absolute Error Tolerance          0.1000E-02
Relative Error Tolerance          0.1000E-02

Limits of Integration             0.0000        1.0000
Low  Order Gauss-Legendre Rule    1
High Order Gauss-Legendre Rule    4
Number of Repetitions             7
```

*****Strip Limits*****		Strip Area	Error
0.0000E+00	0.3125E-01	0.16702531E-01	0.54465760E-03
0.3125E-01	0.6250E-01	0.20089520E-01	0.49172107E-04
0.6250E-01	0.1250E+00	0.44058131E-01	0.12112609E-03
0.1250E+00	0.2500E+00	0.94711137E-01	0.35966429E-03
0.2500E+00	0.3750E+00	0.96258990E-01	0.18593671E-03
0.3750E+00	0.5000E+00	0.93126544E-01	0.10465749E-03
0.5000E+00	0.1000E+01	0.30689542E+00	0.21998188E-03

Solution and Total Error 0.67184228E+00 0.15851962E-02

Results 6.3: Adaptive Gauss-Legendre rules

The basis of the method is that the second rule should be significantly more accurate than the first, so nsp2>>nsp1. Strips that pass the accuracy test are not visited again, but strips that fail the accuracy test are split into two and their areas and error estimates recalculated.

In this example, the more accurate rule is given four sampling points (nsp2= 4) and the less accurate rule, one sampling point (nsp1=1, Midpoint rule). For the purposes of illustration in this example, the error bounds have been made quite large so that the method does not generate too many strips.

The results shown in Results 6.3 indicate that seven strips were generated by the adaptive algorithm across the range, with strip widths varying from 0.03125 to 0.5. The overall result, obtained by summing the areas of each strip (computed using the more accurate rule), is given as 0.6718. An overall error estimate of 0.0016 is also provided, which is obtained by summing the error estimates of each strip. The computed area of each strip will have an estimated error that is smaller than the error bound, given as the larger of the absolute and relative errors. Figure 6.12 shows the locations of the strips following application of the adaptive algorithm.

Fortran95 allows the use of RECURSIVE FUNCTIONs as an alternative to the above approach which can lead to more compact algorithms.

6.5 Special integration rules

The conventional rules considered so far incorporated finite limits of integration, and placed no restriction on the form of the function to be integrated. In this section, some rather more specialized rules are described, which are customized for certain classes of problems that are not conveniently tackled using the conventional approaches.

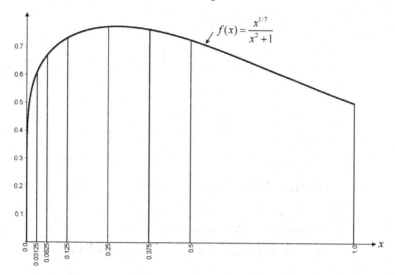

Figure 6.12: Location of strips at convergence from Program 6.3 example.

6.5.0.1 Gauss-Laguerre rules

Gauss-Laguerre rules are specifically designed for integration of an exponentially decaying function of the form

$$\int_0^\infty e^{-x} f(x)\, dx \approx \sum_{i=1}^n w_i f(x_i) \tag{6.60}$$

The weights and sampling points for this rule assume the presence of the exponential term and the semi-infinite range of integration. It should be noted from the right-hand side of equation (6.60) that the sampling points for this rule are substituted only into the function $f(x)$ and not the whole integrand $e^{-x} f(x)$.

The method is so called because the sampling points are the roots of a family of polynomials called Laguerre polynomials which take the form

$$L_n(x) = e^x \frac{d^n}{dx^n}(e^{-x} x^n) = 0 \tag{6.61}$$

A summary of the weights and sampling points for this method up to $n = 5$ is given in Table 6.6. Note how rapidly the weighting coefficients diminish as the order of the method is increased. Gauss-Laguerre rules can be derived using Polynomial substitution as described in a previous section. For example, substitution of $f(x) = 1$ into equation (6.60) yields the following relationships for this rule

$$\sum_{i=1}^{n} w_i = 1$$

$$\sum_{i=1}^{n} w_i x_i = 1$$

(6.62)

TABLE 6.6: Sampling Points and Weights for Gauss-Laguerre rules

n	x_i	w_i
1	1.000000000000000	1.000000000000000
2	0.585786437626905	0.853553390593274
	3.414213562373095	0.146446609406726
3	0.415774556783479	0.711093009929173
	2.294280360279042	0.278517733569241
	6.289945082937479	0.010389256501586
4	0.3225476896193923	0.603154104341634
	1.7457611011583466	0.357418692437800
	4.5366202969211280	0.038887908515005
	9.3950709123011331	0.000539294705561
5	0.2635603197181409	0.521755610582809
	1.4134030591065168	0.398666811083176
	3.5964257710407221	0.075942449681708
	7.0858100058588376	0.003611758679922
	12.6408008442757827	0.000023369972386

Example 6.11

Estimate the value of

$$I = \int_0^\infty e^{-x} \sin x \, dx$$

using the three-point ($n = 3$) Gauss-Laguerre rule.

Solution 6.11

From Table 6.6 with $n = 3$

$$I \approx 0.71109 \sin(0.41577) + 0.27852 \sin(2.29428) + 0.01039 \sin(6.28995)$$
$$= 0.4960 \ (cf. \ exact \ solution \ 0.5)$$

Program 6.4: Gauss-Laguerre rules

```
PROGRAM p64
!---Gauss-Laguerre Rules---
 USE nm_lib; USE precision; IMPLICIT NONE
 INTEGER::i,nsp; REAL(iwp)::area,zero=0.0_iwp
 REAL(iwp),ALLOCATABLE::samp(:,:),wt(:)
 OPEN(10,FILE='nm95.dat'); OPEN(11,FILE='nm95.res')
 READ(10,*)nsp; ALLOCATE(samp(nsp,2),wt(nsp))
 CALL gauss_laguerre(samp,wt); area=zero
 DO i=1,nsp; area=area+wt(i)*f64(samp(i,1)); END DO
 WRITE(11,'(A)')"---Gauss-Laguerre Rules---"
 WRITE(11,'(/,A,I7)')"Gauss-Laguerre's Rule  ",nsp
 WRITE(11,'(A,F12.4)' )"Computed result       ",area
 CONTAINS

FUNCTION f64(x)
  IMPLICIT NONE
  REAL(iwp),INTENT(IN)::x; REAL(iwp)::f64
  f64=SIN(x)
  RETURN
END FUNCTION f64

END PROGRAM p64
```

This program evaluates integrals of the form

$$I = \int_0^\infty e^{-x} f(x) \ dx$$

The user provides the number of sampling points (nsp), and the corresponding sampling points and weights are provided by library subroutine gauss_laguerre.

The demonstration problem is the same as Example 6.11, namely

$$I = \int_0^\infty e^{-x} \sin x \ dx$$

List 6.4:

Scalar integers:
i   simple counter
nsp   number of Gauss-Laguerre points

Scalar reals:
area   holds running total of area from each sampling point
zero   set to 0.0

Dynamic real arrays:
samp   holds sampling points
wt   holds weighting coefficients

---

which has been programmed into f64 at the end of the main program, and will be solved this time with nsp=5 sampling points. Input and output to Program 6.4 are given as Data 6.4 and Results 6.4. With 5 sampling points, the computed result is 0.4989, which represent some improvement over the solution of 0.4960 obtained in Example 6.11 with 3 sampling points. The exact solution in this case is 0.5.

```
Number of sampling points     nsp
                               5
```

Data 6.4: Gauss-Laguerre rules

```
---Gauss-Laguerre Rules---

Gauss-Laguerres Rule          5
Computed result               0.4989
```

Results 6.4: Gauss-Laguerre rules

6.5.1   Gauss-Chebyshev rules

This method is specially designed for integration of functions of the form

$$\int_{-1}^{1} \frac{f(x)}{(1-x)^{\frac{1}{2}}}\, dx \approx \sum_{i=1}^{n} w_i f(x_i) \qquad (6.63)$$

In this case, the "weighting function" is $(1-x)^{-\frac{1}{2}}$ and contains singularities at $x = \pm 1$. As might be expected from the symmetry of the limits of integration in equation (6.62), the weights and sampling points will also be symmetrical about the middle of the range. It can be shown that the general equation is of the form

$$\int_{-1}^{1} \frac{f(x)}{(1-x)^{\frac{1}{2}}} \, dx \approx \frac{\pi}{n} \sum_{i=1}^{n} f\left(\cos\frac{2i-1}{2n}\pi\right) + \frac{2\pi}{2^{2n}(2n)!} f^{2n}(x) \qquad (6.64)$$

and it is left to the reader to compute the required sampling points. The weighting coefficients are seen to involve simple multiples of π.

6.5.2   Fixed weighting coefficients

Using the polynomial substitution technique, many different variations on the methods already described are possible. In the Newton-Cotes approaches, the sampling points were prescribed and the weighting coefficient treated as unknowns. In the Gauss-Legendre approaches, both sampling points and weighting coefficients were treated as unknowns.

A further variation could be to fix the weighting coefficients and treat the sampling points as unknowns. A special case of this type of rule would be where all weighting coefficients are equal to $2/n$, leading to a rule of the form

$$\int_{-1}^{1} f(x) \, dx \approx \frac{2}{n} \sum_{i=1}^{n} f(x_i) \qquad (6.65)$$

It may be noted that when $n = 2$, the rule will be identical to the corresponding Gauss-Legendre rule.

Consider then the case of $n = 3$, hence

$$\int_{-1}^{1} f(x) \, dx \approx \frac{2}{3} [f(x_1) + f(x_2) + f(x_3)] \qquad (6.66)$$

By inspection, $x_1 = -x_3$ and $x_2 = 0$, so only one unknown remains to be found.

Using Polynomial Substitution, let $f(x) = x^2$ giving

$$\int_{-1}^{1} x^2 \, dx = \frac{2}{3} = \frac{2}{3} [x_1^2 + 0 + x_1^2)] \qquad (6.67)$$

thus $x_1^2 = \frac{1}{2}$, $x_1 = -x_3 = -\sqrt{\frac{1}{2}}$ and equation(6.65) becomes

$$\int_{-1}^{1} f(x) \, dx \approx \frac{2}{3} \left[f(-\sqrt{\tfrac{1}{2}}) + f(0) + f(\sqrt{\tfrac{1}{2}}) \right] \qquad (6.68)$$

6.5.3 Hybrid rules

Hybrid rules refer to all other combinations, in which some sampling points and/or weighting coefficients are prescribed, but not others. For example, Lobatto rules always sample at the limits of integration, but allow the intermediate sampling points to be optimally placed. In general, these rules can be written as

$$\int_{-1}^{1} f(x)\,dx \approx w_1 f(-1) + \sum_{i=2}^{n-1} w_i f(x_i) + w_n f(1) \qquad (6.69)$$

It may be noted that when $n = 2$, the rule will be identical to the corresponding Newton-Cotes rule (Trapezoid rule) and for higher order rules of this type, $w_1 = w_n$ and the remaining weights and sampling points will also be symmetrical about the middle of the range.

Using Polynomial Substitution, it is left to the reader to show that the four-point rule ($n = 4$) of this type will be of the form

$$\int_{-1}^{1} f(x)\,dx \approx \frac{1}{6}f(-1) + \frac{5}{6}f(-\sqrt{\frac{1}{5}}) + \frac{5}{6}f(\sqrt{\frac{1}{5}}) + \frac{1}{6}f(1) \qquad (6.70)$$

6.5.4 Sampling points outside the range of integration

All the methods of numerical integration described so far involved sampling points within the range of integration. Although the majority of problems will be of this type, there exists an important class of rule used in the solution of differential equations called "predictors" which require sampling points outside the range of integration.

For example with reference to Figure 6.13, consider a three-point ($n = 3$) integration rule of the type

$$\int_{x_3}^{x_4} f(x)\,dx \approx w_1 f(x_1) + w_2 f(x_2) + w_3 f(x_3) \qquad (6.71)$$

The sampling points include the lower limits of integration and two more points to the left. Since the sampling points are equally spaced at h apart, this rule is of the Newton-Cotes type.

Derivation of this rule by Polynomial Substitution is facilitated by a temporary transformation of the problem as follows

$$\int_{0}^{h} f(x)\,dx \approx w_1 f(-2h) + w_2 f(-h) + w_3 f(0) \qquad (6.72)$$

The three unknown weights can be found by forcing the rule to be exact for $f(x)$ up to degree 2; thus let $f(x) = 1$, x and x^2 to generate three equations leading to

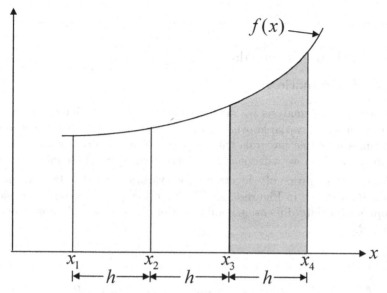

Figure 6.13: Sampling points outside the range of integration.

$$\int_0^h f(x)\, dx \approx \frac{h}{12} \left[5f(-2h) - 16f(-h) + 23f(0) \right] \qquad (6.73)$$

The general form can then be retrieved as

$$\int_{x_3}^{x_4} f(x)\, dx \approx \frac{h}{12} \left[5f(x_1)) - 16f(x_2) + 23f(x_3) \right] \qquad (6.74)$$

Higher order "predictors" of this type can be derived by similar means; for example a four-point ($n = 4$) rule is given by

$$\int_{x_4}^{x_5} f(x)\, dx \approx \frac{h}{24} \left[-9f(x_1)) + 37f(x_2) - 59f(x_3) + 55f(x_4) \right] \qquad (6.75)$$

and is exact for $f(x)$ up to degree 3. Formulas such as those given by e-quations (6.73) and (6.74) are called Adams-Bashforth predictors, and will be encountered again in Chapter 7 which deals with the solution of ordinary differential equations.

6.6   Multiple integrals

6.6.1   Introduction

In engineering analysis we are frequently required to integrate functions of more than one variable over an area or volume. Analytical methods for performing multiple integrals will be possible in a limited number of cases, but in this section we will consider numerical integration techniques.

Consider integration of a function of two variables over the two-dimensional region R as shown in Figure 6.14. The function $f(x, y)$ could be considered to represent a third dimension coming out of the page at right angles over the region R.

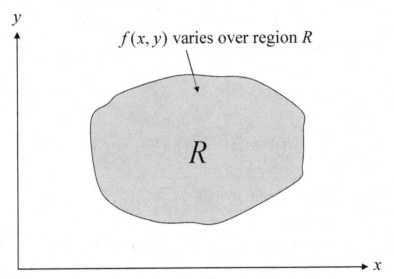

Figure 6.14: Integration over a 2-d region R.

The required integral is

$$I = \int \int_R f(x, y) \, dx \, dy \tag{6.76}$$

By extrapolation of the techniques described previously for a single integral, double integrals involving two variables will lead to integration rules of the form

$$\int\int_R f(x,y)\ dx\ dy \approx \sum_{i=1}^{n} w_i f(x_i, y_i) \tag{6.77}$$

Sampling points now lie in the plane of R and involve coordinates $(x_i, y_i), i = 1 \ldots n$ where the function is to be evaluated. Each function evaluation is weighted by w_i and the sum of the weights must add up to the area of R.

Clearly, a problem arises in defining explicitly the limits of integration for an irregular region such as that shown in Figure 6.14. In practice, it may be sufficient to subdivide the irregularly shaped region into a number of smaller simple subregions (e.g., rectangles), over which numerical integration can be easily performed. The final approximate result over the full region R would then be obtained by adding together the solutions obtained over each subregion. This approach is analogous to the "repeated" rules covered earlier in the chapter.

Initially, we consider integration over rectangular regions which lie parallel to the Cartesian coordinate directions, as this greatly simplifies the definition of the limits. Later on, the concepts are extended to integration over general quadrilateral and triangular regions in the xy plane. All the methods described in this Section are readily extended to triple integrals.

6.6.1.1 Integration over a rectangular region

Consider, in Figure 6.15, integration of a function $f(x,y)$ over
the rectangular region shown. As the boundaries of the rectangle lie parallel to the Cartesian coordinate directions, the variables can be uncoupled, and any of the methods described previously can be applied directly.

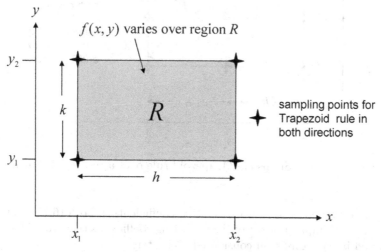

Figure 6.15: Trapezoid rule over a rectangular region.

Newton-Cotes

For example, the Trapezoid rule applied in each direction would lead to four sampling points ($n = 4$), with one at each corner of the rectangle as shown, leading to the rule

$$\int_{y_1}^{y_2} \int_{x_1}^{x_2} f(x,y) \approx \frac{1}{4} hk \left[f(x_1, y_1) + f(x_2, y_1) + f(x_1, y_2) + f(x_2, y_2) \right] \quad (6.78)$$

Simpson's rule applied in each direction would lead to the nine sampling points ($n = 9$) as shown in Figure 6.16, leading to the rule

$$\int_{y_1}^{y_3} \int_{x_1}^{x_3} f(x,y) \approx \frac{1}{9} hk \left[f(x_1, y_1) + 4f(x_2, y_1) + f(x_3, y_1) + \right.$$
$$4f(x_1, y_2) + 16f(x_2, y_2) + 4f(x_3, y_2) +$$
$$\left. f(x_1, y_3) + 4f(x_2, y_3) + f(x_3, y_3) \right] \quad (6.79)$$

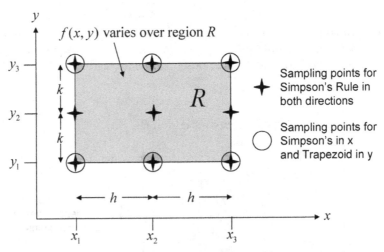

Figure 6.16: Simpson's/Trapezoid rule over a rectangular region.

It should be noted that the weighting coefficients in both (6.77) and (6.78) come from a simple product of the weighting coefficients that would have been obtained in each direction considered separately.

Example 6.12

Estimate the value of

$$I = \int_1^3 \int_1^2 xy(1+x) \, dx \, dy$$

using the Trapezoid rule in each direction.

Solution 6.12

In one-dimension, the Trapezoid rule has 2 sampling points, so in a double integral we will use $n = 2^2 = 4$ sampling points. With reference to Figure 6.16, $h = 1$ and $k = 2$, so from equation (6.77)

$$I \approx \frac{1}{2}hk\left[f(1,1) + f(2,1) + f(1,2) + f(2,2)\right]$$
$$= 16.0000 \ \textit{(cf. exact solution 15.3333)}$$

Example 6.13

Estimate the value of

$$I = \int_1^3 \int_1^2 xy(1+x) \, dx \, dy$$

using Simpson's rule in each direction.

Solution 6.13

In one-dimension, the Simpson's rule has 3 sampling points, so in a double integral we will use $n = 3^2 = 9$ sampling points. With reference to Figure 6.16, $h = 0.5$ and $k = 1$, so from equation (6.78)

$$I \approx \frac{1}{18}[f(1,1) + 4f(1.5,1) + f(2,1) +$$
$$4f(1,2) + 16f(1.5,2) + 4f(2,2) +$$
$$f(1,3) + 4f(1.5,3) + f(2,3)]$$
$$= 15.3333$$

In this instance, it was inefficient to use Simpson's rule in both directions. The exact solution could also have been achieved with Simpson's rule in the x-direction and the Trapezoid rule (say) in the y-direction. In this case we would have used $n = 3 \times 2 = 6$ sampling points (circled in Figure 6.16), leading to

$$I \approx \frac{1}{6}[f(1,1) + 4f(1.5,1) + f(2,1) + f(1,3) + 4f(1.5,3) + f(2,3)]$$
$$= 15.3333$$

Gauss-Legendre

Gauss-Legendre rules can also be applied to multiple integrals of this type, but care must be taken to find the correct locations of the sampling points. One approach would be to perform a coordination transformation so that the limits of integration in each direction become ± 1. This would enable the weights and sampling points from Table 6.4 to be used directly. The general topic of transformation is covered in the next section on integration over general quadrilateral areas.

An alternative approach for rectangular regions is to scale the weights and proportion the sampling points from Table 6.4 about the midpoint of the range in each of the coordinate directions.

Consider two-point Gauss-Legendre integration over the rectangular region shown in Figure 6.17.

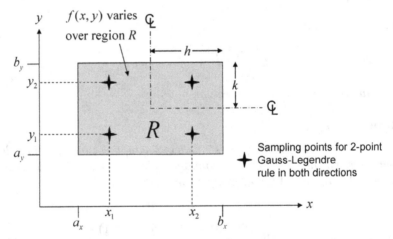

Figure 6.17: Two-point Gauss-Legendre over a rectangular region.

In the x-direction the sampling points will be located at

$$x_1 = a_x + h - \frac{1}{\sqrt{3}}h \quad \text{and} \quad x_2 = a_x + h + \frac{1}{\sqrt{3}}h \qquad (6.80)$$

with corresponding weighting coefficients $w_1 = w_2 = h$. Similarly in the y-direction the sampling points will be located at

$$y_1 = a_y + k - \frac{1}{\sqrt{3}}k \quad \text{and} \quad y_2 = a_y + k + \frac{1}{\sqrt{3}}k \qquad (6.81)$$

with corresponding weighting coefficients $w_1 = w_2 = k$, hence the rule becomes

$$\int_{a_y}^{b_y} \int_{a_x}^{b_x} f(x,y) \, dx \, dy \approx hk \left[f(x_1, y_1) + f(x_2, y_1) + f(x_1, y_2) + f(x_2, y_2) \right]$$

$$(6.82)$$

Example 6.14

Estimate the value of

$$I = \int_{-1}^{0} \int_{0}^{2} x^3 y^4 \, dx \, dy$$

using the 2-point Gauss-Legendre rule in each direction.

Solution 6.14

In this double integral we will use $n = 2^2 = 4$ sampling points. With reference to Figure 6.17, $h = 1$ and $k = 0.5$, so from equations (6.79) and (6.80) the sampling points are located at

$$\begin{aligned} x_1 &= 0.4226 & x_2 &= 1.5774 \\ y_1 &= -0.7887 & x_2 &= -0.2113 \end{aligned}$$

Hence from equation (6.81)

$$\begin{aligned} I \approx 0.5 \, [& f(0.4226, -0.7887) + f(1.5774, -0.7887) \\ & + f(0.4226, -0.2113) + f(1.5774, -0.2113)] \\ = {} & 0.7778 \; (\text{cf. exact solution } 0.8) \end{aligned}$$

In this example, the 2-point rule is capable of exact integration in the x-direction, but not in the y-direction.

By similar reasoning, the three-point Gauss-Legendre rule shown in Figure 6.18 leads to the following sampling points and weighting coefficients in x

$$x_1 = a_x + h - \sqrt{\frac{3}{5}}h \qquad x_2 = a_x + h \qquad x_3 = a_x + h + \sqrt{\frac{3}{5}}h$$

$$(6.83)$$

$$w_1 = \frac{5}{9}h \qquad w_2 = \frac{8}{9}h \qquad w_3 = \frac{5}{9}h$$

and in y

$$y_1 = a_y + k - \sqrt{\frac{3}{5}}k \qquad y_2 = a_y + k \qquad y_3 = a_y + k + \sqrt{\frac{3}{5}}k$$

$$(6.84)$$

$$w_1 = \frac{5}{9}k \qquad w_2 = \frac{8}{9}k \qquad w_3 = \frac{5}{9}k$$

hence the rule becomes

$$\int_{a_y}^{b_y} \int_{a_x}^{b_x} f(x,y)\, dx\, dy \approx \frac{hk}{81} [25f(x_1,y_1) + 40f(x_2,y_1) + 25f(x_3,y_1)$$
$$+ 40f(x_1,y_2) + 64f(x_2,y_2) + 40f(x_3,y_2) \qquad (6.85)$$
$$+ 25\ f(x_1,y_3) + 40f(x_2,y_3) + 25f(x_3,y_3)]$$

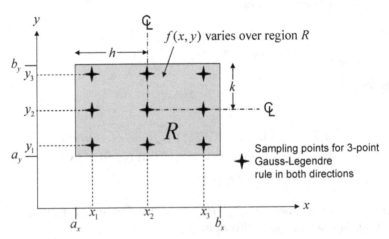

Figure 6.18: Three-point Gauss-Legendre over a rectangular region.

Example 6.15

Estimate the value of

$$I = \int_{-1}^{0} \int_{0}^{2} x^3 y^4 \, dx \, dy$$

using the 3-point Gauss-Legendre rule in each direction.

Solution 6.15

In this double integral we will use $n = 3^2 = 9$ sampling points. With reference to Figure 6.18, $h = 1$ and $k = 0.5$, so from equations (6.82) and (6.83) the sampling points are located at

$$
\begin{array}{lll}
x_1 = & 0.2254 & x_2 = \quad 1 \quad x_3 = \quad 1.7746 \\
y_1 = -0.8873 & y_2 = -0.5 \quad y_3 = -0.1127
\end{array}
$$

hence from equation (6.84), we get $I = 0.8$ which is the exact solution.

An exact solution to the integration of the quartic terms in y was achieved by the three-point rule in this case, although a two-point rule would have been sufficient in the x-direction making a total of six sampling points. Although it is easy to implement different integration rules in different directions, most numerical integration software packages use the same rule in all directions.

6.6.2 Integration over a general quadrilateral area

We now turn our attention to numerical integration of a function $f(x,y)$ over a general quadrilateral area such as that shown in Figure 6.19. Analytical integration of functions over a nonrectangular region is complicated by the variable limits. It should also be noted that the required order of integration to obtain exact solutions over such regions is higher than for regions whose boundaries are parallel to the Cartesian coordinate directions.

The quadrilateral could be broken down into subregions consisting of rectangles and triangles, so consider integration of $f(x,y) = x^m y^n$ over the hatched triangular region in Figure 6.19

$$I = \int_{x_2}^{x_3} \int_{y_2}^{g(x)} x^m y^n \, dy \, dx \tag{6.86}$$

where $g(x)$ is the equation of the straight line between corners 2 and 3.

Let the equation of $g(x)$ be given by

$$g(x) = ax + b \tag{6.87}$$

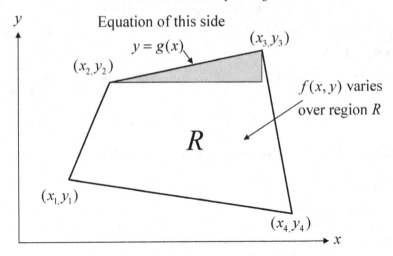

Figure 6.19: Integration over a general quadrilateral region.

Performing the inner integral first with respect to y, and substituting the limits leads to

$$I = \frac{1}{n+1} \int_{x_2}^{x_3} x^m [(ax+b)^{n+1} - y_2^{n+1}] \, dx \qquad (6.88)$$

The remaining outer integral involves an $(m+n+1)^{th}$ order polynomial in x, which indicates the order or the rule required for an exact solution. This may be contrasted with the same function integrated over a rectangular region as considered previously in examples 6.14 and 6.15. In those cases the order of integration required for an exact solution was governed by the polynomial of order m or n (whichever was the greater).

Consider Gauss-Legendre rules for integration of the problem shown in Figure 6.19. Before proceeding, we will perform a coordinate transformation, which replaces the actual problem with integration over the region R, by a normalized problem with integration over a square region with limits of integration in both directions of ± 1 as shown in Figure 6.20, thus

$$I = \int \int_R f(x,y) \, dx \, dy \equiv \int_{-1}^{1} \int_{-1}^{1} g(\xi, \eta) \, d\xi \, d\eta \qquad (6.89)$$

This is analogous to equation (6.52) considered earlier for a single variable.

A one-to-one correspondence between points in the two regions can be achieved by the transformation relationships

$$x(\xi, \eta) = N_1(\xi, \eta)x_1 + N_2(\xi, \eta)x_2 + N_3(\xi, \eta)x_3 + N_4(\xi, \eta)x_4$$
$$y(\xi, \eta) = N_1(\xi, \eta)y_1 + N_2(\xi, \eta)y_2 + N_3(\xi, \eta)y_3 + N_4(\xi, \eta)y_4 \qquad (6.90)$$

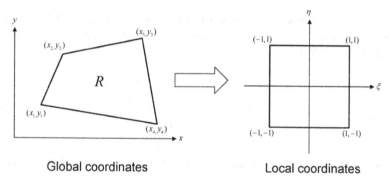

Global coordinates   Local coordinates

Figure 6.20: Transformation from global to local coordinates.

where

$$N_1(\xi, \eta) = \frac{1}{4}(1 - \xi)(1 - \eta)$$

$$N_2(\xi, \eta) = \frac{1}{4}(1 - \xi)(1 + \eta)$$

$$N_3(\xi, \eta) = \frac{1}{4}(1 + \xi)(1 + \eta) \qquad (6.91)$$

$$N_4(\xi, \eta) = \frac{1}{4}(1 + \xi)(1 - \eta)$$

The N-functions are frequently called "shape functions" as used in finite element theory (see Chapter 8), with the properties

$$N_i(\xi_j, \eta_j) = \begin{cases} 1 \text{ if } i = j \\ 0 \text{ if } i \neq j \end{cases} \qquad (6.92)$$

and

$$\sum_{i=1}^{4} N_i = 1 \qquad (6.93)$$

Thus corner (x_1, y_1) is mapped onto $(-1, -1)$ in the transformed space, (x_2, y_2) onto $(-1, 1)$ and so on. In addition, the line joining (x_2, y_2) to (x_3, y_3) is mapped onto the line $\eta = 1$ etc.

The transformation indicated in equation (6.88) can therefore be written as

$$I = \int \int_R f(x, y) \, dx \, dy \equiv \int_{-1}^{1} \int_{-1}^{1} J(\xi, \eta) \, f(x(\xi, \eta), y(\xi, \eta)) \, d\xi \, d\eta \qquad (6.94)$$

where J is called the "Jacobian" and is the scaling factor relating $dx \, dy$ to $d\xi \, d\eta$.

As indicated in equation (6.94), J is a function of position within the transformed region and is computed as the determinant of the "Jacobian matrix" (see Section 3.7.2) by

$$J(\xi,\eta) = \det \begin{vmatrix} \dfrac{\partial x}{\partial \xi} & \dfrac{\partial y}{\partial \xi} \\[2mm] \dfrac{\partial x}{\partial \eta} & \dfrac{\partial y}{\partial \eta} \end{vmatrix} \tag{6.95}$$

The Jacobian matrix is readily computed from equations (6.90), thus

$$\frac{\partial x}{\partial \xi} = \frac{\partial N_1}{\partial \xi}x_1 + \frac{\partial N_2}{\partial \xi}x_2 + \frac{\partial N_3}{\partial \xi}x_3 + \frac{\partial N_4}{\partial \xi}x_4$$

$$\frac{\partial y}{\partial \xi} = \frac{\partial N_1}{\partial \xi}y_1 + \frac{\partial N_2}{\partial \xi}y_2 + \frac{\partial N_3}{\partial \xi}y_3 + \frac{\partial N_4}{\partial \xi}y_4$$

$$\frac{\partial x}{\partial \eta} = \frac{\partial N_1}{\partial \eta}x_1 + \frac{\partial N_2}{\partial \eta}x_2 + \frac{\partial N_3}{\partial \eta}x_3 + \frac{\partial N_4}{\partial \eta}x_4 \tag{6.96}$$

$$\frac{\partial y}{\partial \eta} = \frac{\partial N_1}{\partial \eta}y_1 + \frac{\partial N_2}{\partial \eta}y_2 + \frac{\partial N_3}{\partial \eta}y_3 + \frac{\partial N_4}{\partial \eta}y_4$$

where from equations (6.90)

$$\frac{\partial N_1}{\partial \xi} = -\frac{1}{4}(1-\eta)$$

$$\frac{\partial N_3}{\partial \eta} = \frac{1}{4}(1+\xi) \quad \text{and so on.} \tag{6.97}$$

It should be noted that the shape functions from equation (6.90) are only smooth functions of ξ and η provided all interior angles of the untransformed quadrilateral are less than 180°.

Program 6.5: Multiple integrals by Gauss-Legendre rules

```
PROGRAM p65
!---Multiple Integrals by Gauss-Legendre Rules---
 USE nm_lib; USE precision; IMPLICIT NONE
 INTEGER::i,ndim,nod,nsp; REAL(iwp)::res,zero=0.0_iwp
 REAL(iwp),ALLOCATABLE::coord(:,:),der(:,:),fun(:),samp(:,:),    &
  wt(:)
```

```
OPEN(10,FILE='nm95.dat'); OPEN(11,FILE='nm95.res')
READ(10,*)ndim
IF(ndim==1)nod=2; IF(ndim==2)nod=4; IF(ndim==3)nod=8
ALLOCATE(coord(nod,ndim),der(ndim,nod),fun(nod))
READ(10,*)(coord(i,:),i=1,nod),nsp
ALLOCATE(samp(nsp,ndim),wt(nsp))
CALL gauss_legendre(samp,wt); res=zero
DO i=1,nsp; CALL fun_der(fun,der,samp,i)
  res=res+determinant(MATMUL(der,coord))*wt(i)*            &
    f65(MATMUL(fun,coord))
END DO
WRITE(11,'(A)')"--Multiple Integrals by Gauss-Legendre Rules--"
WRITE(11,'(/A,I5/)')"Number of dimensions      ",ndim
DO i=1,nod
  WRITE(11,'(A,3F12.4)')'Coordinates (x,y[,z])',coord(i,:)
END DO
WRITE(11,'(/A,I5)')"Number of sampling points",nsp
WRITE(11,'(/A,F12.4)')"Computed result       ",res
CONTAINS

FUNCTION f65(point)
  IMPLICIT NONE
  REAL(iwp),INTENT(IN)::point(:)
  REAL(iwp)::x,y,z,f65
  x=point(1); y=point(2); z=point(3)
  f65=x**2*y**2
! f65=x**3-2*y*z
  RETURN
END FUNCTION f65

END PROGRAM p65
```

This program can perform single, double or triple integrals using Gauss-Legendre rules. The user specifies the dimensionality of the problem in the data through the input parameter ndim. Depending on the value of ndim, the following problems can be solved:

$$\text{ndim} = 1 (\text{line}) \qquad I \approx \int f(x)\,dx \qquad \text{nsp} = 1, 2, 3, \ldots$$

$$\text{ndim} = 2 (\text{quadrilateral}) \; I \approx \int\int f(x,y)\,dx\,dy \qquad \text{nsp} = 1, 4, 9, \ldots$$

$$\text{ndim} = 3 (\text{hexahedron}) \quad I \approx \int\int\int f(x,y,z)\,dx\,dy\,dz \quad \text{nsp} = 1, 8, 27, \ldots$$

The user provides the number of sampling points (nsp) and the corresponding sampling points and weights are provided by library subroutine

---

List 6.5:

Scalar integers:
i   simple counter
ndim   number of dimensions
nod   number of corners of domain of integration
nsp   total number of sampling points

Scalar reals:
res   holds running total of result from each sampling point
zero   set to 0.0

Dynamic real arrays:
coord   coordinates defining limits of integration
der   derivatives of shape functions with respect to local coordinates
fun   shape functions
samp   local coordinates of sampling points
wt   weighting coefficients at each sampling point

---

gauss_legendre. It should be noted above that nsp in this context represents the *total* number of Gauss-Legendre sampling points assuming the same rule is being used in each of the (local) coordinate directions. Thus in 2-d, nsp will be a perfect square and in 3-d, a perfect cube.

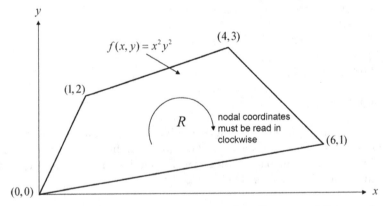

Figure 6.21: Domain of integration in first Program 6.5 example.

The first demonstration problem is a double integral

$$I = \int \int_R x^2 y^2 \ dx \ dy$$

over the quadrilateral region shown in Figure 6.21. The function has been programmed into f65 at the end of the main program.

Input shown in Data 6.5a provide the dimensionality of the problem (ndim), followed by the (x, y) coordinates of the corners of the quadrilateral (coord). It is important that these coordinates are provided in a <u>clockwise</u> sense, since this is the order in which the shape functions have been arranged in the subroutine library. The final data item refers to the number of sampling points required which is set to nsp=9 implying three Gauss-Legendre sampling points in each of the local coordinate directions.

```
ndim
2
Corner coordinates (clockwise)
0.0   0.0
1.0   2.0
4.0   3.0
6.0   1.0

nsp
9
```

Data 6.5a: Multiple Integrals (first example)

```
--Multiple Integrals by Gauss-Legendre Rules--

Number of dimensions          2

Coordinates (x,y[,z])      0.0000        0.0000
Coordinates (x,y[,z])      1.0000        2.0000
Coordinates (x,y[,z])      4.0000        3.0000
Coordinates (x,y[,z])      6.0000        1.0000

Number of sampling points     9

Computed result            267.4389
```

Results 6.5a: Multiple Integrals (first example)

The output given in Results 6.5a gives a result of 267.4389. This is the exact solution as expected in this case. With reference to equation (6.87), exact integration in this case requires integration of a 5th order polynomial which 3-point Gauss-Legendre is able to do.

A second example of the use of Program 6.5 demonstrates a triple integral of the form

$$I = \int_{-4}^{4} \int_{0}^{6} \int_{-1}^{3} (x^3 - 2yz) \, dx \, dy \, dz$$

The integration domain is a cuboid of side lengths 4, 6 and 8 in the x, y and z-directions as shown in Figure 6.22.

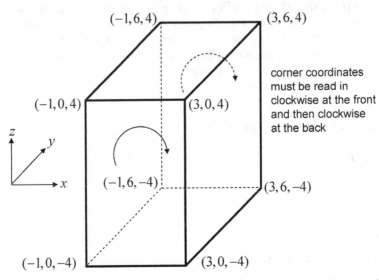

Figure 6.22: Domain of integration in second Program 6.5 example.

```
ndim
3
Corner coordinates (clockwise,front then back)
-1.0  0.0  -4.0
-1.0  0.0   4.0
 3.0  0.0   4.0
 3.0  0.0  -4.0
-1.0  6.0  -4.0
-1.0  6.0   4.0
 3.0  6.0   4.0
 3.0  6.0  -4.0

nsp
8
```

Data 6.5b: Multiple Integrals (second example)

```
---Multiple Integrals by Gauss-Legendre Rules---

Number of dimensions              3
```

```
Coordinates (x,y[,z])      -1.0000      0.0000     -4.0000
Coordinates (x,y[,z])      -1.0000      0.0000      4.0000
Coordinates (x,y[,z])       3.0000      0.0000      4.0000
Coordinates (x,y[,z])       3.0000      0.0000     -4.0000
Coordinates (x,y[,z])      -1.0000      6.0000     -4.0000
Coordinates (x,y[,z])      -1.0000      6.0000      4.0000
Coordinates (x,y[,z])       3.0000      6.0000      4.0000
Coordinates (x,y[,z])       3.0000      6.0000     -4.0000

Number of sampling points    8

Computed result              960.0000
```

Results 6.5b: Multiple Integrals (second example)

This time the program transforms the 3-d domain into normalized local coordinates (ξ, η, ζ) with limits of ± 1 in each direction. Subroutine fun_deriv provides the values of the shape functions and their derivatives for hexahedral shapes with respect to the local coordinates at each of the nsp sampling points. A running total of the estimated integral accumulated from all the sampling points is held in res and printed.

Input and output for this example are shown in Data 6.5b and Results 6.5b respectively. Since the domain of integration is parallel with the coordinate directions, the variables can be considered

separately. The highest order term in $f(x, y, z)$ is the cubic term x^3, thus a two-point Gauss-Legendre rule will give exact solutions with nsp=2^3=8 sampling points. For a three-dimensional problem, ndim=3 and the coordinates of the corners must be read in <u>clockwise</u> for any front plane, followed by clockwise for the corresponding back plane. The exact result of 960 is returned as indicated in Results 6.5b.

6.7 Exercises

1. Calculate the area of a quarter of a circle of radius a by the following methods and compare with the exact result $\pi a^2/4$.

$$\text{Equation of a circle:} x^2 + y^2 = a^2$$

   (a) Rectangle rule
   (b) Trapezoid rule
   (c) Simpson's rule
   (d) 4-point Newton-Cotes

(e) 5-point Newton-Cotes

Answer: (a) a^2, (b) $0.5a^2$, (c) $0.7440a^2$, (d) $0.7581a^2$, (e) $0.7727a^2$

2. Attempt exercise 1 above using the following methods repeated twice.

(a) Rectangle rule
(b) Trapezoid rule
(c) Simpson's rule

Answer: (a) $0.9330a^2$, (b) $0.6830a^2$, (c) $0.7709a^2$

3. Calculate the area of a quarter of an ellipse whose semi-axes are a and b $(a = 2b)$ by the following methods, and compare with the exact result of $\pi b^2 / 2$.

$$\text{Equation of an ellipse:} \frac{x^2}{a^2} + \frac{y^2}{b^2} = 1$$

(a) Rectangle rule
(b) Trapezoid rule
(c) Simpson's rule
(d) 4-point Newton-Cotes
(e) 5-point Newton-Cotes

Answer: (a) $2b^2$, (b) b^2, (c) $1.4880b^2$, (d) $1.5161b^2$, (e) 1.5454^2

4. Attempt exercise 3 above using the following methods repeated twice.

(a) Rectangle rule
(b) Trapezoid rule
(c) Simpson's rule

Answer: (a) $1.8660b^2$, (b) $1.3660b^2$, (c) $1.5418b^2$

5. Determine the weights w_1, w_2 and w_3 in the integration rule

$$\int_0^{2h} f(x)\, dx \approx w_1 f(0) + w_2 f(h) + w_3 f(2h)$$

which ensure that it is exact for all polynomials $f(x)$ of degree 2 or less. Show that the formula is in fact also exact for $f(x)$ of degree 3. Answer: $w_1 = h/3, w_2 = 4h/3, w_3 = h/3$

6. How many repetitions of the Trapezoid rule are necessary in order to compute

$$\int_0^{\pi/3} \sin x\, dx$$

accurate to three decimal places? Answer: 11 repetitions.

7. Compute the volume of a hemisphere by numerical integration using the lowest order Newton-Cotes method that would give an exact solution.
Answer: Simpson's rule gives $2\pi r^3/3$

8. Estimate

$$\int_0^3 (x^3 - 3x^2 + 2)\, dx$$

using (a) Simpson's rule and (b) the Trapezoid rule repeated three times. Which method is the most accurate in this case and why.
Answer: (a) −0.75 (Simpson's rule is exact for cubics), (b) 0.0!

9. From the tabulated data given, estimate the area between the function $y(x)$, and the lines $x = 0.2$ and $x = 0.6$.

x	y
0.2	1.221403
0.3	1.349859
0.4	1.491825
0.5	1.648721
0.6	1.822119

using the following method repeated twice.

(a) Simpson's rule

(b) Trapezoid rule

(c) Midpoint rule

If the exact solution is given by $y(0.6)-y(0.2)$ comment on the numerical solution obtained.
Answer: (a) 0.6007, (b) 0.6027, (c) 0.5997

10. Calculate the area of a quarter of a circle of radius a by the following methods and compare with the exact result of $\pi a^2/4$.

$$\text{Equation of a circle: } x^2 + y^2 = a^2$$

(a) Midpoint rule

(b) 2-point Gauss-Legendre

(c) 3-point Gauss-Legendre

Answer: (a) $0.8660a^2$, (b) $0.7961a^2$, (c) $0.7890a^2$

11. Attempt exercise 10 using the same methods repeated twice.
Answer: (a) $0.8148a^2$, (b) $0.7891a^2$, (c) $0.7867a^2$

12. Use Polynomial Substitution to find the weighting coefficients w_0, w_1 and w_2, and the sampling points x_0, x_1, and x_2 in the Gauss-Legendre formula

$$\int_{-1}^{1} = w_1 f(x_1) + w_2 f(x_2) + w_3 f(x_3)$$

You may assume symmetry of weights and sampling points about the middle of the range.

Answer: $w_1 = w_2 = 5/9$, $w_2 = 8/9$, $x_1 = -x_3 = -1/\sqrt{3}$, $w_2 = 0$

13. Derive the two-point Gauss-Legendre integration rule, and use it to estimate the area enclosed by the ellipse

$$\frac{x^2}{4} + \frac{y^2}{9} = 1$$

Answer: 19.1067, 20.7846 (Exact: 18.8496)

14. Estimate

$$\int_{0.3}^{0.8} e^{-2x} \tan x \, dx$$

using the Midpoint rule repeated once, twice and three times.

Answer: 0.1020, 0.1002, 0.0999 (Exact: 0.0996)

15. Use Gauss-Legendre integration to find the exact value of

$$\int_{0}^{1} (x^7 + 2x^2 - 1) \, dx$$

Answer: 4-point gives -0.2083

16. Estimate the value of

$$\int_{1}^{3} \frac{dx}{(x^4 + 1)^{1/2}}$$

using

(a) Midpoint rule
(b) 2-point Gauss-Legendre
(c) 3-point Gauss-Legendre

Answer: (a) 0.4851, (b) 0.5918, (c) 0.5951 (Exact: 0.5941)

17. Attempt exercise 16 using the same methods repeated twice.
Answer: (a) 0.5641, (b) 0.5947, (c) 0.5942 (Exact: 0.5941)

18. Use Gauss-Legendre integration with four sampling points to estimate the value of

$$(a) \int_{-2}^{2} \frac{dx}{1 + x^2} \quad (a) \int_{0}^{1} x \exp(-3x^2) \, dx$$

Answer: (a) 2.1346 (Exact: 2.2143) (b) 0.1584. (Exact: 0.1584)

19. Estimate the value of

$$\int_0^\infty e^{-x} \cos x \, dx$$

using Gauss-Laguerre rules with

(a) 1-point
(b) 3-points
(c) 5-points

Answer: (a) 0.5403, (b) 0.4765, (c) 0.5005 (Exact: 0.5)

20. Determine approximate values of the integral

$$\int_0^\infty \frac{e^{-x}}{x+4} \, dx$$

by using Gauss-Laguerre rules with one, two and three sampling points.
Answer: 0.2000, 0.2059, 0.2063 (Exact: 0.2063)

21. Use two-point Gauss-Legendre integration to compute

$$\int_1^2 \int_3^4 f(x,y) \, dy \, dx$$

where $f(x,y)=$ (a) xy, (b) x^2y, (c) x^3y, (d) x^4y Check your estimates
against analytical solutions.
Answer: (a) 5.25, (b) 8.1667, (c) 13.125, (d) 21.6806 (approx)

22. Estimate the value of

$$\int_0^1 \int_0^1 e^{-x^2} y^2 \, dy \, dx$$

using Gauss-Legendre rules with (a) 1-point and (b) 2-points.
Answer: (a) 0.1947, (b) 0.2489

23. Use the minimum number of sampling points to find the exact value of

$$\int_1^2 \int_0^3 xy^3 \, dy \, dx$$

Answer: Gauss-Legendre 1-point in x and 2-points in y gives 30.375

24. Estimate of value of

$$\int_{-2}^0 \int_0^3 e^x \sin x \, dx \, dy$$

Answer: 3-point Gauss-Legendre gives -2.4334

25. Find the exact solution to

$$\int\int_R x^2 y \, dx \, dy$$

where R is a region bounded by the (x, y) coordinates $(0,0)$, $(0.5, 1)$, $(1.5, 1.5)$ and $(2, 0.5)$.
Answer: 3-point Gauss-Legendre gives 1.5375

26. Estimate the triple integral

$$\int_0^1 \int_1^2 \int_0^{0.5} e^{xyz} \, dx \, dy \, dz$$

using Gauss-Legendre rules with (a) 1-point and (b) 2-points in each direction.
Answer: (a) 0.6031, (b) 0.6127

27. Use numerical integration with the minimum number of sampling points to exactly integrate the function $f(x, y) = xy^5$ over the region shown in Figure 6.23. *Answer:* 925.75

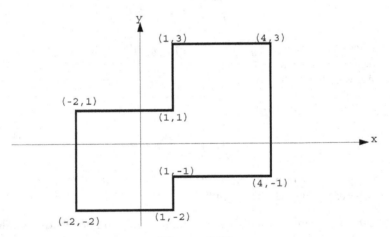

Figure 6.23

28. Derive the optimal values of the unknown weights and sampling point in the following two-point numerical integration rule which samples the function at the lower limit of integration and at one other location.

$$\int_0^1 f(x) dx = w_1 f(0) + w_2 f(x_1)$$

Estimate the value of:

$$\int_{-1}^{2} \frac{1}{1+x^2} dx$$

using both the rule you have just derived and the conventional 2-point Gauss-Legendre rule.
Answer: $w_1 = 1/4$, $w_2 = 3/4$, $x_1 = 2/3$, 1.5, 1.846 (Exact: $\arctan(2) + \pi/4 = 1.8925$)

29. Derive the optimal values of the unknown weights and sampling points in the following four-point numerical integration rule which samples the function at the limits of integration and at two other locations.

$$\int_{-1}^{1} f(x)dx = w_{-2}f(-1) + w_{-1}f(x_{-1}) + w_1f(x_1) + w_2f(1)$$

Use the rule you have just derived to estimate the value of:

$$\int_{-1}^{2} \frac{1}{1+x^2} dx$$

Answer: $w_{-2} = w_2 = 1/6$, $w_{-1} = w_1 = 5/6$, $x_{-1} = -1/\sqrt{5}$, $x_1 = 1/\sqrt{5}$, 1.9168 (Exact: $\arctan(2) + \pi/4 = 1.8925$)

30. Use a 4-point integration rule to estimate the value of:

$$\int_{0}^{2} e^{-x}(x^2 + 3)\,dx$$

Answer: 3.2406 (Exact: $-13e^{-2} + 5 = 3.2406$)

31. Derive the optimal values of the unknown weights and sampling point in the following two-point numerical integration rule

$$\int_{0}^{1} f(x)dx = w_1f(x_1) + w_2f(1)$$

The Seivert integral is defined:

$$S(x, \theta) = \int_{0}^{\theta} e^{-x\sec\phi}\,d\phi$$

Use the rule you have just derived to estimate $S(1.0, \pi/3)$.
Answer: $w_1 = 3/4$, $w_2 = 1/4$, $x_1 = 1/3$, 0.3064 (Exact: 0.3077)

32. Use 2-point Gauss-Legendre integration to estimate the value of

$$\int_{-1.5}^{1} \int_{-1}^{2} e^x \sin y\,dx\,dy$$

Answer: -3.2130 (Exact: -3.2969)

33. Estimate the value of

$$\int_0^1 \int_0^\infty e^{-x} \sin x \sin y \, dx \, dy$$

as accurately as you can using three sampling points in each direction (nine sampling points in total).
Answer: 0.228 (Exact: 0.2298)

34. A seven-point rule for numerically evaluating integrals over the triangular region shown in Figure 6.24 is given by

$$\int_0^1 \int_0^{1-x} f(x,y) \, dy \, dx \approx \sum_1^7 w_i \, f(x_i, y_i)$$

with weights and sampling points as follows:

x_i	y_i	w_i
0.3333	0.3333	0.2250
0.5000	0.5000	0.0667
0.0000	0.5000	0.0667
0.5000	0.0000	0.0667
1.0000	0.0000	0.0250
0.0000	1.0000	0.0250
0.0000	0.0000	0.0250

Use the rule to estimate

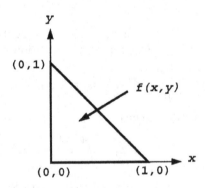

Figure 6.24

$$I = \int_0^1 \int_0^{1-x} (\cos^2 x - \sin^2 y) \, dy \, dx$$

Answer: 0.3550 (Exact: 0.3540)

35. An obscure numerical integration rule called Gauss-Griffiths (Griffiths, 1991) takes the general form

$$\int_0^1 x\, f(x)\, dx = \sum_{i=1}^n w_i\, x_i\, f(x_i)$$

Set up the equations that would enable you to derive the weights and sampling points for the case when $n = 2$, but do not attempt to solve the equations. If it is later shown that:

$x_1 = 0.355051$	$w_1 = 0.512486$
$x_2 = 0.844949$	$w_2 = 0.376403$

use the method to estimate:

$$\int_0^1 x\, \cos x\, dx$$

Answer: 0.38172 (Exact: $\cos(1) + \sin(1) - 1 = 0.38177$)

36. Use any suitable numerical integration rule(s) to estimate the value of:

$$\int_{0.3}^\infty (6x - x^2 - 8)e^{-x}\, dx$$

Answer: -2.1410

Chapter 7

Numerical Solution of Ordinary Differential Equations

7.1 Introduction

The need to solve differential equations arises in a great many problems of engineering analysis where physical laws are expressed in terms of the derivatives of variables rather than just the variables themselves.

The solution to a differential equation essentially involves integration and can sometimes be arrived at analytically. For example, the simplest type of differential equation is of the form

$$\frac{dy}{dx} = f(x) \tag{7.1}$$

where $f(x)$ is a given function of x, and $y(x)$ is the required solution. Provided $f(x)$ can be integrated, the solution to equation (7.1) is of the form

$$y = \int f(x)\, dx + C \tag{7.2}$$

where C is an arbitrary constant. In order to find the value of C, some additional piece of information is required, such as an initial value of y corresponding to a particular value of x.

Differential equations can take many different forms and frequently involve functions of x and y on the right hand side of equation (7.1). Before describing different solution techniques therefore, we need to define some important classes of differential equations, as this may influence our method of tackling a particular problem.

7.2 Definitions and types of ODE

Differential equations fall into two distinct categories depending on the number of independent variables they contain. If there is only one independent variable, the derivatives will be "ordinary", and the equation will be called an "ordinary differential equation". If more than one independent variable exists, the derivatives will be "partial", and the equation will be called a "partial differential equation".

Although ordinary differential equations (ODEs) with only one independent variable may be considered to be a special case of partial differential equations (PDEs), it is best to consider the solution techniques for the two classes quite separately. The remainder of this chapter is devoted to the solution of ODEs while an introduction to the solution of PDEs will be discussed in Chapter 8.

The "order" of an ODE corresponds to the highest derivative that appears in the equation, thus

$$y'' - 3y' + 4 = y \qquad \text{is second order} \tag{7.3}$$

where we use the notation

$$y' = \frac{dy}{dx} \quad ; \quad y'' = \frac{d^2y}{dx^2} \qquad \text{etc.}$$

and

$$\frac{d^4y}{dx^4} + \left(\frac{d^2y}{dx^2}\right)^2 = 1 \qquad \text{is fourth order} \tag{7.4}$$

A "linear" ODE is one which contains no products of the dependent variable or its derivatives, thus equation (7.3) is linear whereas (7.4) is nonlinear, due to the squared term. A general n^{th} order linear equation is given as

$$A_n(x)\frac{d^ny}{dx^n} + A_{n-1}(x)\frac{d^{n-1}y}{dx^{n-1}} + \ldots + A_1(x)\frac{dy}{dx} + A_0(x)y = R(x) \tag{7.5}$$

The "degree" of an ODE is the power to which the highest derivative is raised, thus equations (7.3) and (7.4) are both first degree. A consequence of this definition is that all linear equations are first degree, but not all first degree equations are linear, for example,

$$y'' + 2y' + y = 0 \qquad \text{is second order, first degree, linear, while} \tag{7.6}$$
$$y'' + 2y' + y^2 = 0 \qquad \text{is second order, first degree, nonlinear} \tag{7.7}$$

Nonlinear equations are harder to solve analytically and may have multiple solutions.

The higher the order of a differential equation, the more additional information must be supplied in order to obtain a solution. For example, equations (7.8)-(7.10) are all equivalent statements mathematically, but the second order equation (7.8) requires two additional pieces of information to be equivalent to the first order equation (7.9) which only requires one additional piece of information.

$$y'' = y + x + 1, \qquad y(0) = 0, \;\; y'(0) = 0 \tag{7.8}$$
$$y' = y + x, \qquad\qquad y(0) = 0 \tag{7.9}$$
$$y = e^x - x - 1 \tag{7.10}$$

The third equation (7.10) is purely algebraic and represents the solution to the two differential equations.

In general, to obtain a solution to an nth order ordinary differential equation such as that given in equation (7.5), n additional pieces of information will be required.

The way in which this additional information is supplied greatly influences the method of numerical solution. If all the information is given at the same value of the independent variable, such as in equations (7.8) and (7.9), the problem is termed an "initial value problem" or IVP. If the information is provided at different values of the independent variable, such as in the second order system given by equation (7.11), the problem is termed a "boundary value problem" or BVP

$$y'' + \frac{1}{x}y' - \frac{1}{x^2}y = \frac{6}{x^2}, \quad y(1) = 1, \ y(1.5) = -1 \qquad (7.11)$$

It should also be noted that as first order equations only require one piece of additional information, all first order equations may be treated as initial value problems. Boundary value problems however will be at least of second order.

The numerical solution techniques for initial and boundary value problems differ substantially, so they will be considered separately in this Chapter.

7.3 Initial value problems

We will limit our discussion for now to the numerical solution of first order equations subject to an initial condition of the form

$$\frac{dy}{dx} = f(x, y), \quad \text{with} \ y(x_0) = y_0 \qquad (7.12)$$

where $f(x, y)$ is any function of x and y. The equation is first order and first degree, however it could be linear or nonlinear depending on the nature of the function $f(x, y)$.

It may be noted that if $f(x, y)$ is linear the equation can be solved analytically by separation of variables or by using an "integrating factor". If $f(x, y)$ is nonlinear however, the analytical approach is greatly limited, and may prove impossible. Only a limited number of nonlinear differential equations can be solved analytically.

All numerical techniques for solving equation (7.12) involve starting at the initial condition (x_0, y_0) and stepping along the x-axis. At each step, a new value of y is estimated. As more steps are taken, the form of the required solution $y(x)$ is obtained.

Fundamentally, the *change* in y caused by a *change* in x is obtained by integrating $\dfrac{dy}{dx}$, thus

$$y_{i+1} - y_i = \int_{x_i}^{x_{i+1}} \frac{dy}{dx}\, dx \qquad (7.13)$$

where (x_i, y_i) represents the "old" solution at the beginning of the step, and y_{i+1} is the "new" estimate of y corresponding to x_{i+1}. The step length in x is usually under the user's control and is defined as h, thus

$$x_{i+1} - x_i = h \qquad (7.14)$$

Rearrangement of equation (7.13) leads to

$$y_{i+1} = y_i + \int_{x_i}^{x_{i+1}} \frac{dy}{dx}\, dx \qquad (7.15)$$

or alternatively

$$\begin{bmatrix} \text{"new" value} \\ \text{of } y \end{bmatrix} = \begin{bmatrix} \text{"old" value} \\ \text{of } y \end{bmatrix} + \begin{bmatrix} \text{change in} \\ y \end{bmatrix} \qquad (7.16)$$

Equations (7.15) and (7.16) give the general form of all the numerical solution techniques described in this section for advancing the solution of initial value problems. In the previous chapter we integrated under a curve of y vs. x to compute an "area", but in this chapter we will be integrating under a curve of y' vs. x to compute the *change* in y. Many of the methods of numerical integration described in the previous chapter still apply. The main difference here is that the function to be integrated may depend on both x and y; thus we will need to modify our methods to account for this.

There are two main approaches for performing the integration required by equation (7.15):

(a) One-step methods, which use information from only one preceding point, (x_i, y_i), to estimate the next point (x_{i+1}, y_{i+1}).

(b) Multi-step methods, which use information about several previous points, (x_i, y_i), (x_{i-1}, y_{i-1})... etc. to estimate the next point (x_{i+1}, y_{i+1}). These methods are also sometimes known as "Predictor-Corrector" methods, since each step can make use of two formulas; one that "predicts" the new solution y_{i+1}, and another that refines or "corrects" it.

One-step methods are self-starting, using only the initial condition provided, whereas Multi-step methods require several consecutive initial values of x and y to get started. These additional initial values may be provided by a One-step method if not provided in the initial data.

We will concentrate initially on numerical solution techniques for a single first order equation of the standard form given in equation (7.12).

It will be shown later in this section that higher order ODEs can be broken down into systems of first order equations which can be solved by the same methods.

7.3.1   One-step methods

One-step methods are so called, because information about one previous step only is needed to generate the solution at the next step. This makes One-step methods relatively easy to implement in a computer program. There are many One-step methods of differing levels of complexity. As is typical of many numerical methods, the more work done at each step, the greater the accuracy that is usually obtained. The trade-off to be sought is between increasing the work per step and decreasing the number of steps to span a given range. All the one-step methods covered in this section are eventually implemented in Program 7.1 described later in the Chapter.

7.3.1.1   The Euler method

This method has the advantage of simplicity; however it usually requires small step lengths to achieve reasonable accuracy. The method is widely used in codes that use "explicit" algorithms.

The numerical integration of equation (7.15) is performed using the Rectangle rule (see Section 6.2.2) where the derivative is "sampled" at the beginning of the step.

Starting with the standard problem

$$\frac{dy}{dx} = f(x,y), \quad \text{with } y(x_0) = y_0 \tag{7.17}$$

the Euler method estimates the new value of y using the sequence

$$K_0 = hf(x_0, y_0) \tag{7.18}$$

$$y_1 = y_0 + K_0 \tag{7.19}$$

where

$$h = x_1 - x_0 \tag{7.20}$$

The method is then repeated using the "new" initial conditions (x_1, y_1) to estimate y_2 and so on. In general we get

$$K_0 = hf(x_i, y_i) \tag{7.21}$$

$$y_{i+1} = y_i + K_0 \tag{7.22}$$

where

$$h = x_{i+1} - x_i \tag{7.23}$$

The similarity between equations (7.16) and (7.22) can be noted.

Figure 7.1 shows how the Euler method operates. The numerical approximate solution follows a straight line corresponding to the tangent at the beginning of each step. Clearly an error is introduced by this assumption unless the actual solution happens to be linear. Errors can accumulate quite rapidly in this method unless a small step length h is used, but there is a limit to how small h can be made from efficiency and machine accuracy considerations.

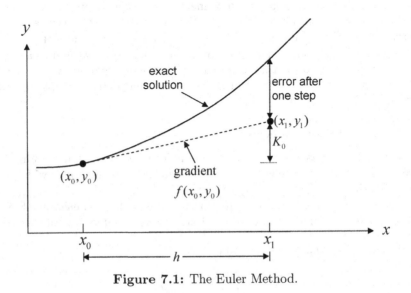

Figure 7.1: The Euler Method.

Example 7.1

Given the equation

$$y' = \frac{x+y}{x} \qquad \text{with} \qquad y(2) = 2$$

estimate $y(2.5)$ using the Euler method. Let (a) $h = 0.25$ and (b) $h = 0.1$

Solution 7.1

For hand calculation purposes in this simple method, the intermediate step involving K_0 has been directly incorporated into equation (7.22).

(a) Two steps will be required if $h = 0.25$

$$y(2.25) = 2 + 0.25 \left(\frac{2+2}{2} \right) = 2.5$$

$$y(2.5) = 2.5 + 0.25 \left(\frac{2.25 + 2.5}{2.25} \right) = 3.028$$

(b) Five steps will be required if $h = 0.1$

$$y(2.1) = 2 + 0.1 \left(\frac{2+2}{2} \right) = 2.2$$

$$y(2.2) = 2.2 + 0.1 \left(\frac{2.1 + 2.2}{2.1} \right) = 2.405$$

$$y(2.3) = 2.405 + 0.1 \left(\frac{2.2 + 2.405}{2.2} \right) = 2.614$$

$$y(2.4) = 2.614 + 0.1 \left(\frac{2.3 + 2.614}{2.3} \right) = 2.828$$

$$y(2.5) = 2.828 + 0.1 \left(\frac{2.4 + 2.828}{2.4} \right) = 3.046$$

The exact solution in this case is given by $y = x[1 + \ln(x/2)]$, hence $y(2.5) = 3.058$.

7.3.1.2   The Modified Euler method

A logical refinement to the Euler method is to use a higher order integration rule to estimate the change in y. The Modified Euler method performs the numerical integration of equation (7.15) using the Trapezoid rule (see Section 6.2.3) where the derivative is "sampled" at the beginning and the end of the step.

Starting with the standard problem

$$\frac{dy}{dx} = f(x, y), \text{ with } y(x_0) = y_0 \tag{7.24}$$

The Modified Euler method estimates the new value of y using the sequence

$$K_0 = hf(x_0, y_0) \tag{7.25}$$
$$K_1 = hf(x_0 + h, y_0 + K_0) \tag{7.26}$$

$$y_1 = y_0 + \frac{1}{2}(K_0 + K_1) \tag{7.27}$$

where K_0 is the change in y based on the slope at the beginning of the step (see Figure 7.1), and K_1 is the change in y based on the slope at the end of the step.

As shown in Figure 7.2, the numerical solution follows the exact solution more closely than the simple Euler method, because the integration of equation (7.15) is performed by sampling the derivative $f(x,y)$ at the beginning *and* end of the step.

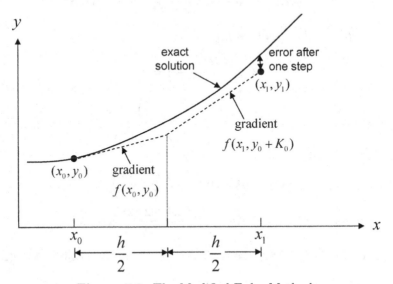

Figure 7.2: The Modified Euler Method.

The calculation of K_0 in equation (7.25) is equivalent to a simple Euler step, which is then used to compute K_1 in equation (7.26) leading to a "modified" Euler approach. The estimated change in y is then a simple average of the two K-values as given in equation (7.27).

Example 7.2

Given the equation

$$y' = \frac{x+y}{x} \qquad with \qquad y(2) = 2$$

estimate $y(2.5)$ using the Modified Euler method. Let $h = 0.25$.

Solution 7.2

Two steps will be required if $h = 0.25$

Step 1:

$$K_0 = 0.25 \left(\frac{2+2}{2} \right) = 0.5$$

$$K_1 = 0.25 \left(\frac{2.25 + 2.5}{2.25} \right) = 0.528$$

$$y(2.25) = 2 + \frac{1}{2}(0.5 + 0.528) = 2.514$$

Step 2:

$$K_0 = 0.25 \left(\frac{2.25 + 2.514}{2.25} \right) = 0.529$$

$$K_1 = 0.25 \left(\frac{2.5 + 3.043}{2.5} \right) = 0.554$$

$$y(2.5) = 2.514 + \frac{1}{2}(0.529 + 0.554) = 3.056$$

As compared with the exact solution of 3.058, this clearly represents a considerable improvement over the simple Euler solution of 3.028 using the same step length of h = 0.25.

7.3.1.3 Mid-Point method

The Mid-Point method performs the numerical integration of equation (7.15) using the Mid-Point rule (or One-point Gauss-Legendre rule, see Section 6.2.4) in which the derivative is "sampled" at the mid-point of the step as shown in Figure 7.3. The method has a similar order of accuracy to the Modified Euler method, and will actually give the same solutions for derivative functions that are linear in both x and y.

Starting with the standard problem

$$\frac{dy}{dx} = f(x, y), \text{ with } y(x_0) = y_0 \tag{7.28}$$

the Mid-Point method estimates the new value of y using the sequence

$$K_0 = hf(x_0, y_0) \tag{7.29}$$

$$K_1 = hf(x_0 + \frac{h}{2}, y_0 + \frac{K_0}{2}) \tag{7.30}$$

$$y_1 = y_0 + K_1 \tag{7.31}$$

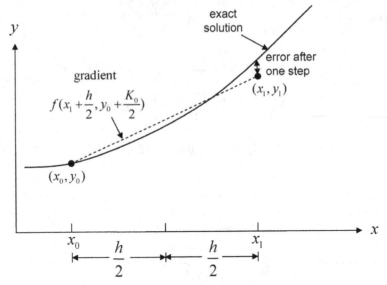

Figure 7.3: Mid-Point Method.

where K_0 as before is a simple Euler step (see Figure 7.1), but in this case K_1 is the change in y based on the slope in the middle of the step.

Example 7.3

Given the equation

$$y' = \frac{x+y}{x} \qquad \text{with} \qquad y(2) = 2$$

estimate $y(2.5)$ using the Mid-Point method. Let $h = 0.25$.

Solution 7.3

Two steps will be required of $h = 0.25$.
Step 1:

$$K_0 = 0.25 \left(\frac{2+2}{2} \right) = 0.5$$

$$K_1 = 0.25 \left(\frac{2.125 + 2.25}{2.125} \right) = 0.515$$

$$y(2.25) = 2 + 0.515 = 2.515$$

Step 2:

$$K_0 = 0.25 \left(\frac{2.25 + 2.515}{2.25} \right) = 0.529$$

$$K_1 = 0.25 \left(\frac{2.375 + 2.779}{2.375} \right) = 0.543$$

$$y(2.5) = 2.515 + 0.543 = 3.057$$

7.3.1.4 Runge-Kutta methods

"Runge-Kutta" methods refer to a family of one-step methods for numerical solution of initial value problems, which include the Euler and Modified Euler methods, which are first and second order methods respectively. The "order" of a method indicates the highest power of h included in the equivalent truncated Taylor series expansion (see Section 7.3.1.5).

The general form of all Runge-Kutta methods for advancing the solution by one step is as follows

$$y_{i+1} = y_i + \frac{\sum_{j=0}^{r-1} W_j K_j}{\sum_{j=0}^{r-1} W_j} \tag{7.32}$$

where the W_j are weighting coefficients and r is the order of the method. The K_j terms are estimates of the change in y evaluated at r locations within the step of width h.

A surprisingly simple fourth order Runge-Kutta method has received widespread use, and is sometimes referred to as *the* Runge-Kutta method.

Starting with the standard problem

$$\frac{dy}{dx} = f(x, y), \quad \text{with } y(x_0) = y_0 \tag{7.33}$$

the fourth order Runge-Kutta method estimates the new value of y using the sequence

$$K_0 = hf(x_0, y_0) \tag{7.34}$$

$$K_1 = hf(x_0 + \frac{h}{2}, y_0 + \frac{K_0}{2}) \tag{7.35}$$

$$K_2 = hf(x_0 + \frac{h}{2}, y_0 + \frac{K_1}{2}) \tag{7.36}$$

$$K_3 = hf(x_0 + h, y_0 + K_2) \tag{7.37}$$

$$y_1 = y_0 + \frac{1}{6}(K_0 + 2K_1 + 2K_2 + K_3) \tag{7.38}$$

The simplicity and accuracy of this method makes it one of the most popular of all one-step methods for numerical solution of first order differential
equations. From equation (7.38) it may be noted that the function is sampled
at the beginning, middle and end of the step in the ratio of $1:4:1$, indicating
a "Simpson-like" rule for estimating the change in y.

Example 7.4

Given the equation

$$y' = \frac{x+y}{x} \quad \text{with} \quad y(2) = 2$$

estimate $y(2.5)$ using the fourth order Runge-Kutta method.

Solution 7.4

*Since we are using a higher order method in this case, a solution will be
attempted using one step of $h = 0.5$*

Step 1:
$$x = 2, \quad y = 2$$
$$K_0 = 0.5\left(\frac{2+2}{2}\right) = 1$$

$$x = 2.25, \quad y = 2.5$$
$$K_1 = 0.5\left(\frac{2.25+2.5}{2.25}\right) = 1.056$$

$$x = 2.25, \quad y = 2.528$$
$$K_2 = 0.5\left(\frac{2.25+2.528}{2.25}\right) = 1.062$$

$$x = 2.5, \quad y = 3.062$$
$$K_3 = 0.5\left(\frac{2.5+3.062}{2.5}\right) = 1.112$$

$$y(2.5) = 2 + \frac{1}{6}[1 + 2(1.056) + 2(1.062) + 1.112] = 3.058$$

which is in agreement with the exact solution to three decimal places.

7.3.1.5   Accuracy of one-step methods

Since One-step methods involve numerical integration, the dominant error
term as discussed in Chapter 6 is again a useful way of comparing methods.

Here we will discuss the dominant error term corresponding to the first two methods discussed previously.

The Euler method is clearly equivalent to the Taylor series expanded about x_0, truncated after the first order term, where

$$y_1 = y_0 + hy_0' + \frac{1}{2}h^2 y_0'' + \dots \tag{7.39}$$

hence the dominant error term is given by $\frac{1}{2}h^2 y_0''$.

The h^2 term implies that if the step length is reduced by a factor of 2, the local error will be reduced by a factor of 4. The global error, however, will only be reduced by a factor of 2 because twice as many of the smaller steps will be required to span a given range.

Interpretation of the dominant error term in the Modified Euler method starts by writing the method in the form

$$y_1 = y_0 + \frac{h}{2}[y_0' + y_1'] \tag{7.40}$$

followed by a simple forward difference approximation of the second derivative (see Section 7.4.1) as

$$y_0'' = \frac{y_1' - y_0'}{h} \tag{7.41}$$

Elimination of y_1' from equations (7.40) and (7.41) leads to

$$y_1 = y_0 + hy_0' + \frac{1}{2}h^2 y_0'' \tag{7.42}$$

which are the first three terms of a truncated Taylor series expanded about x_0. The dominant error term is therefore the first term left out of the series, namely $\frac{1}{6}h^3 y_0'''$.

Halving the step length in this instance will reduce the local error by a factor of 8, but due to a doubling of the number of steps required to span a given range, the global error is only reduced by a factor of 4.

In summary, if the dominant error term in a one-step method based on a truncated Taylor series involves h^{k+1}, then the global error at a particular value of x will be approximately proportional to h^k. Although one-step methods on their own do not give direct error estimates, the dominant error term can be used to estimate the step length needed to achieve a given level of accuracy.

Example 7.5

Given the equation

$$\frac{dy}{dx} = (x+y)^2 \quad \text{with} \quad y(0) = 1$$

estimate y(0.5) using the Modified Euler method with (a) 5 steps of h = 0.1 and (b) 10 steps of h = 0.05. Use these two solutions to estimate the value of h that would be required to give a solution correct to 5 decimal places.

Solution 7.5

Using Program 7.1 for example (see later)

(a) 5 steps of h = 0.1 gives y(0.5) = 2.82541

(b) 10 steps of h = 0.05 gives y(0.5) = 2.88402

The global error by the Modified Euler method is proportional to h^2, hence

$$y_{exact} - 2.82541 = C(0.1)^2$$

$$y_{exact} - 2.88402 = C(0.05)^2$$

where C is a constant of proportionality.

Solving these equations gives C = 7.81477.

In order for the solution to be accurate to five decimal places, the error must not be greater than 0.000005, thus

$$0.000005 = 7.81477h^2 \quad hence \quad h = 0.0008$$

A final run of Program 7.1 with 625 steps of h = 0.0008 gives y(0.5) = 2.90822. The reader can confirm that this result is accurate to five decimal places as compared with the exact solution, which in this case is given by $y = \tan(x + \frac{\pi}{4}) - x$.

A more pragmatic approach to error analysis, and probably the one most often used in engineering practice, is to repeat a particular calculation with a different step size. The sensitivity of the solution to the value of h will then give a good indication of the accuracy of the solution.

7.3.2 Reduction of high order equations

All the examples considered so far have involved the solution of first order equations. This does not prove to be a restriction to solving higher order equations, because it is easily shown that an nth order differential equation can be broken down into an equivalent system of n first order equations. In addition, if the n conditions required to obtain a particular solution to the nth order equation are all given at the same value of the independent variable, then the resulting set of initial value problems can be solved using the same methods as described previously.

Consider a first degree n^{th} order differential equation arranged so that all terms except the nth derivative term are placed on the right-hand side, thus

$$\frac{d^n y}{dx^n} = f\left(x, y, \frac{dy}{dx}, \frac{d^2 y}{dx^2} + \cdots + \frac{d^{n-1} y}{dx^{n-1}}\right) \tag{7.43}$$

with n initial conditions given as

$$y(x_0) = A_1, \quad \frac{dy}{dx} = A_2, \quad \frac{d^2 y}{dx^2} = A_3, \quad \cdots \quad , \frac{d^{n-1} y}{dx^{n-1}} = A_n \tag{7.44}$$

We now replace all terms on the right-hand side of equation (7.43) (except x) with simple variable names, hence let

$$y = y_1, \quad \frac{dy}{dx} = y_2, \quad \frac{d^2 y}{dx^2} = y_3, \quad \cdots \quad , \frac{d^{n-1} y}{dx^{n-1}} = y_n \tag{7.45}$$

By making these substitutions, each variable is itself a *first* derivative of the term that precedes it, thus we can write the system as n first order equations as follows

$$\frac{dy_1}{dx} = y_2, \qquad\qquad y_1(x_0) = A_1$$

$$\frac{dy_2}{dx} = y_3, \qquad\qquad y_2(x_0) = A_2$$

$$\vdots \qquad\qquad\qquad \vdots \tag{7.46}$$

$$\frac{dy_{n-1}}{dx} = y_n, \qquad\qquad y_{n-1}(x_0) = A_{n-1}$$

$$\frac{dy_n}{dx} = f(x, y_1, y_2, \cdots, y_n), \; y_n(x_0) = A_n$$

All the derivatives up to the $n-1^{th}$ are simply treated as dependent variables. This process of reducing a high order equation to a system of first order equations always results in $n-1$ equations with simple right-hand sides, and one final equation resembling the original differential equation.

Equations (7.46) have been arranged in "standard form" whereby the first derivative term of each equation is placed on the left-hand side with all other terms placed on the right.

Example 7.6

Reduce the following third order equation to three first order equations in standard form

$$\frac{d^3y}{dx^3} + 2\frac{dy}{dx} = 2e^x \quad \text{with} \quad \begin{cases} y(x_0) = A \\ \dfrac{dy}{dx}(x_0) = B \\ \dfrac{d^2y}{dx^2}(x_0) = C \end{cases}$$

Solution 7.6

Make the substitutions

$$y = y_1, \quad \frac{dy}{dx} = y_2, \quad \frac{d^2y}{dx^2} = y_3$$

then

$$\frac{dy_1}{dx} = y_2, \qquad y_1(x_0) = A$$

$$\frac{dy_2}{dx} = y_3, \qquad y_2(x_0) = B$$

$$\frac{dy_3}{dx} = 2e^x - 2y_2, \quad y_3(x_0) = C$$

7.3.3   Solution of simultaneous first order equations

Reduction of higher order equations as discussed in the previous section leads to a rather simple system of first order equations where the right-hand side functions in all but one of them consists of a single variable.

In general, a system of n simultaneous first order equations may have fully populated right-hand sides as follows

$$\frac{dy_i}{dx} = f_i(x, y_1, y_2, \ldots, y_n), \quad \text{with } y_i(x_0) = A_i, \quad i = 1, 2, \ldots, n \qquad (7.47)$$

Consider the system of two equations and initial conditions given below

$$\frac{dy}{dx} = f(x, y, z), \quad \text{with } y(x_0) = y_0$$

$$\frac{dz}{dx} = g(x, y, z), \quad \text{with } z(x_0) = z_0 \qquad (7.48)$$

In the interests of clarity in this section, we have called the dependent variables y and z and the right-hand side functions f and g. In programming terminology however, we will be required to use a subscript notation where dependent variables will be y(1), y(2), etc. and functions f(1), f(2), etc.

We may advance the solution of y and z to new values at $x_1 = x_0 + h$ using any of the one-step or Runge-Kutta methods described previously.

In general our solutions will be advanced using expressions of the form

$$y(x_1) = y(x_0) + K$$
$$z(x_1) = z(x_0) + L \tag{7.49}$$

where the values of K and L depend on the method being applied.

The Euler method leads to

$$K = K_0$$
$$L = L_0 \tag{7.50}$$

where

$$K_0 = hf(x_0, y_0, z_0)$$
$$L_0 = hg(x_0, y_0, z_0) \tag{7.51}$$

The Modified Euler method leads to

$$K = \frac{1}{2}(K_0 + K_1)$$
$$L = \frac{1}{2}(L_0 + L_1) \tag{7.52}$$

where

$$K_0 = hf(x_0, y_0, z_0)$$
$$L_0 = hg(x_0, y_0, z_0)$$
$$K_1 = hf(x_0 + h, y_0 + K_0, z_0 + L_0)$$
$$L_1 = hg(x_0 + h, y_0 + K_0, z_0 + L_0) \tag{7.53}$$

The Mid-Point method leads to

$$K = K_1$$
$$L = L_1 \tag{7.54}$$

where

$$K_0 = hf(x_0, y_0, z_0)$$
$$L_0 = hg(x_0, y_0, z_0)$$
$$K_1 = hf(x_0 + \frac{1}{2}h, y_0 + \frac{1}{2}K_0, z_0 + \frac{1}{2}L_0)$$
$$L_1 = hg(x_0 + \frac{1}{2}h, y_0 + \frac{1}{2}K_0, z_0 + \frac{1}{2}L_0) \tag{7.55}$$

and the fourth order Runge-Kutta method leads to

$$K = \frac{1}{6}(K_0 + 2K_1 + 2K_2 + K_3)$$

$$L = \frac{1}{6}(L_0 + 2L_1 + 2L_2 + L_3) \tag{7.56}$$

where

$$K_0 = hf(x_0, y_0, z_0)$$
$$L_0 = hg(x_0, y_0, z_0)$$
$$K_1 = hf(x_0 + \frac{1}{2}h, y_0 + \frac{1}{2}K_0, z_0 + \frac{1}{2}L_0)$$
$$L_1 = hg(x_0 + \frac{1}{2}h, y_0 + \frac{1}{2}K_0, z_0 + \frac{1}{2}L_0) \tag{7.57}$$
$$K_2 = hf(x_0 + \frac{1}{2}h, y_0 + \frac{1}{2}K_1, z_0 + \frac{1}{2}L_1)$$
$$L_2 = hg(x_0 + \frac{1}{2}h, y_0 + \frac{1}{2}K_1, z_0 + \frac{1}{2}L_1)$$
$$K_3 = hf(x_0 + h, y_0 + K_2, z_0 + L_2)$$
$$L_3 = hg(x_0 + h, y_0 + K_2, z_0 + L_2)$$

Example 7.7

A damped oscillator is governed by the differential equation:

$$y'' + 10y' + 500y = 0$$

with initial conditions $y(0) = -0.025$ and $y'(0) = -1$. Estimate $y(0.05)$ using (a) the Modified Euler method with $h = 0.025$ and (b) the fourth order Runge-Kutta method with $h = 0.05$.

Solution 7.7

Firstly reduce the problem to two first order equations in standard form

$$y' = z \qquad \text{with } y(0) = -0.025$$
$$z' = -10z - 500y \quad \text{with } z(0) = -1$$

(a) Modified Euler method ($h = 0.025$)

Step 1: $t = 0, \quad y = -0.025, \quad z = -1$

$$K_0 = 0.025(-1) = -0.025$$
$$L_0 = 0.025(-10(-1) - 500(-0.025)) = 0.5625$$

$$t = 0.025, \quad y = -0.05, \quad z = -0.4375$$

$$K_1 = 0.025(-0.4375) = -0.01094$$
$$L_1 = 0.025(-10(-0.4375) - 500(-0.05)) = 0.73438$$

$$y(0.025) = -0.025 + \frac{1}{2}(-0.025 - 0.01094) = -0.04297$$

$$z(0.025) = -1 + \frac{1}{2}(0.5625 + 0.73438) = -0.35156$$

Step 2: $t = 0.025$, $y = -0.04297$, $z = -0.35156$

$$K_0 = 0.025(-0.35156) = -0.00879$$
$$L_0 = 0.025(-10(-0.35156) - 500(-0.04297)) = 0.62502$$

$$t = 0.05, \quad y = -0.05176, \quad z = 0.27346$$

$$K_1 = 0.025(0.27346) = 0.00684$$
$$L_1 = 0.025(-10(0.27346) - 500(-0.05176)) = 0.57864$$

$$y(0.05) = -0.04297 + \frac{1}{2}(-0.00879 + 0.00684) = -0.04395$$

$$z(0.05) = -0.35156 + \frac{1}{2}(0.62502 + 0.57864) = 0.25027$$

(a) *Fourth order Runge-Kutta* $(h = 0.05)$

Step 1: $t = 0$, $y = -0.025$, $z = -1$

$$K_0 = 0.05(-1) = -0.05$$
$$L_0 = 0.05(-10(-1) - 500(-0.025)) = 1.125$$

$$t = 0.025, \quad y = -0.05, \quad z = -0.4375$$

$$K_1 = 0.05(-0.4375) = -0.02188$$
$$L_1 = 0.05(-10(-0.4375) - 500(-0.05)) = 1.46875$$

$$t = 0.025, \quad y = -0.03594, \quad z = -0.26563$$

$$K_2 = 0.05(-0.26563) = -0.01328$$
$$L_2 = 0.05(-10(-0.26563) - 500(-0.03594) = 1.03132$$

$$t = 0.05, \quad y = -0.03828, \quad z = 0.03132$$

$$K_3 = 0.05(0.03132) = 0.00157$$
$$L_3 = 0.05(-10(0.03132) - 500(-0.03828) = 0.94134$$

$$y(0.05) = -0.025 + \frac{1}{6}(-0.05 + 2(-0.02188 - 0.01328) + 0.00157) = -0.04479$$
$$z(0.05) = -1 + \frac{1}{6}(1.125 + 2(1.46875 + 1.03132) + 0.94134) = 0.17775$$

These results may be compared with the exact solutions of $y(0.05) = -0.04465$ and $y'(0.05) = z(0.05) = 0.19399$.

The results of a more detailed analysis of this problem using Program 7.1 (see later) are shown in Figure 7.4, which contrasts the exact solution given by

$$y = -\frac{9}{760}\sqrt{19}e^{-5x}\sin(5\sqrt{19}x) - \frac{1}{40}e^{-5x}\cos(5\sqrt{19}x)$$

with fourth order Runge-Kutta solutions using step lengths of $h = 0.05$ and $h = 0.1$.

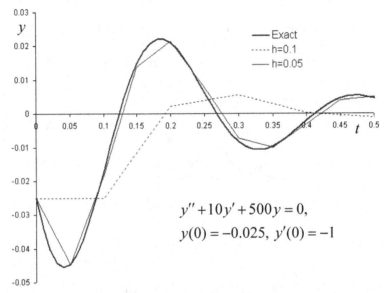

Figure 7.4: Influence of h on accuracy of fourth order Runge-Kutta solution of Example 7.7 using Program 7.1.

The smaller step length clearly leads to a very significant improvement in the numerical solution. While not shown in the figure, further reductions in the step length (with a commensurate increase in the number of computation steps) rapidly converge on the exact solution. For example, an analysis using $h = 0.005$ leads to a numerical solution that is indistinguishable from the exact solution at the resolution shown in Figure 7.4.

Program 7.1: One-step methods for systems of ODEs

```
PROGRAM nm71
!---One-Step Methods for Systems of ODEs---
!   itype= 1 (Euler Method)   itype= 2 (Modified Euler Method)
!   itype= 3 (Mid-Point Method) itype= 4 (4th order R-K Method)
 IMPLICIT NONE; INTEGER,PARAMETER::iwp=SELECTED_REAL_KIND(15,300)
 INTEGER::i,itype,j,n,nsteps; REAL(iwp)::d6=6.0_iwp,h,              &
    two=2.0_iwp,x
 REAL(iwp),ALLOCATABLE::k0(:),k1(:),k2(:),k3(:),y(:)
 OPEN(10,FILE='nm95.dat'); OPEN(11,FILE='nm95.res')
 READ(10,*)itype,n,nsteps,h
 ALLOCATE(k0(n),k1(n),k2(n),k3(n),y(n))
 READ(10,*)x,y
 WRITE(11,'(A)')"---One-Step Methods for Systems of ODEs---"
 SELECT CASE(itype)
 CASE(1)
    WRITE(11,'(/,A)')"*********** EULER METHOD ***********"
    WRITE(11,'(/,A,I2)')"       x              y(i) , i = 1,",n
    DO j=0,nsteps
      WRITE(11,'(10E13.5)')x,y; k0=h*f71(x,y); y=y+k0; x=x+h
    END DO
 CASE(2)
    WRITE(11,'(/,A)')"******* MODIFIED EULER METHOD ********"
    WRITE(11,'(/,A,I2)')"       x              y(i) , i = 1,",n
    DO j=0,nsteps
      WRITE(11,'(10E13.5)')x,y
      k0=h*f71(x,y); k1=h*f71(x+h,y+k0); y=y+(k0+k1)/two; x=x+h
    END DO
 CASE(3)
    WRITE(11,'(/,A)')"********** MID-POINT METHOD ***********"
    WRITE(11,'(/,A,I2)')"       x              y(i) , i = 1,",n
    DO j=0,nsteps
      WRITE(11,'(10E13.5)')x,y
      k0=h*f71(x,y); k1=h*f71(x+h/two,y+k0/two); y=y+k1; x=x+h
    END DO
 CASE(4)
    WRITE(11,'(/,A)')"***** 4TH ORDER RUNGE-KUTTA METHOD ****"
    WRITE(11,'(/,A,I2)')"       x              y(i) , i = 1,",n
    DO j=0,nsteps
      WRITE(11,'(10E13.5)')x,y
      k0=h*f71(x,y); k1=h*f71(x+h/two,y+k0/two)
      k2=h*f71(x+h/two,y+k1/two); k3=h*f71(x+h,y+k2)
```

Numerical Methods for Engineers

```
      y=y+(k0+two*k1+two*k2+k3)/d6; x=x+h
    END DO
  END SELECT
  CONTAINS

  FUNCTION f71(x,y)
   IMPLICIT NONE
   REAL(iwp),INTENT(IN)::x,y(:)
   REAL(iwp)::f71(SIZE(y,1))
   f71(1)=(x+y(1))**2.
!  f71(1)=3._iwp*x*y(2)+4._iwp
!  f71(2)=x*y(1)-y(2)-EXP(x)
!  f71(1)=y(2)
!  f71(2)=2._iwp*y(1)-3._iwp*y(2)+3._iwp*x**2.
   RETURN
  END FUNCTION f71

END PROGRAM nm71
```

List 7.1:

Scalar integers:

i	simple counter
itype	method type to be used
j	simple counter
n	number of equations
nsteps	number of calculations steps

Scalar reals:

d6	set to 6.0
h	calculation step length
two	set to 2.0
x	independent variable

Dynamic real arrays:

k0	estimate of Δy
k1	estimate of Δy
k2	estimate of Δy
k3	estimate of Δy
y	dependent variables

This program allows us to obtain numerical solutions to initial value problems involving systems of n first order ODEs of the form

$$\frac{dy_i}{dx} = f_i(x, y_1, y_2, \ldots, y_n), \text{ with } y_i(x_0) = A_i, \quad i = 1, 2, \ldots, n$$

The user selects the method to be used in the data through the input variable `itype`, where `itype=1` gives the Euler method, `itype=2` gives the Modified Euler method, `itype=3` gives the Mid-Point method and `itype=4` gives the fourth order Runge-Kutta method. The number of equations to be solved is read in as `n`, followed by the number of calculation steps `nsteps` and the step length `h`. The data are completed by the initial value of `x` and the initial value(s) of `y` which is a vector of length `n`. A single first order equation can be solved by simply setting $n = 1$ (see, e.g., Examples 7.5 and 7.7).

The right-hand side functions $f_i(x, y_1, y_2, \ldots, y_n)$ for $i = 1, 2, \ldots, n$ is evaluated by `FUNCTION f71` which must be created by the user and changed from one problem to the next.

A first example to illustrate use of Program 7.1 is the following single (`n=1`) nonlinear equation

$$y' = (x + y)^2 \quad \text{with} \quad y(0) = 1$$

With just one equation, the single dependent variable y is programmed as `y(1)`, so the right-hand side has been entered into `FUNCTION f71` as

```
f71(1)=(x+y(1))**2
```

In this example, $y(0.5)$ is to be estimated with 5 steps of $h = 0.1$ using the fourth order Runge-Kutta method (`itype=4`). The input and output for this example are given in Data 7.1a and Results 7.1a respectively.

One-step method	itype	
	4	
Number of equations	n	
	1	
Number and size of steps	nsteps	h
	5	0.1
Initial value of x and y	x	y(i),i=1,n
	0.0	1.0

Data 7.1a: One-Step Methods (first example)

```
---One-Step Methods for Systems of ODEs---

***** 4TH ORDER RUNGE-KUTTA METHOD ****

        x              y(i) , i = 1, 1
   0.00000E+00  0.10000E+01
   0.10000E+00  0.11230E+01
   0.20000E+00  0.13085E+01
   0.30000E+00  0.15958E+01
   0.40000E+00  0.20649E+01
   0.50000E+00  0.29078E+01
```

Results 7.1a: One-Step Methods (first example)

As shown in Results 7.1a, the method gives the approximate solution $y(0.5) =$ 2.9078 to four decimal places. For comparison, the analytical solution to this problem is given by

$$y = \tan\left(x + \frac{\pi}{4}\right) - x$$

leading to an exact solution of $y(0.5) = 2.9081$.

A second example to illustrate use of Program 7.1 is the pair (n=2) of first order equations

$$\frac{dy}{dx} = 3xz + 4 \qquad \text{with } y(0) = 4$$

$$\frac{dz}{dx} = xy - z - e^x \text{ with } z(0) = 1$$

With two equations, y is programmed as y(1) and z as y(2), and in order to preserve the required precision of calculations, constants such as 3 and 4 in the first function are programmed as 3._iwp and 4._iwp respectively (see Chapter 1). The two right-hand side functions have been entered into FUNCTION f71 as

```
f71(1)=3._iwp*x*y(2)+4._iwp
f71(2)=x*y(1)-y(2)-EXP(x)
```

In this example, $y(0.5)$ and $z(0.5)$ are to be estimated with 5 steps of $h = 0.1$ using the Mid-Point method (itype=3).

The input and output for this example are given in Data 7.1b and Results 7.1b respectively.

```
One-step method                  itype
                                   3

Number of equations                n
                                   2

Number and size of steps         nsteps     h
                                   5        0.1

Initial value of x and y           x      y(i),i=1,n
                                  0.0      4.0   1.0
```

Data 7.1b: One-Step Methods (second example)

```
---One-Step Methods for Systems of ODEs---

********** MID-POINT METHOD **********

     x              y(i) , i = 1, 2
 0.00000E+00   0.40000E+01   0.10000E+01
 0.10000E+00   0.44135E+01   0.82587E+00
 0.20000E+00   0.48473E+01   0.70394E+00
 0.30000E+00   0.52965E+01   0.63663E+00
 0.40000E+00   0.57613E+01   0.62643E+00
 0.50000E+00   0.62471E+01   0.67597E+00
```

Results 7.1b: One-Step Methods (second example)

As shown in Results 7.1b, the method gives the approximate solutions $y(0.5) = 6.2471$ and $z(0.5) = 0.6760$. The user is invited to experiment with different methods and step lengths.

For example the fourth order Runge-Kutta method (itype=4) with step lengths of $h = 0.1$ and $h = 0.05$ both give the same result of $y(0.5) = 6.2494$ and $z(0.5) = 0.6739$, implying solutions accurate to four decimal places.

A third and final example to illustrate use of Program 7.1 involves the second order ODE

$$\frac{d^2y}{dx^2} + 3\frac{dy}{dx} - 2y = 3x^2 \quad \text{with} \quad y(0) = 1 \quad \text{and} \quad \frac{dy}{dx}(0) = 0$$

where $y(0.2)$ and $z(0.2)$ are to be estimated with 4 steps of $h = 0.05$ using the Modified Euler method (itype=2).

By making the substitution

$$y = y_1 \quad \text{and} \quad \frac{dy}{dx} = y_2$$

the equation can be broken down into the following two first order equations

$$\frac{dy_1}{dx} = y_2 \qquad\qquad y_1(0) = 1$$

$$\frac{dy_2}{dx} = 2y_1 - 3y_2 + 3x^2 \qquad y_2(0) = 0$$

The input and output for this example are given in Data 7.1c and Results 7.1c respectively, where the right-hand side functions have been entered into FUNCTION f71 as

```
f71(1)=y(2)
f71(2)=2._iwp*y(1)-3._iwp*y(2)+3._iwp*x**2.
```

```
One-step method                    itype
                                     2

Number of equations                  n
                                     2

Number and size of steps           nsteps      h
                                     4        0.05

Initial value of x and y             x       y(i),i=1,n
                                    0.0      1.0   0.0
```

Data 7.1c: One-Step Methods (third example)

```
---One-Step Methods for Systems of ODEs---

******* MODIFIED EULERS METHOD ********

     x                 y(i) , i = 1, 2
0.00000E+00  0.10000E+01  0.00000E+00
0.50000E-01  0.10025E+01  0.92688E-01
0.10000E+00  0.10093E+01  0.17370E+00
0.15000E+00  0.10199E+01  0.24572E+00
0.20000E+00  0.10339E+01  0.31101E+00
```

Results 7.1c: One-Step Methods (third example)

As shown in Results 7.1c, the method gives the approximate solution $y(0.2)$ = 1.0339 and $z(0.2) = 0.31101$. The user is invited to experiment with different methods and step lengths to converge on the exact solution, which in this case is given to five significant figures of accuracy as $y(0.2) = 1.0336$ and $z(0.2) = 0.31175$.

7.3.4 θ-methods for linear equations

All of the methods described so far for numerical solution of initial value problems have been suitable for both linear or nonlinear equations provided they were first degree and could be arranged in standard form.

If the differential equation is linear, a different one-step approach is possible involving linear interpolation of derivatives between the beginning and end of each step.

7.3.4.1 First order equations

Consider the first order linear equation

$$y' = f(x, y) = k(x) + l(x)y \quad \text{with} \quad y(x_0) = y_0 \qquad (7.58)$$

where $k(x)$ and $l(x)$ are functions of x.

Writing the differential equation at x_0 and x_1, distance h apart, and using an abbreviated notation whereby $k(x_i) = k_i$ and $l(x_i) = l_i$ etc., we get

$$y_0' = k_0 + l_0 y_0 \qquad (7.59)$$
$$y_1' = k_1 + l_1 y_1 \qquad (7.60)$$

We now introduce the dimensionless scaling parameter θ which can be varied in the range

$$0 \le \theta \le 1 \qquad (7.61)$$

and write our one-step method as follows

$$y_1 = y_0 + h[(1 - \theta)y_0' + \theta y_1'] \qquad (7.62)$$

The parameter θ acts as a weighting coefficient on the gradients at the beginning and end of the step. When $\theta = 0$, the simple Euler method is obtained, where only the gradient at x_0 is included.

The most popular choice is $\theta = 0.5$, which gives equal weight to the gradients at x_0 and x_1, and is equivalent to the Trapezoid rule of integration. In the solution of time-dependent systems of differential equations, the use of $\theta = 0.5$ is sometimes referred to as the "Crank-Nicolson" method. These θ-methods are popular because they can be used even when extra coupling of the derivative terms means that the equations cannot easily be reduced to the "standard form" of equations (7.47).

The derivative terms y_0' and y_1' can be eliminated from equations (7.59), (7.60) and (7.62), and since the equation is linear it is easy to make y_1 the subject of the equation as follows

$$y_1 = \frac{y_0 + h[(1 - \theta)(k_0 + l_0 y_0) + \theta k_1]}{1 - h\theta l_1} \tag{7.63}$$

which is an explicit formula for y_1 in terms of h, θ, y_0 and the functions l_0, k_0, l_1 and k_1 evaluated at the beginning and end of the step. For systems of equations the denominator can become a matrix in which case the system is "implicit".

Example 7.8

Given the linear equation

$$\frac{dy}{dx} = 3y - 2x^2 \quad \text{with} \quad y(0) = 0.5$$

estimate $y(0.2)$ using the "θ-method" with two steps of $h = 0.1$. Let $\theta = 0.5$.

Solution 7.8

By comparison with equation (7.58) it is seen that $k(x) = -2x^2$ and $l(x) = 3$.

Step 1: $y_0 = 0.5$, $k_0 = 0$, $k_1 = -2 \times 0.1^2 = -0.02$, $l_0 = 3$, $l_1 = 3$

From equation (7.63)

$$y(0.1) = \frac{0.5 + 0.1[(1 - 0.5)(0 + 3 \times 0.5) + 0.5 \times (-0.02)]}{1 - 0.1 \times 0.5 \times 3} = 0.675$$

Step 2: $y_0 = 0.675$, $k_0 = -0.02$, $k_1 = -2 \times 0.2^2 = -0.08$, $l_0 = 3$, $l_1 = 3$

From equation (7.63)

$$y(0.2) = \frac{0.675 + 0.1[(1 - 0.5)(-0.02 + 3 \times 0.675) + 0.5 \times (-0.08)]}{1 - 0.1 \times 0.5 \times 3} = 0.908$$

which can be compared with the exact solution, which gives $y(0.2) = 0.905$ to three decimal places.

7.3.4.2   Second order equations

Second order linear equations can also be solved by linear interpolation using the parameter θ.

Consider the second order linear equation

$$y'' = k(x) + l(x)y + m(x)y' \quad \text{with} \quad y(x_0) = y_0 \quad \text{and} \quad y'(x_0) = y_0' \quad (7.64)$$

Writing the differential equation at x_0 and x_1, distance h apart, and using an abbreviated notation whereby $k(x_i) = k_i$, $l(x_i) = l_i$ and $m(x_i) = m_i$ etc., we get

$$y_0'' = k_0 + l_0 y + m_0 y_0' \quad (7.65)$$
$$y_1'' = k_1 + l_1 y + m_1 y_1' \quad (7.66)$$

We now obtain the following expressions for y_1 and y_1' using θ to weight the derivatives at x_0 and x_1, hence

$$y_1 = y_0 + h[(1 - \theta)y_0' + \theta y_1'] \quad (7.67)$$
$$y_1' = y_0' + h[(1 - \theta)y_0'' + \theta y_1''] \quad (7.68)$$

The derivative terms y_0'', y_1' and y_1'' can be eliminated from equations (7.65)-(7.68), and since the equation is linear after some rearrangement y_1 and y_1' can be made the subjects of the following two equations

$$y_1 = \frac{y_0(1 - h\theta m_1) + hy_0'[1 - h\theta m_1(1 - \theta)] + h^2\theta[(1 - \theta)(k_0 + l_0 y_0 + m_0 y_0') + \theta k_1]}{1 - h\theta m_1 - h^2\theta^2 l_1}$$
$$y_1' = \frac{y_1 - y_0}{h\theta} - \frac{1 - \theta}{\theta}y_0' \quad (7.69)$$

It may also be noted from the second of equations (7.69) that the method would fail for $\theta = 0$.

For a single second order equation, the fourth order Runge-Kutta method is considerably more accurate than the "θ-method" described in this section. However for large engineering systems involving equations with coupled derivatives, the linear interpolation methods involving θ are still frequently used because of their simplicity.

Program 7.2: Theta-method for linear ODEs

```
PROGRAM nm72
!---Theta-Method for Linear ODEs---
IMPLICIT NONE; INTEGER,PARAMETER::iwp=SELECTED_REAL_KIND(15,300)
INTEGER::j,n,nsteps
REAL(iwp)::h,k0,k1,l0,l1,m0,m1,one=1.0_iwp,theta,x,y1
REAL(iwp),ALLOCATABLE::y(:)
OPEN(10,FILE='nm95.dat'); OPEN(11,FILE='nm95.res')
```

```
READ(10,*)n,nsteps,h,theta; ALLOCATE(y(n)); READ(10,*)x,y
SELECT CASE(n)
CASE(1)
  WRITE(11,'(A)')"--Theta-Method for First Order Linear ODEs--"
  WRITE(11,'(/,A,I2)')"        x                y"
  DO j=0,nsteps; WRITE(11,'(10E13.5)')x,y
    CALL f72(x,k0,l0); CALL f72(x+h,k1,l1)
    y=(y+h*((one-theta)*(k0+l0*y)+theta*k1))/(one-h*theta*l1)
    x=x+h
  END DO
CASE(2)
  WRITE(11,'(A)')"---Theta-Method for Second Order Linear ODEs---"
  WRITE(11,'(/,A,I2)')"        x                y                y'"
  DO j=0,nsteps
    WRITE(11,'(10E13.5)')x,y
    CALL f72(x,k0,l0,m0); CALL f72(x+h,k1,l1,m1)
    y1=y(1)*(one-h*theta*m1)+h*y(2)*(one-h*theta*m1*(1-theta))
    y1=y1+h**2.*theta*((1-theta)*(k0+l0*y(1)+m0*y(2))+theta*k1)
    y1=y1/(one-h*theta*m1-h**2.*theta**2.*l1)
    y(2)=(y1-y(1))/(h*theta)-(one-theta)/theta*y(2)
    y(1)=y1; x=x+h
  END DO
END SELECT
CONTAINS

SUBROUTINE f72(x,k,l,m)
  IMPLICIT NONE
  REAL(iwp),INTENT(IN)::x
  REAL(iwp),INTENT(OUT)::k,l
  REAL(iwp),INTENT(OUT),OPTIONAL::m
  k= x**3.
  l= -2._iwp*x
! k =  0._iwp
! l = -(3._iwp+x**2)
! m =  2._iwp*x
  RETURN
  END SUBROUTINE f72

END PROGRAM nm72
```

This program uses the "θ-method" to obtain numerical solutions to initial value problems governed by first order *linear* ODEs of the form

$$y' = k(x) + l(x)y \quad \text{with} \quad y(x_0) = y_0$$

List 7.2:

Scalar integers:

j	simple counter
n	order of linear equation (1 or 2)
nsteps	number of calculations steps

Scalar reals:

h	calculation step length
k0	$k(x)$ at beginning of step
k1	$k(x)$ at end of step
l0	$l(x)$ at beginning of step
l1	$l(x)$ at end of step
m0	$m(x)$ at beginning of step
m1	$m(x)$ at end of step
one	set to 1.0
theta	weighting parameter
x	independent variable
y1	used to compute y at end of step

Dynamic real arrays:

y	dependent variables y [and y']

---

or second order *linear* ODEs of the form

$$y'' = k(x) + l(x)y + m(x)y' \quad \text{with} \quad y(x_0) = y_0 \quad \text{and} \quad y'(x_0) = y_0'$$

Input consists of the order of the equation where n=1 indicates a first order and n=2 indicates a second order equation. This is followed by the number of calculation steps **nsteps**, the step length **h** and the weighting parameter **theta**. The data are completed by the initial value of x and the initial value(s) of y which is a vector of length n.

The problem-specific functions $k(x)$, $l(x)$ and, if solving a second order equation, $m(x)$, are evaluated by SUBROUTINE f as k, l and m respectively.

The first example to illustrate use of Program 7.2 is the single (n=1) linear equation

$$y' = -2xy + x^3 \quad \text{with} \quad y(0) = 1$$

where $y(1.5)$ is to be estimated using 5 steps of $h = 0.1$ with $\theta = 0.5$.

By comparison with the standard form from equation (7.58), we see that in this case $k(x) = x^3$ and $l(x) = -2x$.

The input and output for this example are given in Data 7.2a and Results 7.2a respectively, and the functions of x have been entered into SUBROUTINE f72 as

```
k= x**3.
l= -2._iwp*x
```

Order of equation	n
	1

Number and size of steps	nsteps	h	theta
and weighting parameter	5	0.1	0.5

Initial value of x and y	x	y(i),i=1,n
	0.0	1.0

Data 7.2a: Theta-Method for Linear ODEs (first example)

--Theta-Method for First Order Linear ODEs--

x	y
0.00000E+00	0.10000E+01
0.10000E+00	0.99015E+00
0.20000E+00	0.96147E+00
0.30000E+00	0.91649E+00
0.40000E+00	0.85918E+00
0.50000E+00	0.79454E+00

Results 7.2a: Theta-Method for Linear ODEs (first example)

As shown in Results 7.2a, the method gives the approximate solution $y(0.5) = 0.7945$ to four decimal places. For comparison, the exact solution to this problem is given by

$$y = \frac{2}{3}x^2 + \frac{4}{9}x + \frac{19}{54}e^{3x} + \frac{4}{27}$$

leading to $y(0.5) = 0.7932$.

A second example to illustrate use of Program 7.2 is the second order n=2 linear equation

$$y'' = 2xy' - (3 + x^2)y \quad \text{with} \quad y(0) = 2 \quad \text{and} \quad y'(0) = 0$$

where $y(1)$ and $y'(1)$ are to be estimated using 5 steps of $h = 0.2$ with $\theta = 0.5$.

By comparison with the standard form from equation (7.64) we see that in this case $k(x) = 0$, $l(x) = -(3 + x^2)$ and $m(x) = 2x$.

The input and output for this example are given in Data 7.2b and Results 7.2b respectively, and the functions of x have been entered into SUBROUTINE f72 as,

```
k =   0._iwp
l = -(3._iwp+x**2)
m =   2._iwp*x
```

```
Order of equation                 n
                                  2

Number and size of steps       nsteps      h        theta
and weighting parameter          5        0.2        0.5

Initial value(s) of x and y       x       y(i),i=1,n
                                 0.0       2.0    0.0
```

Data 7.2b: Theta-Method for Linear ODEs (second example)

```
---Theta-Method for Second Order Linear ODEs---

       x              y               y'
 0.00000E+00   0.20000E+01    0.00000E+00
 0.20000E+00   0.18780E+01   -0.12197E+01
 0.40000E+00   0.15044E+01   -0.25161E+01
 0.60000E+00   0.85728E+00   -0.39555E+01
 0.80000E+00  -0.95809E-01   -0.55754E+01
 0.10000E+01  -0.13880E+01   -0.73467E+01
```

Results 7.2b: Theta-Method for Linear ODEs (second example)

As shown in Results 7.2b, the method gives the approximate solution $y(1) = -1.388$ and $y'(1) = -7.347$ respectively to three decimal places. For comparison, the exact solution to this problem is given by

$$y = 2e^{\frac{1}{2}x^2} \cos(2x) \quad \text{and} \quad y' = 2xe^{\frac{1}{2}x^2} \cos(2x) - 4xe^{\frac{1}{2}x^2} \sin(2x)$$

leading to exact solutions of $y(1) = -1.372$ and $y'(1) = -7.369$.

7.3.5   Predictor-corrector methods

Predictor-Corrector methods use information from several previous known points to compute the next as indicated in Figure 7.5.

A disadvantage of the methods is that they are not self-starting and may need a one-step method to generate a few points in order to get started. The attraction of the methods is that more efficient use is made of existing information in order to advance to the next step. This is in contrast to the fourth order Runge-Kutta method for example, where at each step, four function evaluations are required which are never used again.

Predictor-Corrector methods make use of two formulas; a predictor formula that extrapolates existing data to estimate the next point, and a corrector formula that improves on this estimate.

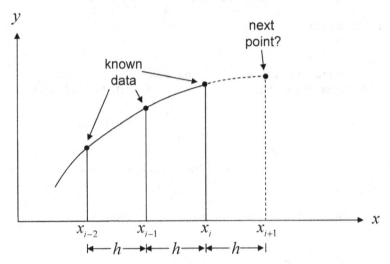

Figure 7.5: Predictor-Corrector Methods.

It is possible to apply the corrector formula repeatedly until some convergence criterion is satisfied; however this option is not implemented in any of the methods described in this section.

Predictor formulas estimate the new value of y_{i+1} by integrating under the curve of y' vs. x using sampling points at x_i, x_{i-1}, x_{i-2} etc. Any numerical integration formula which does *not* require a prior estimate of y'_{i+1} is suitable for use as a predictor. Formulas of this type have already been discussed in Section 6.4.6, where the sampling points were outside the range of integration.

Corrector formulas improve on the predicted value y_{i+1} by again integrating under the curve of y' vs. x, but this time using sampling points x_{i+1}, x_i, x_{i-1} etc. The corrector formula is able to sample at x_{i+1} because a value of y'_{i+1} is now available from the predictor stage. Any numerical integration formula which requires a prior estimate of y'_{i+1} is suitable for use as a corrector.

The Modified Euler method described earlier in this chapter is a type of Predictor-Corrector. Given a first order differential equation in standard form, $y' = f(x,y)$ with $y(x_0) = y_0$, the method starts with the Euler method, which is a predictor using the Rectangle rule

$$y_{i+1}^p = y_i + hf(x_i, y_i) \tag{7.70}$$

This is followed by a corrector that uses the Trapezoid rule

$$y_{i+1}^c = y_i + \frac{1}{2}h[f(x_i, y_i) + f(x_{i+1}, y_{i+1}^p)] \tag{7.71}$$

Note that the predictor does not require a prior estimate of y'_{i+1} while the corrector does.

The best known Predictor-Corrector methods however use formulas that have the same order of accuracy in both parts of the algorithm. As will be shown, this facilitates a convenient way of estimating the error of the corrected term based on the *difference* between the predicted and corrected terms.

7.3.5.1 The Milne-Simpson method

This method uses a formula due to Milne as a predictor, and the familiar Simpson's rule as a corrector. The method is fourth order, i.e., the dominant error term in both the predictor and corrector includes h^5, and requires four initial values of y to get started. Note the smaller error term associated with the corrector formula as compared with the predictor, which is to be expected since a predictor involves the less precise process of extrapolation.

Given a first order differential equation, $y' = f(x, y)$ with four initial conditions $y(x_{i-3}) = y_{i-3}$, $y(x_{i-2}) = y_{i-2}$, $y(x_{i-1}) = y_{i-1}$ and $y(x_i) = y_i$, the method starts with Milne's Predictor

$$y_{i+1}^p = y_{i-3} + \frac{4h}{3}[2f(x_{i-2}, y_{i-2}) - f(x_{i-1}, y_{i-1}) + 2f(x_i, y_i)] + \frac{28}{90}h^5 y^{(iv)}(\xi)$$
$$(7.72)$$

followed by Simpson's Corrector

$$y_{i+1}^c = y_{i-1} + \frac{h}{3}[f(x_{i-1}, y_{i-1}) + 4f(x_i, y_i) + f(x_{i+1}, y_{i+1}^p)] - \frac{1}{90}h^5 y^{(iv)}(\xi)$$
$$(7.73)$$

Milne's Predictor integrates under the curve of y' vs. x between limits of x_{i-3} and x_{i+1} using three centrally placed sampling points as shown in Figure 7.6.

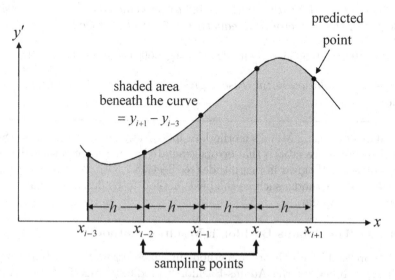

Figure 7.6: Milnes Predictor.

Example 7.9

Given $y' = 2x^2 + 2y$ with initial conditions

$$y(-0.6) = 0.1918$$
$$y(-0.4) = 0.4140$$
$$y(-0.2) = 0.6655$$
$$y(0) = 1.0000$$

use the Milne-Simpson Predictor-Corrector method to estimate $y(0.2)$.

Solution 7.9

Four initial conditions are provided with a step length of $h = 0.2$. To facilitate hand calculation, make a table of the initial conditions with a counter, and compute the value of the right-hand side derivative function $f(x, y)$ at each point.

	x	y	$f(x,y)$
$i-3$	-0.6	0.1918	
$i-2$	-0.4	0.4140	1.1480
$i-1$	-0.2	0.6655	1.4110
i	0.0	1.0000	2.0000

First apply the predictor from equation (7.72).

$$y^P(0.2) = 0.1918 + \frac{4(0.2)}{3}[2(1.1480) - (1.4110) + 2(2.0000)] = 1.4945$$

This enables the derivative at $x_{i+1} = 0.2$ to be predicted as $f_{i+1}(0.2, 1.4945) = 3.0689$ and the corrector from equation (7.73) to be applied.

$$y^c(0.2) = 0.6655 + \frac{0.2}{3}[(1.4110) + 4(2.0000) + (3.0689)] = 1.4975$$

The exact solution in this case is given by $y(0.2) = 1.4977$ to four decimal places.

A danger in using Milne's method, or indeed any method which uses Simpson's rule as a corrector, is that errors generated at one stage of the calculation may subsequently grow in magnitude (see Section 7.3.4). For this reason other fourth order methods, such as the next method to be described, have tended to be more popular.

7.3.5.2   The Adams-Bashforth-Moulton method

A more stable fourth order method, in which errors do not tend to grow so fast, is based on the Adams-Bashforth predictor, together with Adams-Moulton Corrector.

Given a first order differential equation, $y' = f(x, y)$ with four initial conditions $y(x_{i-3}) = y_{i-3}$, $y(x_{i-2}) = y_{i-2}$, $y(x_{i-1}) = y_{i-1}$ and $y(x_i) = y_i$ the method starts with Adams-Bashforth's Predictor

$$y_{i+1}^p = y_i + \frac{h}{24}[-9f(x_{i-3}, y_{i-3}) + 37f(x_{i-2}, y_{i-2})$$
$$-59f(x_{i-1}, y_{i-1}) + 55f(x_i, y_i)] + \frac{251}{720}h^5 y^{(iv)}(\xi) \qquad (7.74)$$

followed by Adams-Moulton's Corrector

$$y_{i+1}^c = y_i + \frac{h}{24}[f(x_{i-2}, y_{i-2}) - 5f(x_{i-1}, y_{i-1})$$
$$+19f(x_i, y_i) + 9f(x_{i+1}, y_{i+1}^p)] - \frac{19}{720}h^5 y^{(iv)}(\xi) \qquad (7.75)$$

The Adams-Bashforth-Moulton method has larger error terms than Milne-Simpson, although the dominant error terms still indicate that the corrector is considerably more accurate than the predictor.

The improved stability is also obtained at a cost of some additional work, since both formulas require four sampling points, as opposed to three in the Milne-Simpson method. As shown in Figure 7.7, the Adams-Bashforth Predictor uses four sampling points to integrate under the curve of y' vs. x between limits of x_i and x_{i+1}. The Adams-Moulton corrector is similar, but with the sampling points shifted one step to the right.

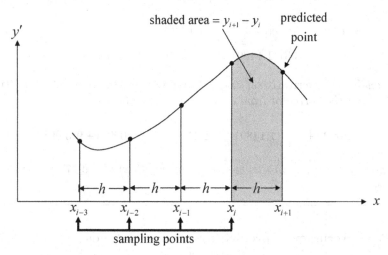

Figure 7.7: Adams-Bashforth Predictor.

Example 7.10

Given $y' = 2x^2 + 2y$ with initial conditions

$$y(-0.6) = 0.1918$$
$$y(-0.4) = 0.4140$$
$$y(-0.2) = 0.6655$$
$$y(0) = 1.0000$$

use the Adams-Bashforth-Moulton Predictor-Corrector method to estimate $y(0.2)$.

Solution 7.10

Four initial conditions are provided with a step length of $h = 0.2$. To facilitate hand calculation, make a table of the initial conditions with a counter, and compute the value of the right-hand side derivative function $f(x,y)$ at each point.

	x	y	$f(x,y)$
$i-3$	−0.6	0.1918	1.1036
$i-2$	−0.4	0.4140	1.1480
$i-1$	−0.2	0.6655	1.4110
i	0.0	1.0000	2.0000

First apply the predictor from equation (7.74).

$$y^p(0.2) = 1.0000 + \frac{0.2}{24}[-9(1.1036)$$
$$+37(1.1480) - 59(1.4110) + 55(2.0000)] = 1.4941$$

This enables the derivative at $x_{i+1} = 0.2$ to be predicted as $f_{i+1}(0.2, 1.4941) = 3.0682$ and the corrector from equation (7.75) to be applied.

$$y^c(0.2) = 1.0000 + \frac{0.2}{24}[(1.1480) - 5(1.4110) + 19(2.0000) + 9(3.0682)] = 1.4976$$

The exact solution in this case is given by $y(0.2) = 1.4977$ to four decimal places.

7.3.5.3   Accuracy of predictor-corrector methods

An attractive feature of Predictor-Corrector methods in which both formulas have the same order of accuracy is that they can make use of their dominant error terms to give a simple error estimate of the corrected value.

At a typical step, let y_{i+1}^p and y_{i+1}^c represent the approximate values computed by the predictor and the corrector formulas respectively, and let $y(x_{i+1})$ represent the (unknown) exact solution.

The errors in the Milne Predictor and Simpson Corrector formulas can be written respectively as

$$y(x_{i+1}) - y_{i+1}^p = \frac{28}{90} h^5 y^{(iv)}(\xi) \tag{7.76}$$

$$y(x_{i+1}) - y_{i+1}^c = -\frac{1}{90} h^5 y^{(iv)}(\xi) \tag{7.77}$$

After subtracting one equation from the other and eliminating $y(x_{i+1})$ we can write

$$y_{i+1}^c - y_{i+1}^p = \frac{29}{90} h^5 y^{(iv)}(\xi) \tag{7.78}$$

hence

$$h^5 y^{(iv)}(\xi) = \frac{90}{29}(y_{i+1}^c - y_{i+1}^p) \tag{7.79}$$

Substituting this back into equation (7.77) gives

$$y(x_{i+1}) - y_{i+1}^c = -\frac{1}{29}(y_{i+1}^c - y_{i+1}^p) \tag{7.80}$$

Similar operations on the Adams-Bashforth-Moulton Predictor-Corrector formulas leads to

$$y(x_{i+1}) - y_{i+1}^c = -\frac{1}{14}(y_{i+1}^c - y_{i+1}^p) \tag{7.81}$$

Equations (7.80) and (7.81) indicate that the error in the corrected value y_{i+1}^c is approximately proportional to the difference between the predicted and corrected values.

Program 7.3 to be described next allows solution of a first order ODE by either of the two fourth order Predictor-Corrector methods discussed in this section. The program uses a constant step length h; however the ability to estimate the error at each step of the solution process means that a computer program could take account of the error, and adjust the step size accordingly. If the error was too big, one strategy would be to keep halving the step size until the accuracy criterion was met. At a subsequent stage of the calculation, the previously obtained step length might become excessively small, in which case it could be increased systematically.

It should be noted however, that if the step length is changed during the calculation, it will be necessary to recall some one-step starting procedure in order to generate enough points at the new step length for the predictor-corrector algorithm to proceed. An "adaptive" approach such as this was described in Chapter 6 in relation to repeated numerical integration rules and implemented in Program 6.3.

Program 7.3: Fourth order predictor-corrector methods

```
PROGRAM nm73
!---Fourth Order Predictor-Corrector Methods---
 IMPLICIT NONE; INTEGER,PARAMETER::iwp=SELECTED_REAL_KIND(15,300)
 INTEGER::i,itype,j,nsteps
 REAL(iwp)::d3=3.0_iwp,d4=4.0_iwp,d5=5.0_iwp,d9=9.0_iwp,         &
   d14=14.0_iwp,d19=19.0_iwp,d24=24.0_iwp,d29=29.0_iwp,          &
   d37=37.0_iwp,d55=55.0_iwp,d59=59.0_iwp,e,h,two=2.0_iwp,       &
   x(-3:1),y(-3:1),y1
 OPEN(10,FILE='nm95.dat'); OPEN(11,FILE='nm95.res')
 READ(10,*)itype,nsteps,h; READ(10,*)(x(i),y(i),i=-3,0)
 SELECT CASE(itype)
 CASE(1)
   WRITE(11,'(A)')"---Milne-Simpson 4th Order P-C Methods---"
   WRITE(11,'(/,A)')"      x             y           Error"
   WRITE(11,'(2E13.5)')(x(i),y(i),i=-3,0)
   DO j=1,nsteps
     x(1)=x(0)+h
     y1=y(-3)+d4*h/d3*(two*f73(x(-2),y(-2))-f73(x(-1),y(-1))+     &
       two*f73(x(0),y(0)))
     y(1)=y(-1)+h/d3*(f73(x(-1),y(-1))+d4*f73(x(0),y(0))+         &
       f73(x(1),y1))
     e=-(y(1)-y1)/d29; WRITE(11,'(3E13.5)')x(1),y(1),e
     y(-3:0)=y(-2:1); x(-3:0)=x(-2:1)
   END DO
 CASE(2)
   WRITE(11,'(A)')"---Adams-Bashforth-Moulton 4th Order P-C &
     &Methods---"
   WRITE(11,'(/,A)')"      x             y           Error"
   WRITE(11,'(2E13.5)')(x(i),y(i),i=-3,0)
   DO j=1,nsteps
     x(1)=x(0)+h
     y1=y(0)+h/d24*(-d9*f73(x(-3),y(-3))+d37*f73(x(-2),y(-2))-    &
       d59*f73(x(-1),y(-1))+d55*f73(x(0),y(0)))
     y(1)=y(0)+h/d24*(f73(x(-2),y(-2))-d5*f73(x(-1),y(-1))+       &
       d19*f73(x(0),y(0))+d9*f73(x(1),y1))
     e=-(y(1)-y1)/d14; WRITE(11,'(3E13.5)')x(1),y(1),e
     y(-3:0)=y(-2:1); x(-3:0)=x(-2:1)
   END DO
 END SELECT
 CONTAINS
```

```
FUNCTION f73(x,y)
 IMPLICIT NONE
 REAL(iwp),INTENT(IN)::x,y; REAL(iwp)::f73
 f73=x*y**2+2._iwp*x**2
 RETURN
END FUNCTION f73

END PROGRAM nm73
```

List 7.3:

Scalar integers:
i	simple counter
itype	method type to be used
j	simple counter
nsteps	number of calculations steps

Scalar reals:
d3	set to 3.0
d4	set to 4.0
d5	set to 5.0
d9	set to 9.0
d14	set to 14.0
d19	set to 19.0
d24	set to 24.0
d29	set to 29.0
d37	set to 37.0
d55	set to 55.0
d59	set to 59.0
e	error estimate
h	calculation step length
two	set to 2.0
y1	used to compute y at end of step

Real arrays:
x	holds four previous values of x
y	holds four previous values of y

This program allows us to obtain numerical solutions to an initial value problem involving a first order ODE of the form

$$\frac{dy}{dx} = f(x,y)$$

with initial values provided from four previous steps as

$$y(x_i) = y_i, \quad i = -3, -2, -1, 0$$

The user selects the method to be used in the data through the input variable itype, where itype=1 gives the Milne-Simpson method and itype=2 gives the Adams-Bashforth-Moulton method.

The example to illustrate use of Program 7.3 is the nonlinear equation

$$y' = xy^2 + 2x^2$$

with the four initial values

$$y(1.00) = 3.61623$$
$$y(0.95) = 2.99272$$
$$y(0.90) = 2.55325$$
$$y(0.85) = 2.22755$$

The function has been entered into FUNCTION f73 as

f73=x*y**2.+2._iwp*x**2.

The program is to be used to continue the sequence of solutions to $y(0.6)$ using 5 steps of $h = -0.05$ using the Adams-Bashforth-Moulton method (itype=2). This example demonstrates that h can be positive or negative, depending on the direction along the x-axis for which new solutions are required. The input and output for this example are given in Data 7.3 and Results 7.3 respectively.

```
Predictor-Corrector method        itype
                                   2

Number and size of steps          nsteps    h
                                   5       -0.05

Initial values of x and y          x         y
                                   1.00    3.61623
                                   0.95    2.99272
                                   0.90    2.55325
                                   0.85    2.22755
```

Data 7.3: Fourth Order Predictor-Corrector Methods

```
---Adams-Bashforth-Moulton 4th Order P-C Methods---
```

x	y	Error
0.10000E+01	0.36162E+01	
0.95000E+00	0.29927E+01	
0.90000E+00	0.25532E+01	
0.85000E+00	0.22275E+01	
0.80000E+00	0.19767E+01	0.71356E-03
0.75000E+00	0.17798E+01	0.30774E-03
0.70000E+00	0.16224E+01	0.12696E-03
0.65000E+00	0.14948E+01	0.72215E-04
0.60000E+00	0.13907E+01	0.38560E-04

Results 7.3: Fourth Order Predictor-Corrector Methods

As shown in Data 7.3, the four initial values are provided in the order dictated by the direction in which additional solutions are required. Thus with a negative h in this case, the initial conditions are given in descending values of x.

Results 7.3 give the approximate solution $y(0.6) = 1.3907$ to four decimal places. The user is invited to show that the Milne-Simpson method for the same problem (itype=1) gives $y(0.6) = 1.3912$. For comparison, the exact solution to this problem is given as 1.3917.

The error term provided in the output file is approximate, and is best used as a qualitative guide for comparing the different methods. The most convincing way of assessing accuracy is for the user to run the problem several times with decreasing values of the step length h. In this way the user can assess the sensitivity of results to the step size used. A disadvantage of multi-step methods as mentioned previously is that if the step length is changed, the initial values must also be adjusted to the new step length.

7.3.6   Stiff equations

Certain types of differential equations do not lend themselves to numerical solution by the techniques described so far in this chapter. Problems can occur if the solution to the system of equations contains components with widely different "time scales".

For example, the solution to a second order differential equation might be of the form

$$y(x) = C_1 e^{-x} + C_2 e^{-100x} \tag{7.82}$$

where the second term decays very much more rapidly than the first. Such a system of equation is said to be "stiff" and solutions are unreliable when treated by traditional methods. Any stepping method used to tackle such a problem numerically must have a step length small enough to account for the "fastest-changing" component of the solution, and this step size must be maintained even after the "fast" component has died out. As has been discussed previously, very small step lengths can have disadvantages from efficiency and accuracy consideration.

7.3.7   Error propagation and numerical stability

All numerical methods are approximate in nature, and some of the sources of these errors were discussed in Chapter 1.

When solving differential equations by repetitive algorithms, where errors introduced at one stage are carried on into subsequent calculations, the questions arises as to whether these errors will propagate with increased magnitude or remain within acceptable limits.

An "unstable" process in the context of numerical methods is one in which a small perturbation introduced into the calculation will grow spontaneously. Several sources of "instability" can occur.

One source of "instability" can be in the differential equation itself rather than the numerical method being used to solve it. For example, the second order equation,

$$y'' - 3y' - 10y = 0 \quad \text{with} \quad y(0) = 1 \quad \text{and} \quad y'(0) = -2 \qquad (7.83)$$

has the exact solution $y = e^{-2x}$ which decays as x increases.

If a small perturbation, ϵ, is introduced into one of the initial conditions, i.e.,

$$y(0) = 1 + \epsilon \qquad (7.84)$$

then the new solution is given by

$$y = e^{-2x} + \frac{\epsilon}{7}(2e^{5x} + 5e^{-2x}) \qquad (7.85)$$

which tends to infinity as x becomes very large.

The small perturbation has caused a huge change in the solution for large x, hence this differential equation is said to be unstable. When tackling "unstable" problems such as this by numerical methods, any of the sources of error described in Chapter 1 could contribute to an initial perturbation of this type and lead to completely erroneous solutions.

Another source of instability in the solution of differential equations can come from the difference formula itself. As mentioned earlier in this chapter, Simpson's corrector formula from equation (7.73) can lead to instability. It can be shown that this formula does not cope well with differential equations of the form

$$y' = cy \quad \text{where } c \text{ is negative} \qquad (7.86)$$

In these cases, a spurious solution causes the errors to grow exponentially. Spurious solutions can occur whenever the order of the difference formula is higher than that of the differential equation being solved.

In some methods the spurious solutions can cause problems, whereas in others their effects die away. For example, the Adams-Bashforth-Moulton family of predictor-corrector formulas do not suffer from instability problems.

7.3.8   Concluding remarks on initial value problems

No single method for obtaining numerical solutions to initial value problems can be recommended for all occasions. Fourth order methods are to be recommended on the grounds of accuracy without undue complexity. The question still remains as to whether to use one-step or predictor-corrector methods.

The following points should be considered before deciding on a particular method.

(a) Predictor-corrector methods are not self-starting, and must actually rely on a one-step method in order to generate enough points to get started. If a change in step size is made during the solution process, a temporary reversion to the one-step method is usually required. Changes in the step size are very easy to implement in a one-step method.

(b) One-step methods are of comparable accuracy to predictor-corrector methods of the same order. However, the predictor-corrector methods provide a simple error estimate at each step, enabling the step size to be adjusted for maximum accuracy and efficiency. No such estimate is usually available with a one-step method, hence the step size h is often made smaller than necessary to be conservative.

(c) To advance one step using the fourth order Runge-Kutta method requires four evaluations of the function $f(x, y)$ which are used just once and then discarded. To advance one step using a fourth order predictor-corrector method usually requires just one new function evaluation. In addition, function evaluations in a predictor-corrector method can be stored and used again, as only whole intervals of x are required.

The fourth order methods of Runge-Kutta and Adams-Bashforth-Moulton are the preferred one-step and predictor-corrector approaches. The fourth order Runge-Kutta method has the advantages of being simple and self-starting, and is probably to be recommended for most engineering applications.

Many methods exist for numerical solution of differential equations, including newer hybrid methods which take advantage of both the simplicity of the one-step methods and the error-estimates of the predictor-corrector methods. These refinements are only obtained at the cost of greater complexity and more function evaluations per step. The reader is referred to more advanced texts on numerical analysis to learn of such developments.

7.4   Boundary value problems

When we attempt to solve ordinary differential equations of second order
or higher with information provided at different values of the independent
variable, we must use different numerical methods from those described in
the previous section.

The initial value problems covered previously often involved "time" as the
independent variable, and solution techniques required us to "march" along
in steps until the required solution was reached. The domain of the solution
in such cases is not finite, because in principle we can step along indefinitely
in either the positive or negative direction.

Boundary value problems involve a finite solution domain, and solutions
are required within that domain. The independent variable in such problems
is usually a coordinate measuring distance in space. A typical second order
boundary value problem might be of the form

$$y'' = f(x, y, y') \quad \text{with} \quad y(A) = y_A \quad \text{and} \quad y(B) = y_B \qquad (7.87)$$

The domain of the solution (assuming $B > A$) is given by values of x in the
range $A \leq x \leq B$. We are interested in finding the values of y corresponding
to this range of x.

Numerical solution methods that will be considered in this chapter fall into
three categories

(a) Techniques that replace the original differential equation by its finite
difference equivalent.

(b) "Shooting methods", which attempt to replace the boundary value problem by an equivalent initial value problem.

(c) Methods of "weighted residuals", where a trial solution satisfying the
boundary conditions is guessed, and certain parameters within that solution
adjusted in order to minimise errors.

7.4.1   Finite difference methods

In this method, derivative terms in the differential equation are replaced
by finite difference approximations (see Chapter 5). As will be shown, if the
differential equation is linear this process leads to a system of linear simultaneous equations, which can be solved using the methods described in Chapter
2.

Initially, we need to define some finite difference approximations to regularly
encountered derivatives. Consider the solution curve in Figure 7.8 in which
the x-axis is subdivided into regular grid points, distance h apart.

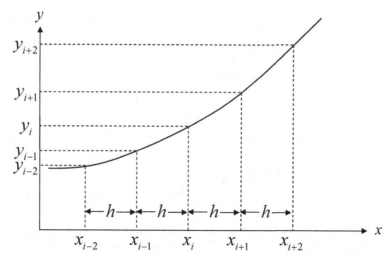

Figure 7.8: Finite difference grid points.

We may express the first derivative of y with respect to x evaluated at x_i, in any of the following ways

$$y_i' \approx \frac{y_{i+1} - y_i}{h} \qquad \text{"forward" difference} \qquad (7.88)$$

$$y_i' \approx \frac{y_i - y_{i-1}}{h} \qquad \text{"backward" difference} \qquad (7.89)$$

$$y_i' \approx \frac{y_{i+1} - y_{i-1}}{2h} \qquad \text{"central" difference} \qquad (7.90)$$

These finite difference formulas are approximating the first derivative by computing the slope of a straight line joining points in the vicinity of the function at x_i. The forward and backward formulas have bias, in that they include points only to the right or left of x_i. The central difference form takes equal account of points to the left and right of x_i.

Higher derivatives can also be approximated in this way. For example, the second derivative at x_i could be estimated by taking the backward difference of first derivatives as follows

$$y_i'' \approx \frac{y_i' - y_{i-1}'}{h} \qquad (7.91)$$

If we then substitute forward difference formulas for the first derivatives as

$$y_i' \approx \frac{y_{i+1} - y_i}{h} \text{ and } y_{i-1}' \approx \frac{y_i - y_{i-1}}{h} \qquad (7.92)$$

into equation (7.91), we can write a central difference formula for y_i'' as

$$y_i'' \approx \frac{y_{i-1} - 2y_i + y_{i+1}}{h^2} \qquad (7.93)$$

Similarly a central difference formula for the third derivative at x_i can be given by

$$y_i''' \approx \frac{y_i'' - y_{i-1}''}{h}$$

$$= \frac{-y_{i-2} + 2y_{i-1} - 2y_{i+1} + y_{i+2}}{2h^3} \qquad (7.94)$$

and for a fourth derivative at x_i by

$$y_i^{iv} \approx \frac{y_{i-1}'' - 2y_i'' + y_{i+1}''}{h^2}$$

$$= \frac{y_{i-2} - 4y_{i-1} + 6y_i - 4y_{i+1} + y_{i+2}}{h^4} \qquad (7.95)$$

Forward and backward difference versions of these higher derivatives are also readily obtained, and by including more points in the difference formulas, greater accuracy can be achieved. A summary of the coefficients for forward, central and backward difference formulas was presented in Section 5.3.

Backward difference formulae are simply mirror images of the forward difference formulas, except for odd numbered derivatives (i.e., y', y''', etc.) where the signs must be reversed.

For example, a four-point forward difference formula for a first derivative at x_0 including the dominant error term from Table 5.2 is

$$y_0' = \frac{1}{6h}(-11y_0 + 18y_1 - 9y_2 + 2y_3) - \frac{1}{4}h^3 y^{iv}(\xi) \qquad (7.96)$$

while the corresponding backward difference formula from Table 5.4 is

$$y_0' = \frac{1}{6h}(-2y_{-3} + 9y_{-2} - 18y_{-1} + 11y_0) + \frac{1}{4}h^3 y^{iv}(\xi) \qquad (7.97)$$

The solution to a "two-point" boundary value problem such as that given by equation (7.87), involves splitting the range of x for which a solution is required into n equal parts, each of width h. As shown in Figure 7.9 with $n = 4$, if the boundary conditions are given at, $x = A$ and $x = B$, we let $x_i = x_0 + ih$, $i = 1, 2, \ldots, n$, where $x_0 = A$ and $x_n = B$.

The differential equation is then written in finite difference form at each of the internal points $i = 1, 2, \ldots, n - 1$. If the differential equation is linear, this leads to $n - 1$ simultaneous linear equations in the unknown values y_i, $i = 1, 2, \ldots, n - 1$, where y_i represents the solution at x_i. The more subdivisions made, the greater the detail and accuracy of the solution but the more simultaneous equations that must be solved.

Example 7.11

Given $y'' = 3x + 4y$ with boundary conditions $y(0) = 0$ and $y(1) = 1$, solve the equation in the range $0 \leq x \leq 1$ by finite differences using $h = 0.2$.

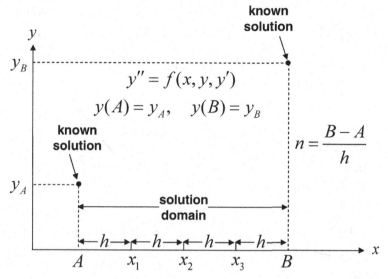

Figure 7.9: Two-point boundary value problem ($n = 4$).

Solution 7.11

Firstly we write the differential equation in finite difference form. From Tables 5.2-5.4 we have a choice of formulas for the required second derivative. It is usual to use the lowest order central difference form, unless there is a particular reason to use the less accurate forward or backward difference forms, hence from equation (7.93)

$$y_i'' \approx \frac{y_{i-1} - 2y_i + y_{i+1}}{h^2}$$

and the differential equation can be written as

$$\frac{1}{h^2}(y_{i-1} - 2y_i + y_{i+1}) = 3x_i + 4y_i$$

The solution domain is split into five equal strips, each of width $h = 0.2$ as shown in Figure 7.10.

The finite difference equation is then written at each of the grid points for which a solution is required. Known boundary conditions are introduced where needed leading to the four equations given in Table 7.1.

These (nonsymmetric) equations can be solved using any suitable program from Chapter 2. Table 7.2 compares the numerical solutions with the exact solutions given by

$$y = \frac{7(e^{2x} - e^{-2x})}{4(e^2 - e^{-2})} - \frac{3}{4}x$$

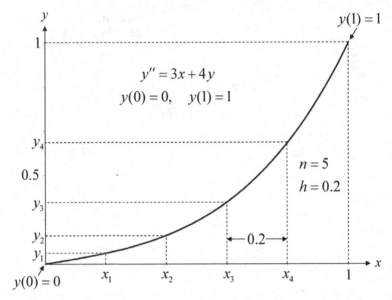

Figure 7.10: Finite difference subdivisions (Example 7.11).

TABLE 7.1: Difference equations from Example 7.11

i	x_i	Finite difference equation
1	0.2	$(y_2 - 2y_1 + 0)/0.04 = 0.6 + 4y_1$
2	0.4	$(y_3 - 2y_2 + y_1)/0.04 = 1.2 + 4y_2$
3	0.6	$(y_4 - 2y_3 + y_2)/0.04 = 1.8 + 4y_3$
4	0.8	$(1 - 2y_4 + y_3)/0.04 = 2.4 + 4y_4$

TABLE 7.2: Exact and finite difference solutions from Example 7.11

x	y_{exact}	y_{FD}
0.0	0.0	0.0
0.2	0.0482	0.0495
0.4	0.1285	0.1310
0.6	0.2783	0.2814
0.8	0.5472	0.5488
1.0	1.0	1.0

Clearly the accuracy of the finite difference solution could have been further improved by taking more subdivisions in the range $0 \le x \le 1$, at the expense of solving more equations. It should be noted that the equation coefficients in this class of problem have a narrow bandwidth (see Section 2.4).

Example 7.11 highlights a problem that would be encountered if one of the higher order finite difference representations of y'' had been used. For example, if we had used the five-point central different formula for y'' (see Table 5.3), values of y outside the solution domain would have been required in order to express y_1'' and y_4''. Further information would then be required in order to solve the system, because there would be more unknowns than equations.

The need to introduce points outside the solution domain, at least temporarily, is encountered quite frequently and can be resolved by incorporating the appropriate boundary conditions as shown in the next example.

Consider a boundary value problem where

$$y'' = f(x, y, y') \quad \text{with} \quad y(A) = y_A \quad \text{and} \quad y'(B) = y_B' \qquad (7.98)$$

In this case, $y(B)$ remains unknown, so the finite difference version of the differential equation will need to be written at $x = B$. Even when using the simplest central difference formula for y'' given by equation (7.93), the unknown y_5, corresponding to the "solution" at x_5, will be introduced as shown in Figure 7.11. In this case, we have 5 unknowns but only 4 equations.

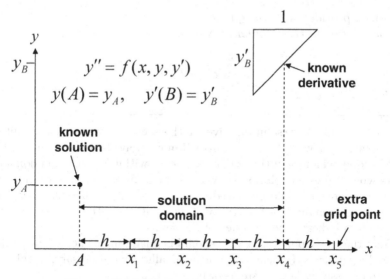

Figure 7.11: Derivative boundary condition.

The fifth equation comes from the derivative boundary condition which can also be written in finite difference form, i.e., using central differences

$$y_B' = \frac{y_5 - y_3}{2h} \qquad (7.99)$$

Example 7.12

Given $x^2 y'' + 2xy' - 2y = x^2$ with boundary conditions $y(1) = 1$ and $y'(1.5) = -1$. Solve the equation in the range $1 \leq x \leq 1.5$ by finite difference using $h = 0.1$.

Solution 7.12

The differential equation is written in finite difference form using central differences for both the derivative terms as follows

$$\frac{x_i^2}{h^2}(y_{i+1} - 2y_i + y_{i-1}) + \frac{x_i}{h}(y_{i+1} - y_{i-1}) - 2y_i = x_i^2$$

and applied at the five x-values for which solutions are required as shown in Figure 7.12.

The resulting equations are given in Table 7.3 and it can be noted from Figure 7.12 that a sixth unknown $y_6 = y(1.6)$ outside the solution domain has been introduced by the fifth equation. The derivative boundary condition in central difference form

$$\frac{y_6 - y_4}{0.2} = -1$$

is used to provide the sixth equation.

The solution to these six linear equations, together with the exact

$$y = \frac{135}{86}\frac{1}{x^2} - \frac{141}{172}x + \frac{1}{4}x^2$$

is given in Table 7.4.

The finite difference solutions given in this section have performed quite well with relatively few grid points. This will not always be the case, especially if quite sudden changes in the derivatives occur within the solution domain. In cases where no exact solution is available for comparison, it is recommended that the problem should be solved using two or three different gradations of the solution domain. The sensitivity of the solutions to the grid size parameter h will often indicate the accuracy of the solution.

From the user's point of view, h should be made small enough to enable the desired accuracy to be achieved, but no smaller than necessary, as this would lead to excessively large systems of banded equations.

7.4.2   Shooting methods

Shooting methods attempt to solve boundary value problems as if they were initial value problems. Consider the following second order equation

$$y'' = f(x, y, y') \quad \text{with} \quad y(A) = y_A \quad \text{and} \quad y(B) = y_B \tag{7.100}$$

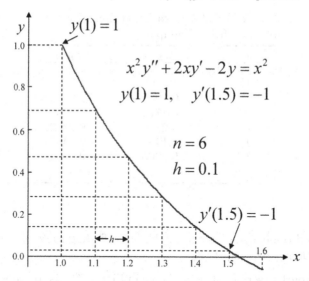

Figure 7.12: Finite difference subdivisions (Example 7.12).

TABLE 7.3: Difference equations from Example 7.12

i	x_i	Finite difference equation
1	1.1	$\dfrac{1.1^2}{0.1^2}(y_2 - 2y_1 + 1) + \dfrac{1.1}{0.1}(y_2 - 1) - 2y_1 = 1.1^2$
2	1.2	$\dfrac{1.2^2}{0.1^2}(y_3 - 2y_2 + y_1) + \dfrac{1.2}{0.1}(y_3 - y_1) - 2y_2 = 1.2^2$
3	1.3	$\dfrac{1.3^2}{0.1^2}(y_4 - 2y_3 + y_2) + \dfrac{1.3}{0.1}(y_4 - y_2) - 2y_3 = 1.3^2$
4	1.4	$\dfrac{1.4^2}{0.1^2}(y_5 - 2y_4 + y_3) + \dfrac{1.4}{0.1}(y_5 - y_3) - 2y_4 = 1.4^2$
5	1.5	$\dfrac{1.5^2}{0.1^2}(y_6 - 2y_5 + y_4) + \dfrac{1.5}{0.1}(y_6 - y_4) - 2y_5 = 1.5^2$

In a shooting method we will solve a sequence of initial value problems of the form

$$y'' = f(x, y, y') \quad \text{with} \quad y(A) = y_A \quad \text{and} \quad y_i'(A) = a_i \tag{7.101}$$

By varying the initial gradient a_i in a methodical way, we can eventually reach a solution at $x = B$ that is sufficiently close to the required boundary value y_B. There are several different strategies for converging on the required

TABLE 7.4: Exact and finite difference solutions from Example 7.12

x	y_{exact}	y_{FD}
1.0	1.0000	1.0000
1.1	0.7981	0.7000
1.2	0.4774	0.4797
1.3	0.2857	0.2902
1.4	0.1432	0.1487
1.5	0.0305	0.0379
1.6	-0.0584	-0.0513

solution, and these are similar to methods for finding roots of nonlinear algebraic equations (see Chapter 3).

The approach described here as shown in Figure 7.13 is to choose two initial gradients $y_0'(A) = a_0$ and $y_1'(A) = a_1$ which, following application of a one-step method for each, give values of $y_0(B) = y_0$ and $y_1(B) = y_1$ respectively that straddle the required boundary condition of y_B, i.e., $y_0 < y_B < y_1$ (or $y_1 < y_B < y_0$).

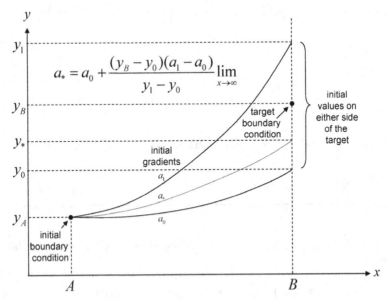

Figure 7.13: Shooting method for a nonlinear equation.

By linear interpolation, an improved estimate of the initial gradient is then given by

$$a_* = a_0 + \frac{(y_B - y_0)(a_1 - a_0)}{y_1 - y_0} \qquad (7.102)$$

which leads to $y(B) = y_*$.

One of the initial gradients is then replaced by a_* according to the following test

If $(y_* - y_B)(y_0 - y_B) > 0$

then overwrite y_0 by y_* and a_0 by a_* else

overwrite y_1 by y_* and a_1 by a_*

The goal throughout the iterative process is to retain one initial gradient that overestimates the target boundary condition, and one that underestimates it. As the iterations proceed, y_* tends to the target value and calculations are stopped when a convergence tolerance is satisfied according to the criterion

$$\left| \frac{y_* - y_B}{y_B} \right| < \text{tolerance} \qquad (7.103)$$

The iterative process is essentially the "False Position Method" described in Chapter 3.

Program 7.4: Shooting method for second order ODEs

```
PROGRAM nm74
!---Shooting Method for Second Order ODEs---
 IMPLICIT NONE; INTEGER,PARAMETER::iwp=SELECTED_REAL_KIND(15,300)
 INTEGER::i,iters,j,limit,nsteps
 REAL(iwp)::astar,a0(2),d6=6.0_iwp,h,k0(2),k1(2),k2(2),k3(2),    &
    tol,two=2.0_iwp,x,xa,xb,y(2),ya,yb,zero=0.0_iwp
 REAL(iwp),ALLOCATABLE::y0(:,:),ystar(:)
 OPEN(10,FILE='nm95.dat'); OPEN(11,FILE='nm95.res')
 READ(10,*)nsteps,xa,ya,xb,yb,a0,tol,limit
 ALLOCATE(y0(0:nsteps,2),ystar(0:nsteps))
 WRITE(11,'(A)')"---Shooting Method for Second Order ODEs---"
 h=(xb-xa)/nsteps
 DO j=1,2
   x=xa; y(1)=ya; y(2)=a0(j)
   DO i=0,nsteps
     y0(i,j)=y(1); k0=h*f74(x,y); k1=h*f74(x+h/two,y+k0/two)
     k2=h*f74(x+h/two,y+k1/two); k3=h*f74(x+h,y+k2)
     y=y+(k0+two*k1+two*k2+k3)/d6; x=x+h
   END DO
```

```
END DO
IF((y0(nsteps,1)-yb)*(y0(nsteps,2)-yb)>zero)                    &
  WRITE(11,'(A)')"Try new gradients....?"
iters=0
DO
  iters=iters+1
  astar=a0(1)+(yb-y0(nsteps,1))*(a0(2)-a0(1))/                    &
    (y0(nsteps,2)-y0(nsteps,1))
  x=xa; y(1)=ya; y(2)=astar
  DO i=0,nsteps
    ystar(i)=y(1)
    k0=h*f74(x,y); k1=h*f74(x+h/two,y+k0/two)
    k2=h*f74(x+h/two,y+k1/two); k3=h*f74(x+h,y+k2)
    y=y+(k0+two*k1+two*k2+k3)/d6; x=x+h
  END DO
  IF(ABS((ystar(nsteps)-yb)/yb)<tol)THEN
    WRITE(11,'(/,A,I2)')"       x               y"
    WRITE(11,'(2E13.5)')(xa+i*h,ystar(i),i=0,nsteps)
    WRITE(11,'(/,A,/,I5)')"Iterations to Convergence",iters
    EXIT
  END IF
  IF((ystar(nsteps)-yb)*(y0(nsteps,1)-yb)>zero)THEN
    y0(:,1)=ystar; a0(1)=astar; ELSE; y0(:,2)=ystar; a0(2)=astar
  END IF
END DO
CONTAINS

FUNCTION f74(x,y)
 IMPLICIT NONE
 REAL(iwp),INTENT(IN)::x,y(2)
 REAL(iwp)::f74(2)
 f74(1)=y(2)
 f74(2)=3._iwp*x**2+4._iwp*y(1)
! f74(1)=y(2)
! f74(2)=-2._iwp*y(2)**2/y(1)
 RETURN
END FUNCTION f74

END PROGRAM nm74
```

This program obtains numerical solutions to second order ODEs with two point boundary conditions of the form

$$y'' = f(x,y,y') \quad \text{with} \quad y(A) = y_A \quad \text{and} \quad y(B) = y_B$$

The second order equation must first be reduced to two first order equations

---

List 7.4:

Scalar integers:

i	simple counter
iters	iteration counter
j	simple counter
limit	iteration ceiling
nsteps	number of calculations steps

Scalar reals:

astar	interpolated initial gradient
d6	set to 6.0
h	calculation step length
tol	convergence tolerance
two	set to 2.0
x	independent variable
xa	x-coordinate of first boundary condition
xb	x-coordinate of second boundary condition
ya	y-coordinate of first boundary condition
yb	y-coordinate of second boundary condition
zero	set to 0.0

Real arrays:

a0	holds initial gradients a_0 and a_1
k0	estimate of Δy
k1	estimate of Δy
k2	estimate of Δy
k3	estimate of Δy
y	holds y and y' during one-step process

Dynamic real arrays:

y0	holds values of y_0 and y_1 on either side of target
ystar	holds interpolated values of y

---

as

$$y' = z \qquad \text{with} \quad y(A) = y_A$$

$$z' = f(x, y, z) \text{ with} \quad z(A) = a_i$$

where two values of the initial gradient a_0 and $a(1)$ must be provided by the user. The program then iteratively adjusts the initial gradient until the boundary condition at $y(B) = y_B$ is satisfied within a specified tolerance.

The first item to be read in is **nsteps**, the number of steps within the solution domain $A \le x \le B$. Since the boundaries are known in this problem, the step length is calculated internally as **h=(B-A)/nsteps**. The next data are the boundary conditions read as **xa, ya** at the beginning of the range, and **xb, yb** at the end. The two initial values of the gradient are then read

into the vector a0. For the algorithm to proceed, these must result in values
of y at $x = B$ that lie either side of the required boundary condition y_B. A
warning is printed if this is not the case. Data are completed by a convergence
tolerance tol and an iteration ceiling limit.

The first example to illustrate use of Program 7.4 is the linear equation

$$y'' = 3x^2 + 4y \quad \text{with} \quad y(0) = 0 \quad \text{and} \quad y(1) = 1$$

for which a solution is required in the range $0 \le x \le 1$.

The equation is decomposed into two first order equations (see Section 7.3.2)
as follows

$$y' = z \qquad \text{with} \quad y(0) = 0$$

$$z' = 3x^2 + 4y \text{ with } z(0) = a_i$$

and the right-hand-side functions entered into FUNCTION f74 as

```
f74(1)=y(2)
f74(2)=3._iwp*x**2+4._iwp*y(1)
```

Number of steps	nsteps	
	5	

Boundary conditions	(xa, ya)	(xb, yb)
	0.0 0.0	1.0 1.0

Initial gradients	a0(1)	a0(2)
	0.0	1.0

Convergence data	tol	limit
	0.0001	25

Data 7.4a: Shooting Method (first example)

```
---Shooting Method for Second Order ODEs---

        x                y
 0.00000E+00  0.00000E+00
 0.20000E+00  0.81315E-01
 0.40000E+00  0.18147E+00
 0.60000E+00  0.33130E+00
 0.80000E+00  0.57939E+00
 0.10000E+01  0.10000E+01
```

```
Iterations to Convergence
    1
```

Results 7.4a: Shooting Method (first example)

The input and output for this example are given in Data 7.4a and Results 7.4a. As shown in Data 7.4a, five steps ($h = 0.2$) are to be computed within the range and this is followed by the boundary conditions at the beginning and end of the range. The two initial gradients are read in as 0.0 and 1.0. For linear equations such as in this case, any two initial guesses will lead to a converged solution in a single iteration. If the equation is nonlinear however, convergence is slower and may fail entirely if the initial guesses lead to solutions on the same side of the target boundary condition. In this case the program prints a warning to the output file suggesting that a different pair of initial conditions should be tried. The data in this case are completed by a convergence tolerance of `tol=0.0001` and an iteration ceiling of `limit=25`. The iteration ceiling is there to avoid an "infinite loop" in the event of a data error or a poorly posed problem.

The output given in Results 7.4a is in good agreement with the exact solution. For example, the numerical solution at $x = 0.6$ is given as $y = 0.33130$ compared with the exact solution, which can be shown to be $y = 0.33123$ to five decimal places.

The second example to illustrate use of Program 7.4 is the nonlinear equation

$$y'' + \frac{2(y')^2}{y} = 0 \quad \text{with} \quad y(0) = 1 \quad \text{and} \quad y(1) = 2$$

for which a solution is required in the range $0 \le x \le 1$.

The equation is decomposed into two first order equations (see Section 7.3.2) as follows

$$y' = z \quad \text{with} \quad y(0) = 1$$
$$z' = \frac{-2z^2}{y} \quad \text{with} \quad z(0) = a_i$$

and the right-hand side functions entered into FUNCTION f74 as,

```
f74(1)=y(2)
f74(2)=-2._iwp*y(2)**2/y(1)
```

Number of steps   nsteps
                           10

Boundary conditions   (xa, ya)   (xb, yb)
                          0.0   1.0   1.0   2.0

```
Initial gradients        a0(1)    a0(2)
                          1.0      3.0


Convergence data         tol           limit
                         0.000001       25
```

Data 7.4b: Shooting Method (second example)

```
---Shooting Method for Second Order ODEs---


            x                y
    0.00000E+00   0.10000E+01
    0.10000E+00   0.11932E+01
    0.20000E+00   0.13386E+01
    0.30000E+00   0.14579E+01
    0.40000E+00   0.15604E+01
    0.50000E+00   0.16509E+01
    0.60000E+00   0.17324E+01
    0.70000E+00   0.18069E+01
    0.80000E+00   0.18757E+01
    0.90000E+00   0.19399E+01
    0.10000E+01   0.20000E+01


Iterations to Convergence
    7
```

Results 7.4b: Shooting Method (second example)

The input and output for this example are given in Data 7.4b and Results 7.4b. As shown in Data 7.4a, ten steps ($h = 0.1$) are to be computed within the range. The two initial gradients are read in as 1.0 and 3.0 which were shown by trial and error to give solutions on either side of the target boundary condition. A tighter convergence tolerance of tol=0.000001 is used in this example.

As shown in Results 7.4b, the shooting method took 7 iterations to converge to the required tolerance. The user is invited to demonstrate the good agreement between the numerical solution and the exact solution, which in this case is given by $y = \sqrt[3]{7x + 1}$.

7.4.3 Weighted residual methods

These methods have been the subject of a great deal of research in recent years. The methods can form the basis of the finite element method (see Chapter 8) which is now the most widely used numerical method for the solution of large boundary value problems.

The starting point for weighted residual methods is to guess a solution to the differential equation which satisfies the boundary conditions. This "trial solution" will contain certain parameters which can be adjusted to minimize the errors. Several different methods are available for minimizing the error or "residual", such that the trial solution is as close to the exact solution as possible.

Consider the second order boundary value problem

$$y'' = f(x, y, y') \quad \text{with} \quad y(A) = y_A \quad \text{and} \quad y(B) = y_B \tag{7.104}$$

The differential equation can be rearranged as

$$R = y'' - f(x, y, y') \tag{7.105}$$

where R is the "residual" of the equation. Only the exact solution $y(x)$ will satisfy the boundary conditions and cause R to equal zero for all x.

Consider a trial solution of the form

$$\tilde{y} = F(x) + \sum_{i=1}^{n} C_i \psi_i(x) \quad \text{with} \quad \tilde{y}(A) = y_A \quad \text{and} \quad \tilde{y}(B) = y_B \tag{7.106}$$

where $F(x)$ can equal zero if appropriate.

The trial solution must satisfy the boundary conditions of the differential equation, and is made up of trial functions $\psi_1(x)$, $\psi_2(x)$, etc., each of which is multiplied by an "undetermined parameter" C_1, C_2, etc. The trial solution from equation (7.106) is differentiated twice, and substituted into (7.105) to give

$$R = \tilde{y}'' - f(x, \tilde{y}, \tilde{y}') \tag{7.107}$$

The goal is now to minimize the residual R in some sense over the solution domain $A \leq x \leq B$ by selecting the "best" values of the undetermined parameters C_1, C_2, etc. Various methods for minimizing R will be described.

Clearly the choice of a suitable trial solution is crucial to the whole process. Ideally, the trial functions $\psi_1(x)$, $\psi_2(x)$, etc. will be simple polynomials such as 1, x, x^2, etc., although transcendental functions such as $\sin x$, $\ln x$, etc., can also be used. Sometimes the engineering application underlying the differential equations points to a "sensible" choice of trial functions.

From equation (7.106), there is no upper limit on n, the number of terms in the trial solution. In related methods, such as the finite element method (see Chapter 8), it is usual to split the solution domain into smaller pieces or elements, and use rather simple trial solutions over each. The general ethos of the finite element method is that the more terms included in the trial solution, the more accurate each element will be, and hence fewer will be needed over the solution domain to achieve a desired level of accuracy. A compromise is usually reached between complexity and repetition.

In this chapter, we will only consider the weighted residual method applied to a single trial solution over the whole solution domain.

Example 7.13

Given the boundary value problem

$$y'' = 3x + 4y \quad \text{with} \quad y(0) = 0 \quad \text{and} \quad y(1) = 1$$

obtain an expression for the Residual using a trial solution with one undetermined parameter.

Solution 7.13

In order to find a trial solution which satisfies the boundary condition, we could derive a Lagrangian polynomial (see Section 5.2.1) which passes through the points

x	y
0	0
1/2	a
1	1

We have introduced a variable a at the midpoint of the range that will be related to our single undetermined parameter. The resulting Lagrangian polynomial is second order and of the form

$$\tilde{y} = x^2(2 - 4a) - x(1 - 4a)$$

which can be arranged in the form given by equation (7.106) as

$$\tilde{y} = F(x) + C_1\psi_1(x)$$

$$\text{where} \quad F(x) = 2x^2 - x$$
$$C_1 = -4a$$
$$\psi_1(x) = x^2 - x$$

The residual of the original differential equation is given by

$$R = \tilde{y}'' - 3x - 4\tilde{y}$$

Differentiation of the trial solution twice gives

$$\tilde{y}' = 2x(2 - 4a) - (1 - 4a)$$
$$\tilde{y}'' = 2(2 - 4a)$$

which can be substituted into the expression for the residual. In this example there is no first derivative, but after rearrangement and substitution for C_1 we get

$$R = -4x^2(2 + C_1) + x(1 + 4C_1) + 2(2 + C_1)$$

Example 7.14

Obtain a trial solution to the problem of Example 7.13 involving two undetermined parameters.

Solution 7.14

Derive a Lagrangian polynomial which passes through the following points

x	y
0	0
1/3	a
2/3	b
1	1

We have introduced two variables a and b at the one-third points of the range that will be related to our two undetermined parameters. The resulting Lagrangian polynomial is cubic and of the form

$$\tilde{y} = \frac{1}{2}[x^3(27a - 27b + 9) - x^2(45a - 36b + 9) + x(18a - 9b + 2)]$$

which can be arranged in the form given by equation (7.106) as

$$\tilde{y} = F(x) + C_1\psi_1(x) + C_2\psi_2(x)$$

$$\text{where } F(x) = \frac{1}{2}(9x^3 - 9x^2 + 2x)$$

$$C_1 = \frac{27}{2}(a - b)$$

$$\psi_1(x) = x^3 - x^2$$

$$C_2 = -\frac{9}{2}(2a - b)$$

$$\psi_2(x) = x^2 - x$$

Four of the best-known methods for minimizing the residual R are now covered briefly. In each case, the trial solution and residual from Example 7.13, i.e.,

$$\tilde{y} = (2x^2 - x) + C_1(x^2 - x) \tag{7.108}$$

and

$$R = -4x^2(2 + C_1) + x(1 + 4C_1) + 2(2 + C_1) \tag{7.109}$$

are operated on.

7.4.3.1   Collocation

In this method, the residual is made equal to zero at as many points in the solution domain as there are unknown C_i. Hence, if we have n undetermined parameters, we write

$$R(x_i) = 0 \quad \text{for} \quad i = 1, 2, \ldots, n \qquad (7.110)$$

where $x_1, x_2, \ldots, x_n$ are n "collocation points" within the solution domain. This leads to n simultaneous equations in the unknown C_i.

Example 7.15

Use collocation to find a solution to the differential equation of Example 7.13.

Solution 7.15

Using the trial solution and residual from Solution 7.13 we have one undetermined parameter C_1. It seems logical to collocate at the midpoint of the solution domain, $x = 0.5$, hence from equation (7.109)

$$R(0.5) = -4(0.5)^2(2 + C_1) + 0.5(1 + 4C_1) + 2(2 + C_1) = 0$$

which can be solved to give $C_1 = -\dfrac{5}{6}$.

The trial solution is therefore given by

$$\tilde{y} = (2x^2 - x) - \frac{5}{6}(x^2 - x)$$

which can be simplified as

$$\tilde{y} = \frac{x}{6}(7x - 1)$$

Note: If our trial solution had two undetermined parameters such as that given in Example 7.14, we would have needed to collocate at two locations such as $x = \frac{1}{3}$ and $x = \frac{2}{3}$ to give two equations in the unknown C_1 and C_2.

7.4.3.2   Subdomain

In this method, the solution domain is split into as many "subdomains" or parts of the domain as there are unknown C_i. We then integrate the residual over each subdomain and set the result to zero. Hence, for n undetermined parameters we will integrate over n subdomains as follows

$$\int_{x_0}^{x_1} R\,dx = 0, \quad \int_{x_1}^{x_2} R\,dx = 0, \ldots, \int_{x_{n-1}}^{x_n} R\,dx = 0 \qquad (7.111)$$

where the solution domain is in the range $x_0 \leq x \leq x_n$ and $x_1, x_2, \ldots, x_{n-1}$ are points within that range. This leads to n simultaneous equations in the unknown C_i.

Example 7.16

Use the subdomain method to find a solution to the differential equation of Example 7.13.

Solution 7.16

Using the trial solution and residual from Solution 7.13 we have one undetermined parameter C_1. In this case we require only one "subdomain" which is the solution domain itself, hence integrating R from equation (7.109) we get

$$\int_{x_0}^{x_1} R\,dx = \int_0^1 \left[-4x^2(2+C_1) + x(1+4C_1) + 2(2+C_1)\right]dx$$

$$= \left[-\frac{4}{3}x^3(2+C_1) + \frac{1}{2}x^2(1+4C_1) + 2x(2+C_1)\right]_0^1 = 0$$

which can be solved to give $C_1 = -\dfrac{11}{16}$.

The trial solution is therefore given by

$$\tilde{y} = (2x^2 - x) - \frac{11}{16}(x^2 - x)$$

which can be simplified as

$$\tilde{y} = \frac{x}{16}(21x - 5)$$

Note: If our trial solution had two undetermined parameters such as that given in Example 7.14, we would have needed to integrate over two subdomains such as

$$\int_0^{0.5} R\,dx = 0 \quad \text{and} \quad \int_{0.5}^1 R\,dx = 0$$

to give two equations in the unknown C_1 and C_2.

7.4.3.3   Least Squares

In this method we integrate the square of the residual over the full solution domain. We then differentiate with respect to each of the undetermined parameters in turn, and set the result to zero. This has the effect of minimizing the integral of the square of the residual.

Noting that in general

$$\frac{\partial}{\partial C_i}\int_A^B R^2\,dx = 2\int_A^B R\frac{\partial R}{\partial C_i}\,dx \tag{7.112}$$

where the solution domain is $A \le x \le B$.

For n undetermined parameters, we can therefore write

$$\int_A^B R\frac{\partial R}{\partial C_1}\,dx = 0, \int_A^B R\frac{\partial R}{\partial C_2}\,dx = 0, \ldots \int_A^B R\frac{\partial R}{\partial C_n}\,dx = 0 \qquad (7.113)$$

leading to n simultaneous equations in the unknown C_i.

Example 7.17

Use the Least Squares method to find a solution to the differential equation of Example 7.13.

Solution 7.17

Using the trial solution and residual from Solution 7.13 we have one undetermined parameter C_1, hence from equation (7.113) write

$$\int_0^1 R\frac{\partial R}{\partial C_1}\,dx = \int_0^1 [-4x^2(2+C_1) + x(1+4C_1) + 2(2+C_1)][-4x^2 + 4x + 2]\,dx$$

$$= \left[\frac{16}{5}(2+C_1)x^5 - (9+8C_1)x^4 - \frac{28}{3}x^3\right.$$

$$\left. +(9+8C_1)x^2 + 4(2+C_1)x\right]_0^1 = 0$$

which can be solved to give $C_1 = -\frac{19}{27}$.

The trial solution is therefore given by

$$\tilde{y} = (2x^2 - x) - \frac{19}{27}(x^2 - x)$$

which can be simplified as

$$\tilde{y} = \frac{x}{27}(35x - 8)$$

Note: If our trial solution had two undetermined parameters such as that given in Example 7.14, we would have written

$$\int_0^1 R\frac{\partial R}{\partial C_1}\,dx = 0 \quad \text{and} \quad \int_0^1 R\frac{\partial R}{\partial C_2}\,dx = 0$$

to give two equations in the unknown C_1 and C_2.

7.4.3.4   Galerkin's method

In this method, which is the most popular in finite element applications, we "weight" the residual by the trial functions and set the integrals to zero.

Hence for n undetermined parameters, we have in general

$$\int_A^B R\psi_1(x)\,dx = 0, \int_A^B R\psi_2(x)\,dx = 0, \ldots, \int_A^B R\psi_n(x)\,dx = 0 \quad (7.114)$$

where the solution domain is $A \leq x \leq B$.

This leads to n simultaneous equations in the unknown C_i.

Example 7.18

Use Galerkin's method to find a solution to the differential equation of Example 7.13.

Solution 7.18

Using the trial solution and residual from Solution 7.13 we have one undetermined parameter C_1, hence from equation (7.114) write

$$\int_0^1 R\psi_1(x)\,dx = \int_0^1 [-4x^2(2+C_1) + x(1+4C_1) + 2(2+C_1)][x^2 - x]\,dx$$

$$= \left[-\frac{4}{5}(2+C_1)x^5 + \frac{1}{4}(9+8C_1)x^4 \right.$$

$$\left. + \frac{1}{3}(3-2C_1)x^3 - (2+C_1)x^2 \right]_0^1 = 0$$

which can be solved to give $C_1 = -\frac{3}{4}$.

The trial solution is therefore given by

$$\tilde{y} = (2x^2 - x) - \frac{3}{4}(x^2 - x)$$

which can be simplified as

$$\tilde{y} = \frac{x}{4}(5x - 1)$$

Note: If our trial solution had two undetermined parameters such as that given in Example 7.14, we would have written

$$\int_0^1 R\psi_1(x)\,dx = 0 \quad \text{and} \quad \int_0^1 R\psi_2(x)\,dx = 0$$

to give two equations in the unknown C_1 and C_2.

7.4.3.5 Concluding remarks on weighted residual examples

The differential equation with boundary conditions

$$y'' = 3x + 4y \quad \text{with} \quad y(0) = 0 \quad \text{and} \quad y(1) = 1$$

has been solved using a trial solution with just one undetermined parameter. Four different weighted residual methods were demonstrated, and a summary of those results, together with the exact solution are given below

$$\text{Collocation} \quad \tilde{y} = \frac{x}{6}(7x - 1)$$

$$\text{Subdomain} \quad \tilde{y} = \frac{x}{16}(21x - 5)$$

$$\text{Least Squares} \quad \tilde{y} = \frac{x}{27}(35x - 8) \qquad (7.115)$$

$$\text{Galerkin} \quad \tilde{y} = \frac{x}{4}(5x - 1)$$

$$\text{Exact} \quad y = \frac{7(e^{2x} - e^{-2x})}{4(e^2 - e^{-2})} - \frac{3}{4}x$$

The error associated with each of the trial solutions is shown in Figure 7.14. It is clear that, for this example, no single method emerges as the "best." Indeed each of the methods is the "most accurate" at some point within the solution domain $0 \le x \le 1$.

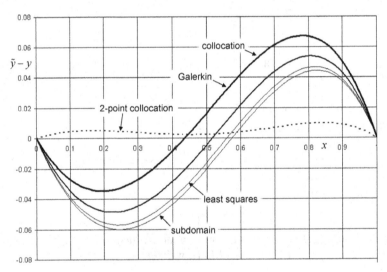

Figure 7.14: Error due to various weighted residual methods.

For simple problems such as the one demonstrated here, the Collocation method is the easiest to apply, because it does not require any integration. For finite element methods however, the Galerkin method is most often employed, because it can be shown that use of the trial functions to weight the residual leads to desirable properties such as symmetry of simultaneous equation coefficient matrices.

For comparison, the trial solution with two undetermined parameters developed in Example 7.14 has been derived using two-point collocation. It is left to the reader to show that the resulting trial solution becomes

$$\tilde{y} = \frac{1}{2}(9x^3 - 9x^2 + 2x) - \frac{108}{31}(x^3 - x^2) + \frac{576}{806}(x^2 - x) \qquad (7.116)$$

leading to considerably reduced errors, also shown in Figure 7.14.

All the four methods described in this section for optimizing the accuracy of a trial solution employ the philosophy that weighted averages of the residual should vanish. The difference between methods, therefore, lies only in the way the residual is weighted. In general, if we have n undetermined parameters, we set

$$\int_A^B W_i(x)R\,dx = 0 \quad \text{for} \quad i = 1, 2, \ldots, n \qquad (7.117)$$

where the solution domain is $A \le x \le B$.

In the previous examples with one undetermined parameter, the weighting function W_1 took the following form over the solution domain of $0 \le x \le 1$ for each of the methods

$$\text{Collocation} \begin{cases} W_1(x) = 0 & \text{for } x \neq \frac{1}{2} \\ W_1(x) = 1 & \text{for } x = \frac{1}{2} \end{cases}$$

$$\text{Subdomain } W_1(x) = 1 \qquad (7.118)$$

$$\text{Least Squares } W_1(x) = -4x^2 + 4x + 2$$

$$\text{Galerkin } W_1(x) = x^2 - x$$

These weighting functions are summarized graphically in Figure 7.15.

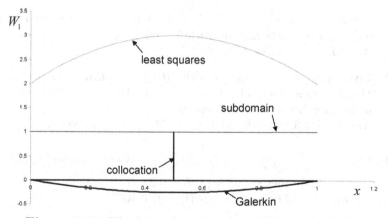

Figure 7.15: Weighting functions from different methods.

7.5   Exercises

1. Given $y' = xy \sin y$ with $y(1) = 1$, estimate $y(1.5)$ using the Euler method with (a) $h = 0.25$, (b) $h = 0.1$.
   Answer: (a) 1.5643, (b) 1.6963

2. Given $y' = (x^2 + y^2)/y$ with $y(1) = 1$, estimate $y(0.5)$ using the Modified Euler method with (a) $h = -0.1$, (b) $h = -0.05$.
   Answer: (a) 0.1869, (b) 0.1923 (Exact: 0.1939)

3. Rework exercise 2 using the Mid-Point rule.
   Answer: (a) 0.1952, (b) 0.1943

4. Rework exercise 2 using the fourth order Runge-Kutta method with two steps of $h = -0.25$.
   Answer: 0.1941

5. Use the Euler method in the following cases

   (a) $y' = y$ with $y(0) = 1$ find $y(0.2)$ using $h = 0.05$
   (b) $y' = x + y$ with $y(0) = 0$ find $y(0.5)$ using $h = 0.1$
   (c) $y' = 3x - y$ with $y(1) = 0$ find $y(0.5)$ using $h = -0.1$

   Answer: (a) 1.2155 (Exact: 1.2214), (b) 0.1105 (Exact: 0.1487), (c) -1.5 (Exact: -1.5)

6. Since the equations in exercise 5 are linear, rework them using the "θ-method" with $h = 0.5$.
   Answer: (a) 1.2215, (b) 0.1494, (c) -1.5

7. Given $y' = (x + y)^2$ with $y(0) = 1$, estimate $y(0.2)$ using the fourth order Runge-Kutta method with $h = 0.2$
   Answer: 1.3085 (Exact: 1.3085)

8. Given $y' = x + y + xy + 1$ with $y(0) = 2$, estimate $y(0.4)$ using the Modified Euler method with (a) $h = 0.1$, (b) $h = 0.2$, and use these to make a further improved estimate.
   Answer: (a) 3.8428, (b) 3.8277, 3.8478 (Exact: 3.8482)

9. Given $\dfrac{y''}{[1 + (y')^2]^{3/2}} = 1 - x$ with $y(0) = y'(0) = 0$, estimate $y(0.5)$ using the Mid-Point method with (a) $h = 0.25$, (b) $h = 0.05$.
   Answer: (a) 0.1124, (b) 0.1087 (Exact: 0.1085)

10. Given $y'' + y + (y')^2 = 0$ with $y(0) = -1$ and $y'(0) = 1$, estimate $y(0.5)$ and $y'(0.5)$ using the fourth order Runge-Kutta method with $h = 0.25$.
    Answer: -0.5164, 0.9090

11. Given $y'' - 2y' + y = 2x$ with $y(1) = 0$ and $y'(1) = 0$, estimate $y(0.4)$ using the fourth order Runge-Kutta method with $h = -0.15$.
    Answer: 0.1900 (Exact: 0.2020)

12. The angular acceleration of a pendulum is governed by the equation, $\ddot{\theta} = -\dfrac{g}{L}\sin\theta$ with $\theta(0) = \pi/4$ and $\dot{\theta}(0) = 0$. If $g = 9.91\mathrm{m/s^2}$ and $L = 10\mathrm{m}$ use the fourth order Runge-Kutta method with $h = 0.25$ to estimate the angular position of the pendulum after 0.5 seconds.
    Answer: 0.7000

13. Given $y'' + y = x$ with $y(0) = 1$ and $y'(0) = 0$, estimate $y(0.5)$ using the fourth order Runge-Kutta method with $h = 0.25$.
    Answer: 0.8982 (Exact: 0.8982)

14. Given $y'' = -\dfrac{x}{y'}$ with $y(0) = y'(0) = 1$, estimate $y'(0.2)$ using the Modified Euler method with $h = 0.05$.
    Answer: 0.9798 (Exact: 0.9798)

15. Given $y'' - 2xy' + (3 + x^2)y = 0$ with $y(0) = 2$ and $y'(0) = 0$, estimate $y(0.1)$ using a single step of the fourth order Runge-Kutta method.
    Answer: 1.9700 (Exact: 1.9700)

16. The differential equation $y' = \sec x + y\tan x$ has the following solutions in the range $0 \le x \le 0.5$.

x	y
0.0	0.0
0.1	0.1005
0.2	0.2041
0.3	0.3140
0.4	0.4343
0.5	0.5697

    Continue the solution to estimate $y(0.7)$ using the Adams-Bashforth-Moulton predictor-corrector method.
    Answer: 0.9153 (Exact: 0.9152)

17. Rework exercise 16 using the Milne-Simpson predictor-corrector method.
    Answer: 0.9152

18. Use polynomial substitution to derive the weighting coefficients of the predictor formula

$$y_{i+1} = y_{i-3} + h \sum_{j=i-2}^{i} w_j y'_j$$

    where the sampling points are equally spaced at h apart, and use the formula to estimate $y(0.4)$ given the differential equation $y' = \tan(xy)$, with initial values

x	y
0.0	0.0000
0.1	0.1005
0.2	0.1020
0.3	0.1046

Answer: $w_{i-2} = \dfrac{8}{3}$, $w_{i-1} = -\dfrac{4}{3}$, $w_i = \dfrac{8}{3}$, 0.1083

19. Given the boundary value problem $y'' = \dfrac{2}{x}y' + \dfrac{2}{x^2}y - \sin x$ with $y(1) = 1$ and $y(2) = 2$, use a shooting method with $h = 0.1$ to estimate $y(1.5)$. *Answer:* 1.3092 (Exact: 1.3092)

20. Given the nonlinear boundary value problem $2yy'' - (y')^2 + 4y^2 = 0$ with $y(\dfrac{\pi}{6}) = 0.75$ and $y(\dfrac{5\pi}{12}) = 0.0670$, try initial gradients $y'(\dfrac{\pi}{6}) = -0.5$ and -0.9 to estimate $y(\dfrac{\pi}{3})$ using a step length of $h = \dfrac{\pi}{12}$. *Answer:* 0.2514 (Exact: 0.25)

21. Solve the following problems using a shooting method

    (a) Given $y'' = 6y^2$ with $y(1) = 1$ and $y(2) = 0.25$, estimate $y(1.5)$ using $h = 0.1$. Try initial gradients $y'(1) = -1.8$ and -2.1.

    (b) Given $y'' = -e^{-2y}$ with $y(1) = 0$ and $y(1.5) = 0.4055$, estimate $y(1.25)$ using $h = 0.05$. Try initial gradients $y'(1) = 0.8$ and 1.2. *Answer:* (a) 0.4445 (Exact: 0.4444), (b) 0.2232 (Exact: 0.2232)

22. A cantilever of unit length and flexural rigidity supports a unit load at its free end. Use a finite difference scheme with a step length of (a) 0.25 and (b) 0.125 to estimate the free end deflection. The governing equation is $w'' = M(x) = 1 - x$ where x is the distance along the beam from the supported end, $M(x)$ is the bending moment at that section and w is the beam deflection. Note: Use a program from Chapter 2 to solve the equations. *Answer:* (a) 0.344, (b) 0.336 (Exact: 0.3333)

23. A simply supported beam of length L $(0 \le x \le L)$ and flexural rigidity EI rests on an elastic foundation of modulus k. The beam supports a uniform load of q. If w is the deflection of the beam, the governing equation in dimensionless form is given by

$$\frac{d^4\phi}{dz^4} + K\phi = 1 \text{ with } \phi(0) = \phi(1) = 0 \text{ and } \phi''(0) = \phi''(1) = 0$$

where $\phi = \dfrac{w\,EI}{qL^4}$, $z = \dfrac{x}{L}$ and $K = \dfrac{kL^4}{EI}$

Using a finite difference scheme with a dimensionless step length of $h = 0.5$ show that the dimensionless midpoint deflection is given as $\phi(0.5) = \dfrac{1}{64 + K}$. If the elastic foundation is removed by putting $K = 0$, what is the percentage error of the finite difference solution if the exact displacement in this case is given as $\phi(0.5) = \dfrac{5}{384}$?

Answer: 20%

24. Rework exercise 23 to obtain the percentage error using a smaller step length of $h = 0.25$.

    Answer: 5%

25. Given $y'' + y = 2$ with $y(0) = 1$ and $y'(1) + y(1) = 3.682941$, estimate y at intervals of 0.2 in the range $0 \le x \le 1$ using a finite difference scheme. Note: Use a program from Chapter 2 to solve the equations. *Answer:* $y(0.2) = 1.220$, $y(0.4) = 1.471$, $y(0.6) = 1.743$, $y(0.8) = 2.2025$, $y(1.0) = 2.307$ (Exact: $y(0.2) = 1.219$, $y(0.4) = 1.468$, $y(0.6) = 1.739$, $y(0.8) = 2.021$, $y(1.0) = 2.301$)

26. Estimate the value of the cantilever tip deflection from Exercise 22 using a method of weighted residuals with the trial solution, $\tilde{w} = C_1 x^2 + C_2 x^3$. *Answer:* All methods give the exact solution of 0.3333

27. Given $y'' + \pi^2 y = x$ with $y(0) = 1$ and $y(1) = -0.8987$, estimate $y(0.5)$ using the trial solution

$$y = 1 - 1.8987x + \sum_{i=1}^{2} C_i x^i (1 - x)$$

    with (a) Collocation and (b) Galerkin.
    Answer: (a) 0.0508, (b) 0.0495

28. Apply the Galerkin method to

$$y'' + \lambda y = 0 \text{ with } y(0) = y(1) = 0$$

    using the trial solution

$$y = C_1 x (1 - x) + C_2 x^2 (1 - x)^2$$

    and hence estimate the smallest eigenvalue λ.
    Answer: 9.86975

29. Rework Exercise 28 using finite differences with a step length of (a) 0.3333 and (b) 0.25. It may be assumed that y will be symmetrical about the point $x = 0.5$.
    Answer: (a) 9, (b) 9.37

30. The Lorenz equations originating from models of atmospheric physics are given as follows:

$$\frac{dx}{dt} = -10x + 10y$$

$$\frac{dy}{dt} = 28x - y - xz$$

$$\frac{dz}{dt} = -2.666667z + xy$$

with initial conditions $x(0) = y(0) = z(0) = 5$.

Use one step of the Modified Euler Method to estimate the values of x, y and z when $t = 0.1$.
Answer: $x(0.1) = 10.5$, $y(0.1) = 15.1583$, $z(0.1) = 8.7611$

31. Given the differential equations and initial conditions

$$\dot{x} = x - y + 2t - t^2 - t^3 \quad x(0) = 1$$
$$\dot{y} = x + y - 4t^2 + t^3 \qquad y(0) = 0$$

use the 4th order Runge-Kutta method ($h = 0.5$) to estimate $x(0.5)$ and $y(0.5)$.
Answer: $x(0.5) = 1.700$, $y(0.5) = 0.665$

32. A damped oscillator is governed by the differential equation:

$$2y'' + 5y' + 1600y = 0$$

with initial conditions $y(0) = -0.025$ and $y'(0) = 0$. Use the Mid-Point method with a time step of $\Delta t = 0.02$ to estimate the time taken for the oscillator to first pass through its equilibrium position corresponding to $y = 0$.
Answer: $t = 0.054$ by linear interpolation

33. A typical "predator-prey" model is based on the pair of differential equations

$$\frac{dx}{dt} = ax - bxy$$

$$\frac{dy}{dt} = -cy + dxy$$

Using the parameters, $a = 1.2$, $b = 0.6$, $c = 0.8$, and $d = 0.3$, with initial conditions $x(0) = 2$ and $y(0) = 1$, use one step of the Runge-Kutta 4th order method to estimate $x(0.5)$ and $y(0.5)$.
Answer: $x(0.5) = 2.7263$, $y(0.5) = 0.9525$

34. A predictor/corrector method has been proposed of the form

$$y_{i+1} = y_i + \frac{h}{12}\left(5y'_{i-2} - 16y'_{i-1} + 23y'_i\right)$$

$$y_{i+1} = y_i + w_{i-1}y'_{i-1} + w_i y'_i + w_{i+1}y'_{i+1}$$

Use polynomial substitution to show that the weights in the corrector formula are given by $w_{i-1} = -\frac{h}{12}$, $w_i = \frac{8h}{12}$ and $w_{i+1} = \frac{5h}{12}$, and then use the formulas to estimate $y(0.7)$, given the differential equation $y' = \sec x + y \tan x$ with initial values:

x	y
0.1	0.1005
0.3	0.3140
0.5	0.5697

Answer: 0.9166

35. Given the differential equation

$$\frac{dx}{dt} - 2t^2x^2 - x = 0$$

with initial conditions:

t	x
0.00	0.5000
0.25	0.6461
0.50	0.8779
0.75	1.4100

use the Milne/Simpson predictor-corrector method to estimate $x(1)$.
Answer: 3.9222 (Exact: 4.8239)

Chapter 8

Introduction to Partial Differential Equations

8.1 Introduction

A partial differential equation (PDE) contains derivatives involving two or more independent variables. This is in contrast to ordinary differential equations (ODE) as described in Chapter 7, which involve only one independent variable.

Many phenomena in engineering and science are described by PDEs. For example, a dependent variable, such as a pressure or a temperature, may vary as a function of time (t) and space (x, y, z).

Two of the best known numerical methods for solving PDEs are the Finite Difference and Finite Element methods, both of which will be covered in this chapter. Nothing more than an introductory treatment is attempted here, since many more advanced texts are devoted to this topic, including one by the authors themselves on the Finite Element Method (Smith and Griffiths 2004). This and other references on the subject are included in the Bibliography at the end of the text.

The aim of this chapter is to familiarize the student with some important classes of PDE, and to give insight into the types of physical phenomena they describe. Techniques for solving these problems will then be described through simple examples.

8.2 Definitions and types of PDE

Consider the following second order PDE in two independent variables

$$a\frac{\partial^2 u}{\partial x^2} + b\frac{\partial^2 u}{\partial x \partial y} + c\frac{\partial^2 u}{\partial y^2} + d\frac{\partial u}{\partial x} + e\frac{\partial u}{\partial y} + fu + g = 0 \qquad (8.1)$$

Note the presence of "mixed" derivatives such as that associated with the coefficient b.

If $a, b, c, \ldots, g$ are functions of x and y only, the equation is "linear", but if these coefficients contain u or its derivatives, the equation is "nonlinear".

The degree of a PDE is the power to which the highest derivative is raised, thus equation (8.1) is first degree. Only first degree equations will be considered in this chapter.

A regularly encountered shorthand notation in the study of second order PDEs is the "Laplace operator" ∇^2 where

$$\nabla^2 u = \frac{\partial^2 u}{\partial x^2} + \frac{\partial^2 u}{\partial y^2} \qquad \text{in 2-dimensions, and} \qquad (8.2)$$

$$\nabla^2 u = \frac{\partial^2 u}{\partial x^2} + \frac{\partial^2 u}{\partial y^2} + \frac{\partial^2 u}{\partial z^2} \quad \text{in 3-dimensions} \qquad (8.3)$$

Another is the "Biharmonic operator" ∇^4 where

$$\nabla^4 u = \frac{\partial^4 u}{\partial x^4} + 2\frac{\partial^4 u}{\partial x^2 \partial y^2} + \frac{\partial^4 u}{\partial y^4} \quad \text{in 2-dimensions} \qquad (8.4)$$

Table 8.1 summarizes some commonly encountered PDEs and the types of engineering applications they represent.

TABLE 8.1: Common types of PDEs in engineering analysis

Name	Equation	Application
Laplace's Eq.	$\nabla^2 u = 0$	Steady heat/fluid flow
Poisson's Eq.	$\nabla^2 u = f(x, y)$	Steady heat/fluid flow with sources or sinks, torsion
Diffusion Eq.	$\nabla^2 u = c\dfrac{\partial u}{\partial t}$	Transient heat/fluid flow
Wave Eq.	$\nabla^2 u = \dfrac{1}{c^2}\dfrac{\partial u}{\partial t}$	Vibration, wave propagation
Biharmonic Eq.	$\nabla^4 u = f(x, y)$	Deformation of thin plates

8.3 First order equations

Although the majority of engineering applications involve second order PDEs, we start with a consideration of first order equations, as this will lead to a convenient introduction to the "method of characteristics".

Consider the first order equation

$$a\frac{\partial u}{\partial x} + b\frac{\partial u}{\partial y} = c \tag{8.5}$$

where a, b and c may be functions of x, y and u, but not derivatives of u.

The following substitutions can be made to simplify the algebra

$$p = \frac{\partial u}{\partial x}$$

$$q = \frac{\partial u}{\partial y} \tag{8.6}$$

hence $\qquad ap + bq = c \tag{8.7}$

A "solution" to equation (8.5) will be an estimate of the value of u at any point within the x,y-plane. In order to find values of u in this solution domain, some "initial conditions" will be required.

As shown in Figure 8.1, let initial values of u be known along the line $\mathcal{I}$ in the solution domain. We now consider an arbitrary line $\mathcal{C}$ which intersects the initial line $\mathcal{I}$.

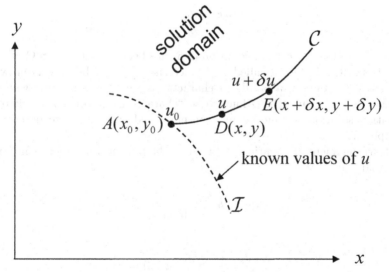

Figure 8.1: Solution domain and "initial conditions" line $\mathcal{I}$.

Consider a small change in u along the line $\mathcal{C}$ between points D and E.

This leads to the equation

$$\delta u = \frac{\partial u}{\partial x}\delta x + \frac{\partial u}{\partial y}\delta y \tag{8.8}$$

which in the limit as $\delta x \rightarrow 0$ and $\delta y \rightarrow 0$ becomes

$$du = \frac{\partial u}{\partial x}dx + \frac{\partial u}{\partial y}dy \tag{8.9}$$

or, from equations (8.6)

$$du = p\,dx + q\,dy \tag{8.10}$$

Elimination of p between equations (8.7) and (8.10) and rearrangement leads to

$$q(a\,dy - b\,dx) + (c\,dx - a\,du) = 0 \tag{8.11}$$

This statement would be true for any line C, but if we choose C so that at all points along its length, the following condition is satisfied

$$a\,dy - b\,dx = 0 \tag{8.12}$$

then q can also be eliminated from equation (8.7) which becomes

$$c\,dx - a\,du = 0 \tag{8.13}$$

Combining equations (8.12) and (8.13) leads to

$$\frac{dx}{a} = \frac{dy}{b} = \frac{du}{c} \tag{8.14}$$

thus the original PDE from equation (8.5) has been reduced to an ODE along the chosen line C which is called a "characteristic". It can be shown that within the solution domain, a family of characteristic lines exists along which the PDE reduces to an ODE. Along these "characteristics" ordinary differential equation solution techniques may be employed such as those described in Chapter 7.

Rearrangement of equation (8.14) gives the pair of first order ordinary differential equations

$$\frac{dy}{dx} = \frac{b}{a} \quad \text{with } y(x_0) = y_0$$

$$\tag{8.15}$$

$$\frac{du}{dx} = \frac{c}{a} \quad \text{with } u(x_0) = u_0$$

where the initial conditions (x_0, y_0) and (x_0, u_0) correspond to values of x, y and u at the point of intersection of the characteristic line C with the initial conditions line $\mathcal{I}$ as shown in Figure 8.1.

Thus, if we know the value of u_0 at any initial point (x_0, y_0), solution of the ordinary differential equations given by equation (8.15), using Program 7.1 for example, leads to values of u along the characteristic line $\mathcal{C}$ passing through (x_0, y_0).

Example 8.1

Given the first order linear PDE

$$\frac{\partial u}{\partial x} + 3x^2 \frac{\partial u}{\partial y} = x + y$$

estimate $u(3, 19)$ given the initial condition $u(x, 0) = x^2$

Solution 8.1

The equation is linear so an analytical solution is possible in this case. Along the characteristics

$$\frac{dx}{1} = \frac{dy}{3x^2} = \frac{du}{x + y}$$

hence

$$\frac{dy}{dx} = 3x^2$$

$$\frac{du}{dx} = x + y$$

Integration of the first equation leads to $y = x^3 + k$, which represents the family of characteristic lines. We are interested in the characteristic passing through the point $(3, 19)$, thus

$$19 = 27 + k$$
$$k = -8$$
$$y = x^3 - 8$$

which is the equation of line $\mathcal{C}$.

Substituting for y in the second equation gives

$$\frac{du}{dx} = x + x^3 - 8$$

hence

$$u = \frac{1}{2}x^2 + \frac{1}{4}x^4 - 8x + K$$

The initial value line $\mathcal{I}$ coincides with the x-axis as shown in Figure 8.2 and intersects $\mathcal{C}$ at $(2,0)$ where $u = 4$, thus

$$4 = 2 + 4 - 16 + K$$
$$K = 14$$
$$u = \frac{1}{2}x^2 + \frac{1}{4}x^4 - 8x + 14$$

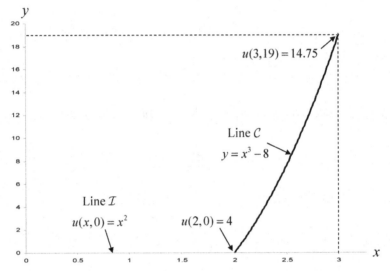

Figure 8.2: Lines $\mathcal{C}$ and $\mathcal{I}$ in Example 8.1.

Hence, when $x = 3$, $u = 14.75$.

Example 8.2

Given the first order nonlinear PDE

$$\sqrt{x}\frac{\partial u}{\partial x} + u\frac{\partial u}{\partial y} = -u^2$$

find the value of u when $x = 1.1$ on the characteristic passing through the point $(1,0)$ given the initial condition $u(x,0) = 1$.

Solution 8.2

Along the characteristics

$$\frac{dx}{\sqrt{x}} = \frac{dy}{u} = -\frac{du}{u^2}$$

hence

$$\frac{dy}{dx} = \frac{u}{\sqrt{x}} \quad \text{with } y(1) = 0$$

$$\frac{du}{dx} = -\frac{u^2}{\sqrt{x}} \quad \text{with } u(1) = 1$$

For nonlinear equations such as this, a numerical solution may be more convenient. We require $u(1.1)$ along this characteristic so any of the methods described in Chapter 7 for initial value problems could be used. Using Program 7.1 and the fourth order Runge-Kutta method with a step length of $h = 0.1$ we get $u(1.1) = 0.9111$ and $y(1.1) = 0.0931$.

8.4   Second order equations

The majority of partial differential equations that are likely to be encountered in engineering analysis are second order. The concept of characteristic lines introduced in the preceding section is a useful starting point, because it leads to an important means of classifying second order equations.

Consider the general equation (8.1), but without first derivative terms, or terms including u, thus

$$a\frac{\partial^2 u}{\partial x^2} + b\frac{\partial^2 u}{\partial x \partial y} + c\frac{\partial^2 u}{\partial y^2} + g = 0 \tag{8.16}$$

We assume for now that the equation is linear, that is the terms a, b, c and g are functions of x and y only and not of u or its derivatives.

The following substitutions can be made to simplify the algebra

$$p = \frac{\partial u}{\partial x}, \quad q = \frac{\partial u}{\partial y}, \quad r = \frac{\partial^2 u}{\partial x^2}, \quad s = \frac{\partial^2 u}{\partial x \partial y}, \quad t = \frac{\partial^2 u}{\partial y^2}$$

hence

$$ar + bs + ct + g = 0 \tag{8.17}$$

Considering small changes in p and q with respect to x and y, we can write

$$dp = \frac{\partial p}{\partial x}dx + \frac{\partial p}{\partial y}dy = r\,dx + s\,dy$$

$$\tag{8.18}$$

$$dq = \frac{\partial q}{\partial x}dx + \frac{\partial q}{\partial y}dy = s\,dx + t\,dy$$

Eliminating r and t from equations (8.17) and (8.18) gives

$$\frac{a}{dx}(dp - s\,dy) + bs + \frac{c}{dy}(dq - s\,dx) + g = 0 \qquad (8.19)$$

which after multiplication by $\dfrac{dy}{dx}$ leads to

$$s\left[a\left(\frac{dy}{dx}\right)^2 - b\left(\frac{dy}{dx}\right) + c\right] - \left[a\frac{dp}{dx}\frac{dy}{dx} + c\frac{dq}{dx} + g\frac{dy}{dx}\right] = 0 \qquad (8.20)$$

The only remaining partial derivative s can be eliminated by choosing curves or characteristic lines in the solution domain satisfying

$$a\left(\frac{dy}{dx}\right)^2 - b\left(\frac{dy}{dx}\right) + c = 0 \qquad (8.21)$$

Depending on the roots of equation (8.21) we can classify three different types of PDEs as follows:

(a) $b^2 - 4ac < 0$ Equation is ELLIPTIC
No real characteristic lines exist.

$$\text{e.g., Laplace's equation } \frac{\partial^2 u}{\partial x^2} + \frac{\partial^2 u}{\partial y^2} = 0$$

$$\text{Poisson's equation } \frac{\partial^2 u}{\partial x^2} + \frac{\partial^2 u}{\partial y^2} = f(x, y)$$

In both cases $b = 0$ and $a = c = 1$ hence $b^2 - 4ac = -4$

(b) $b^2 - 4ac = 0$ Equation is PARABOLIC
Characteristic lines are coincident.

$$\text{e.g., Heat diffusion or soil consolidation equation } c_y\frac{\partial^2 u}{\partial y^2} - \frac{\partial u}{\partial t} = 0$$

In this case $a = c_y$ and $b = c = 0$ hence $b^2 - 4ac = 0$

(c) $b^2 - 4ac > 0$ Equation is HYPERBOLIC
Two families of characteristic lines exist.

$$\text{e.g., Wave equation } \frac{\partial^2 u}{\partial y^2} - \frac{1}{k^2}\frac{\partial^2 u}{\partial t^2} = 0$$

In this case $a = 1$, $b = 0$ and $c = -\dfrac{1}{k^2}$ hence $b^2 - 4ac = \dfrac{4}{k^2}$

The characteristics method can be used for hyperbolic partial differential equations, where essentially the same techniques of "integration along the characteristics" as used for first order equations can be employed again.

The most widely used methods for numerical solution of PDEs, irrespective of their classification, are the finite difference and finite element methods.

The finite difference method has already been discussed in Chapter 7 in relation to the solution of ordinary differential equations, and will be extended in the next Section to partial differential equations.

The finite element method is a more powerful method, but it is also more complicated. Section 8.6 of this text therefore gives only a brief introduction to the method, but the interested reader is encouraged to consult some of the numerous texts dedicated to the method (e.g. Smith and Griffiths 2004).

8.5 Finite difference method

In the finite difference method, derivatives that occur in the governing PDE are replaced by "finite difference" approximations (see Section 7.4.1). Since PDEs involve at least two independent variables, the finite difference expressions now involve various combinations of the unknown variable occurring at "grid points" surrounding the location at which a derivative is required.

Consider in Figure 8.3 a two-dimensional Cartesian solution domain, split into a regular rectangular grid with a spacing of h in the x-direction and k in the y-direction. The single dependent variable u in this case is a function of the two independent variables x and y. The point at which a derivative is required is given the subscript i, j, where i counts in the x-direction and j counts in the y-direction.

Any partial derivative may be approximated using finite difference expressions such as those given in Tables 5.2, 5.3 and 5.4. For example,

$$\left(\frac{\partial u}{\partial x}\right)_{i,j} \approx \frac{1}{2h}(u_{i+1,j} - u_{i-1,j})$$

$$\left(\frac{\partial u}{\partial y}\right)_{i,j} \approx \frac{1}{2k}(u_{i,j+1} - u_{i,j-1})$$

(8.22)

would be central difference formulas for first derivatives. These formulas are sometimes conveniently expressed as computational "molecules," where, for

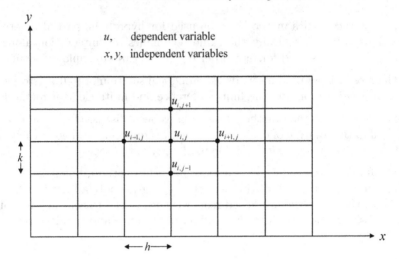

Figure 8.3: Two-dimensional finite difference grid.

example,

$$\left(\frac{\partial u}{\partial x}\right)_{i,j} \approx \frac{1}{2h}\left[\,\text{(-1)}\!-\!\text{(0)}\!-\!\text{(1)}\,\right]$$

$$(8.23)$$

$$\left(\frac{\partial u}{\partial y}\right)_{i,j} \approx \frac{1}{2k}\begin{bmatrix}\text{(1)}\\\text{(0)}\\\text{(-1)}\end{bmatrix}$$

Combining these two first derivatives, and letting $h = k$, we get

$$\left(\frac{\partial u}{\partial x}+\frac{\partial u}{\partial y}\right)_{i,j} \approx \frac{1}{2h}\begin{bmatrix}&\text{(1)}&\\\text{(-1)}\!-\!\text{(0)}\!-\!\text{(1)}\\&\text{(-1)}&\end{bmatrix}$$

$$(8.24)$$

Similarly, for second derivatives

$$\left(\frac{\partial^2 u}{\partial x^2}\right)_{i,j} \approx \frac{1}{h^2}\left[\,\text{(1)}\!-\!\text{(-2)}\!-\!\text{(1)}\,\right]$$

$$(8.25)$$

$$\left(\frac{\partial u^2}{\partial y^2}\right)_{i,j} \approx \frac{1}{k^2}\begin{bmatrix}\text{(1)}\\\text{(-2)}\\\text{(1)}\end{bmatrix}$$

which for the special case of $h = k$ leads to the Laplacian "molecule"

$$\nabla^2 u_{i,j} \approx \frac{1}{h^2} \begin{bmatrix} & ① & \\ ①-④-① \\ & ① & \end{bmatrix} \qquad (8.26)$$

Central difference "molecules" for third and fourth derivatives (see Table 5.3) can be written in the form

$$\left(\frac{\partial^3 u}{\partial x^3}\right)_{i,j} \approx \frac{1}{2h^3} [\,①-②-⓪-②-①\,] \qquad (8.27)$$

$$\left(\frac{\partial^4 u}{\partial x^4}\right)_{i,j} \approx \frac{1}{h^4} [\,①-④-⑥-④-①\,] \qquad (8.28)$$

and "mixed" derivatives $(h = k)$ as

$$\frac{\partial}{\partial x}\left(\frac{\partial u}{\partial y}\right)_{i,j} = \left(\frac{\partial^2 u}{\partial x \partial y}\right)_{i,j} \approx \frac{1}{4h^2} \begin{bmatrix} ⒈-⓪-① \\ ⓪-⓪-⓪ \\ ①-⓪-⒈ \end{bmatrix} \qquad (8.29)$$

and the biharmonic operator $(h = k)$ as

$$\nabla^4 u_{i,j} \approx \frac{1}{h^4} \begin{bmatrix} & & ① & & \\ & ②-⑧-② & \\ ①-⑧-⑳-⑧-① \\ & ②-⑧-② & \\ & & ① & & \end{bmatrix} \qquad (8.30)$$

All these examples used the lowest order central difference form from Table 5.3. Clearly higher order versions could be devised including forward and backward differences if required. It is advisable to use central difference formulas when possible, as they give the greatest accuracy in relation to the number of terms included.

When using these central difference formulas, the $(i,j)^{th}$ grid point always lies at the middle of "molecule". This is the point at which a derivative is to be approximated, and the "molecule" can be visualized as an overlay to the grid drawn on the two-dimensional solution domain.

In summary, the finite difference approach to the solution of linear partial differential equations can be considered in four steps:

1. Divide the solution domain into a grid. The shape of the grid should reflect the nature of the problem and the boundary conditions.

2. Obtain the finite difference formula that approximately represents the governing PDE. This may be conveniently written as a computational "molecule".

3. Overlay the "molecule" over each grid point at which a solution is required taking account of boundary conditions.

4. Elliptic systems, such as those governed by Laplace's equation, lead to linear, symmetric, banded systems of simultaneous equations which can be solved using the techniques described in Chapter 2. As will be seen, in the case of parabolic and hyperbolic systems, spatial discretization by finite differences reduces the problem to a system of ODEs in time which can be solved using the techniques from Chapter 7. Temporal discretization by finite differences can be performed using either implicit or explicit approaches. Implicit approaches will also lead to the solution of simultaneous equations at each time step, whereas explicit approaches lead to simpler time stepping algorithms without the need to solve simultaneous equations, but at the expense of being only conditionally stable (i.e., not subject to unbounded errors, see Section 7.3.7).

8.5.1 Elliptic systems

Problems governed by elliptic systems of PDEs are characterized by closed solution domains as shown in Figure 8.4. For example, Laplace's equation (in two-dimensions) is given by

$$k_x \frac{\partial^2 u}{\partial x^2} + k_y \frac{\partial^2 u}{\partial y^2} = 0 \tag{8.31}$$

For a general anisotropic ($k_x \neq k_y$) material with a rectangular ($h \neq k$) grid, the finite difference equation that would be applied at a typical internal grid point within the solution domain from equations (8.25) is given by

$$\frac{k_x}{h^2}(u_{i-1,j} - 2u_{i,j} + u_{i+1,j}) + \frac{k_y}{k^2}(u_{i,j-1} - 2u_{i,j} + u_{i,j+1}) = 0 \tag{8.32}$$

For isotropic materials with square grids, the formula is greatly simplified as

$$4u_{i,j} - u_{i-1,j} - u_{i,j+1} - u_{i+1,j} - u_{i,j-1} = 0 \tag{8.33}$$

which leads to the molecule shown in equation (8.26).

Solution of an elliptic problem proceeds by applying the finite difference equations (or molecules) at each unknown grid point within the solution domain. The majority of grid points will typically be internal and surrounded by other grid points, thus the formula given by equation (8.33) (or 8.34) will

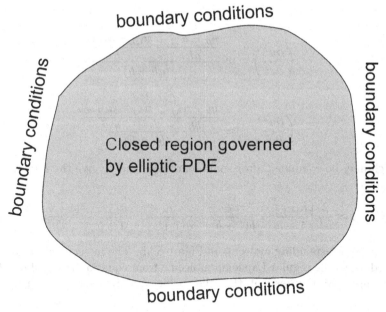

Figure 8.4: Closed solution domain for elliptic problems.

be used in the majority of cases. When boundary grid points have known values, they should be included directly into the finite difference equations.

The problem is eventually reduced to a system of linear, symmetric, banded equations that can be solved using the techniques described in Chapter 2.

8.5.1.1 Boundary conditions

It may be noted that the rectangular layout of the "molecules" described so far implies that the solution domains must also be rectangular. Problems with irregular boundaries can be tackled using finite differences; however the finite difference formulas and molecules must be modified to account for the actual location of the boundary between the regular grid points. Consider the case of a Laplacian molecule applied to the irregular boundary shown in Figure 8.5a. The factors $a < 1$ and $b < 1$ refer to the proportion of the regular grid spacing h required to reach the boundary from the center of the molecule in the x- and y-directions respectively.

Taking an average of the distance between grid points, we can write the second derivatives as

$$\left(\frac{\partial u^2}{\partial x^2}\right)_{i,j} \approx \frac{\frac{u_a - u_{i,j}}{ah} - \frac{u_{i,j} - u_{i-1,j}}{h}}{\frac{1}{2}(ah + h)}$$

(8.34)

$$\left(\frac{\partial u^2}{\partial y^2}\right)_{i,j} \approx \frac{\frac{u_b - u_{i,j}}{bh} - \frac{u_{i,j} - u_{i,j-1}}{h}}{\frac{1}{2}(bh + h)}$$

which may be combined after some rearrangement to give the Laplacian formula as

$$\nabla^2 u_{i,j} \approx \frac{2}{h^2}\left[\frac{u_{i-1,j}}{1+a} + \frac{u_b}{b(1+b)} + \frac{u_a}{a(1+a)} + \frac{u_{i,j-1}}{1+b} - \frac{a+b}{ab}u_{i,j}\right] \quad (8.35)$$

with the corresponding molecule in Figure 8.5b. The molecule is easily shown to reduce to the regular Lapacian molecule from equation (8.27) if $a = b = 1$. Note also from Figures 8.5b that the coefficients always sum to zero.

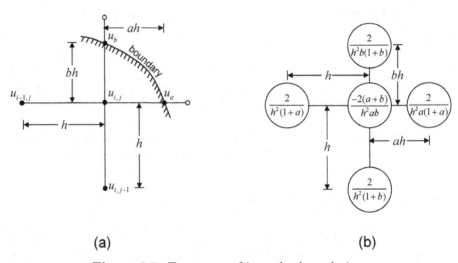

(a)   (b)

Figure 8.5: Treatment of irregular boundaries.

Although irregular boundaries can be included using the interpolation method described above, it is not conveniently generalized since each boundary molecule may have different a and b values in different locations.

Even when dealing with rectangular grids, other types of conditions may be encountered involving impermeable (or insulated) boundaries. The derivation of these molecules involves the assumption of steady state flow and the concept of "half flow channels" parallel to impermeable surfaces. Consider for example steady flow around an external corner of an impermeable material as shown in Figure 8.6.

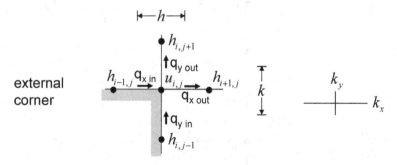

Figure 8.6: Derivation of FD formulas for impermeable boundaries.

The total head values are indicated on the grid points and the x- and y-components of flow indicated in the positive directions. For now we will assume a rectangular grid ($h \neq k$) and an anisotropic material ($k_x \neq k_y$). From Darcy's law for 1-d flow (i.e., $q = kia$, where k=permeability, i=hydraulic gradient and a=area), and using forward difference approximations to the hydraulic gradient between grid points, we get

$$q_{x_{in}} = -k_x \frac{h_{i,j} - h_{i-1,j}}{h} \frac{k}{2}$$

$$q_{y_{in}} = -k_y \frac{h_{i,j} - h_{i,j-1}}{k} \frac{h}{2}$$

$$q_{x_{out}} = -k_x \frac{h_{i+1,j} - h_{i,j}}{h} k$$

$$q_{y_{out}} = -k_y \frac{h_{i,j+1} - h_{i,j}}{k} h$$

(8.36)

Since this is a steady-state condition

$$q_{x_{in}} + q_{y_{in}} = q_{x_{out}} + q_{y_{out}} \tag{8.37}$$

and if we let $h = k$ and $k_x = k_y$ we get

$$3h_{i,j} - \frac{1}{2}h_{i-1,j} - h_{i,j+1} - h_{i+1,j} - \frac{1}{2}h_{i-1,j} = 0 \tag{8.38}$$

which leads to the "molecule" given in Figure 8.7b. Molecules for two other types of boundary conditions derived using this approach are also given in Figures 8.7a and c.

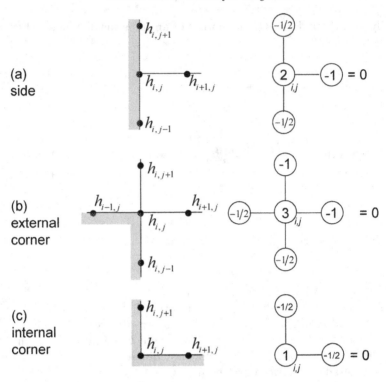

Figure 8.7: Summary of molecules for impermeable boundary condition (isotropic material with square FD grid).

It should be noted that different orientations of the boundary types shown in Figures 8.7 are easily derived by a simple rotation of the terms in the molecules.

In Examples 8.3 and 8.4 to follow, the value of ϕ will be known around the entire boundary. Derivative boundary conditions may also apply and can be tackled by the introduction of hypothetical grid points outside the solution domain as will be demonstrated in Example 8.5. A quite commonly encountered derivative boundary condition corresponds to impermeable or "no-flow" boundary conditions. In this case, the derivative normal to the impermeable boundary is set to zero as in $\dfrac{\partial \phi}{\partial n} = 0$ where n represents the normal direction. As in the Example 8.5, this boundary condition can be introduced through the use of hypothetical grid points; however the "molecules" described in Figure 8.7 offer a more convenient approach.

Example 8.3

The isotropic square plate shown in Figure 8.8 is subjected to constant boundary temperatures such that the top and left sides are fixed at 50° and the bottom and right side are fixed at 0°. Use a finite difference scheme to estimate the steady state temperatures at the internal grid points.

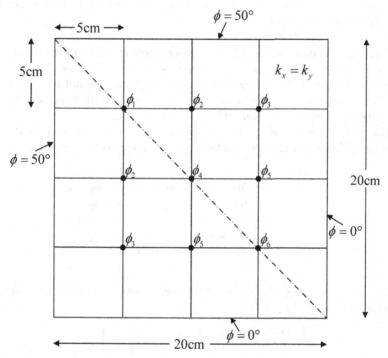

Figure 8.8: Finite difference grid from Example 8.3.

Solution 8.3

Using the symbol ϕ to signify temperature, the system is governed by Laplace's equation

$$\frac{\partial^2 \phi}{\partial x^2} + \frac{\partial^2 \phi}{\partial y^2} = 0$$

and since the finite difference grid is square, the molecule is given by

$$\begin{bmatrix} & \boxed{-1} & \\ \boxed{-1} & \boxed{4} & \boxed{-1} \\ & \boxed{-1} & \end{bmatrix} = 0$$

Note that since the Laplacian molecule is set equal to zero, we have changed the signs of the coefficients from those shown in equation 8.26, so that the central term is positive. While this change will not affect the solution, it leads to positive numbers on the diagonal of the coefficient matrix which can be computationally preferable.

A line of symmetry passing from the top left corner of the plate to the bottom right is indicated on Figure 8.8. This reduces the number of unknowns in the problem from 9 to 6, as indicated by the subscripts of the unknown ϕ values written next to each grid point. If symmetries are not exploited in the solution of a problem such as this, the same results will be achieved, but at the expense of greater computational effort.

The center of the molecule is now placed at each unknown grid point, taking account of known boundary conditions, to give the following equations

Point	
1	$-50 - 50 - \phi_2 - \phi_2 + 4\phi_1 = 0$
2	$-50 - \phi_1 - \phi_4 - \phi_3 + 4\phi_2 = 0$
3	$-50 - \phi_2 - \phi_5 - 0 + 4\phi_3 = 0$
4	$-\phi_2 - \phi_2 - \phi_5 - \phi_5 + 4\phi_4 = 0$
5	$-\phi_3 - \phi_4 - \phi_6 - 0 + 4\phi_5 = 0$
6	$-\phi_5 - \phi_5 - 0 - 0 + 4\phi_6 = 0$

which, after collecting terms and simplification, can be written in the standard matrix form of $[\mathbf{A}]\{\mathbf{x}\} = \{\mathbf{b}\}$ as

$$\begin{bmatrix} 2 & -1 & 0 & 0 & 0 & 0 \\ -1 & 4 & -1 & -1 & 0 & 0 \\ 0 & -1 & 4 & 0 & -1 & 0 \\ 0 & -1 & 0 & 2 & -1 & 0 \\ 0 & 0 & -1 & -1 & 4 & -1 \\ 0 & 0 & 0 & 0 & -1 & 2 \end{bmatrix} \begin{Bmatrix} \phi_1 \\ \phi_2 \\ \phi_3 \\ \phi_4 \\ \phi_5 \\ \phi_6 \end{Bmatrix} = \begin{Bmatrix} 50 \\ 50 \\ 50 \\ 0 \\ 0 \\ 0 \end{Bmatrix}$$

It should be noted that Laplacian problems such as this with the sign change noted above always lead to a positive definite, symmetric coefficient matrix ($[\mathbf{A}] = [\mathbf{A}]^T$). It can also be noted that a feature of finite difference methods, and indeed all "grid" methods including the finite element method, is that the discretization leads to a banded structure for $[\mathbf{A}]$.

An appropriate solution technique from Chapter 2 might therefore be Program 2.4 for symmetric, banded, positive definite systems. The input matrix $[\mathbf{A}]$ in vector "skyline" form would be

$$\begin{bmatrix} 2 & -1 & 4 & -1 & 4 & -1 & 0 & 2 & -1 & -1 & 4 & -1 & 2 \end{bmatrix}^T$$

with diagonal location vector

$$\begin{bmatrix} 1 & 3 & 5 & 8 & 11 & 13 \end{bmatrix}^T$$

The $\{\mathbf{b}\}$ vector would be

$$\begin{bmatrix} 50 & 50 & 50 & 0 & 0 & 0 \end{bmatrix}^T$$

leading to the solution

$$\begin{Bmatrix} \phi_1 \\ \phi_2 \\ \phi_3 \\ \phi_4 \\ \phi_5 \\ \phi_6 \end{Bmatrix} = \begin{Bmatrix} 42.86 \\ 35.71 \\ 25.00 \\ 25.00 \\ 14.29 \\ 7.14 \end{Bmatrix}$$

Example 8.4

Repeat the previous example assuming an anisotropic material in which the horizontal conductivity is two times bigger than the vertical conductivity.

Solution 8.4

With an anisotropic material where $k_x = 2k_y$, the finite difference formula from equation (8.32) can be written as

$$2(u_{i-1,j} - 2u_{i,j} + u_{i+1,j}) + (u_{i,j-1} - 2u_{i,j} + u_{i,j+1}) = 0$$

and since the finite difference grid is square, the molecule is given by

The geometry is no longer symmetric, so the finite difference grid has 9 unknowns as shown in Figure 8.9.

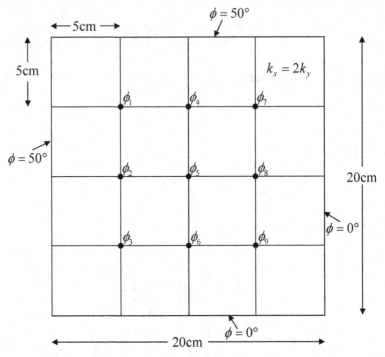

Figure 8.9: Finite difference grid from Example 8.4.

The center of the molecule is now placed at each unknown grid point, taking account of known boundary conditions, to give the following equations

Point					
1	-100	-50	$-2\phi_4$	$-\phi_2$	$+6\phi_1 = 0$
2	-100	$-\phi_1$	$-2\phi_5$	$-\phi_3$	$+6\phi_2 = 0$
3	-100	$-\phi_2$	$-2\phi_6$	-0	$+6\phi_3 = 0$
4	$-2\phi_1$	-50	$-2\phi_7$	$-\phi_5$	$+6\phi_4 = 0$
5	$-2\phi_2$	$-\phi_4$	$-2\phi_8$	$-\phi_6$	$+6\phi_5 = 0$
6	$-2\phi_3$	$-\phi_5$	$-2\phi_9$	-0	$+6\phi_6 = 0$
7	$-2\phi_4$	-50	-0	$-\phi_8$	$+6\phi_7 = 0$
8	$-2\phi_5$	$-\phi_7$	-0	$-\phi_9$	$+6\phi_8 = 0$
9	$-2\phi_6$	$-\phi_8$	-0	-0	$+6\phi_9 = 0$

which, after collecting terms and simplification, can be written in the standard *matrix form of* $[\mathbf{A}]\{\mathbf{x}\} = \{\mathbf{b}\}$ *as*

$$
\begin{bmatrix}
6 & -1 & 0 & -2 & 0 & 0 & 0 & 0 & 0 \\
-1 & 6 & -1 & 0 & -2 & 0 & 0 & 0 & 0 \\
0 & -1 & 6 & 0 & 0 & -2 & 0 & 0 & 0 \\
-2 & 0 & 0 & 6 & -1 & 0 & -2 & 0 & 0 \\
0 & -2 & 0 & -1 & 6 & -1 & 0 & -2 & 0 \\
0 & 0 & -2 & 0 & -1 & 6 & 0 & 0 & -2 \\
0 & 0 & 0 & -2 & 0 & 0 & 6 & -1 & 0 \\
0 & 0 & 0 & 0 & -2 & 0 & -1 & 6 & -1 \\
0 & 0 & 0 & 0 & 0 & -2 & 0 & -1 & 6
\end{bmatrix}
\begin{Bmatrix}
\phi_1 \\ \phi_2 \\ \phi_3 \\ \phi_4 \\ \phi_5 \\ \phi_6 \\ \phi_7 \\ \phi_8 \\ \phi_9
\end{Bmatrix}
=
\begin{Bmatrix}
150 \\ 100 \\ 100 \\ 50 \\ 0 \\ 0 \\ 50 \\ 0 \\ 0
\end{Bmatrix}
$$

leading to the solution

$$
\begin{Bmatrix}
\phi_1 \\ \phi_2 \\ \phi_3 \\ \phi_4 \\ \phi_5 \\ \phi_6 \\ \phi_7 \\ \phi_8 \\ \phi_9
\end{Bmatrix}
=
\begin{Bmatrix}
42.44 \\ 36.76 \\ 28.15 \\ 33.93 \\ 25.00 \\ 16.07 \\ 21.85 \\ 13.24 \\ 7.56
\end{Bmatrix}
$$

Example 8.5

The isotropic plate shown in Figure 8.10 has temperatures prescribed on three sides as indicated and a derivative boundary condition on the fourth side. Find the steady state temperature distribution using finite differences.

Solution 8.5

By symmetry only the top half of the plate will be analyzed. There are four internal grid points with temperatures labeled ϕ_1 thru ϕ_4. To incorporate derivative boundary condition, two hypothetical grid points have been included outside the plate, with temperatures given by ϕ_5 and ϕ_6.

The distribution is governed by Laplace's equation, but in this case the grid is not square since $h = 10$cm and $k = 5$cm. Equation (8.32) leads to the following Laplacian finite difference equation

$$
\frac{1}{100}(\phi_{i-1,j} - 2\phi_{i,j} + \phi_{i+1,j}) + \frac{1}{25}(\phi_{i,j-1} - 2\phi_{i,j} + \phi_{i,j+1}) = 0
$$

giving the molecule

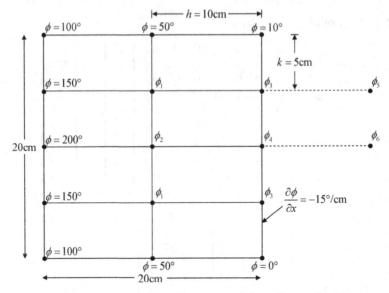

Figure 8.10: Finite difference grid from Example 8.5.

The center of the molecule is now placed at each unknown grid point, taking account of known boundary conditions, to give the following equations

Point
$$1 \qquad -150 - 200 - \phi_3 - 4\phi_2 + 10\phi_1 = 0$$
$$2 \qquad -200 - 4\phi_1 - \phi_4 - 4\phi_1 + 10\phi_2 = 0$$
$$3 \qquad -\phi_1 - 40 - \phi_5 - 4\phi_4 + 10\phi_3 = 0$$
$$4 \qquad -\phi_2 - 4\phi_3 - \phi_6 - 4\phi_3 + 10\phi_4 = 0$$

Two final equations are provided by the derivative boundary condition expressed in central finite difference form as

$$\frac{\phi_5 - \phi_1}{20} = -15$$

$$\frac{\phi_6 - \phi_2}{20} = -15$$

After substituting for ϕ_5 and ϕ_6 in the four Laplacian equations above and some rearrangement, we get the standard matrix form of $[\mathbf{A}]\{\mathbf{x}\} = \{\mathbf{b}\}$ as

$$
\begin{bmatrix}
20 & -8 & -2 & 0 \\
-8 & 10 & 0 & -1 \\
-2 & 0 & 10 & -4 \\
0 & -1 & -4 & 5
\end{bmatrix}
\begin{Bmatrix}
\phi_1 \\
\phi_2 \\
\phi_3 \\
\phi_4
\end{Bmatrix}
=
\begin{Bmatrix}
700 \\
200 \\
-260 \\
-150
\end{Bmatrix}
$$

It may be noted that the symmetrical form given above was obtained by multiplying the first equation by 2 and dividing the last equation by 2.

The input matrix $[\mathbf{A}]$ in vector "skyline" form for Program 2.4 would be

$$\begin{bmatrix} 20 & -8 & 10 & -2 & 0 & 10 & -1 & -4 & 5 \end{bmatrix}^T$$

with diagonal location vector

$$\begin{bmatrix} 1 & 3 & 6 & 9 \end{bmatrix}^T$$

The $\{\mathbf{b}\}$ vector would be

$$\begin{bmatrix} 700 & 200 & -260 & -150 \end{bmatrix}^T$$

leading to the solution

$$\begin{Bmatrix} \phi_1 \\ \phi_2 \\ \phi_3 \\ \phi_4 \end{Bmatrix} = \begin{Bmatrix} 55.90 \\ 60.34 \\ -32.34 \\ -43.80 \end{Bmatrix}$$

Example 8.6

Steady flow is taking place through the system shown in Figure 8.11 where hatched lines represent impermeable boundaries. The up- and down-stream boundaries are fixed to 100 and 0 respectively. Compute the total head at all the internal grid points. The porous material through which flow is occurring is isotropic.

Solution 8.6

The problem includes several different types of impermeable boundary. Since the problem is isotropic and has a square finite difference grid, the molecules from Figure 8.7 can be used directly to give the following equations

Point					
1	-100	-50	$-2h_4$	$-h_2$	$+6h_1 = 0$
2	-100	$-h_1$	$-2h_5$	$-h_3$	$+6h_2 = 0$
3	-100	$-h_2$	$-2h_6$	-0	$+6h_3 = 0$
4	$-2h_1$	-50	$-2h_7$	$-h_5$	$+6h_4 = 0$
5	$-2h_2$	$-h_4$	$-2h_8$	$-h_6$	$+6h_5 = 0$
6	$-2h_3$	$-h_5$	$-2h_9$	-0	$+6h_6 = 0$
7	$-2h_4$	-50	-0	$-h_8$	$+6h_7 = 0$
8	$-2h_5$	$-h_7$	-0	$-h_9$	$+6h_8 = 0$
9	$-2h_6$	$-h_8$	-0	-0	$+6h_9 = 0$
10	$-2h_6$	$-h_8$	-0	-0	$+6h_9 = 0$
11	$-2h_6$	$-h_8$	-0	-0	$+6h_9 = 0$
12	$-2h_6$	$-h_8$	-0	-0	$+6h_9 = 0$
13	$-2h_6$	$-h_8$	-0	-0	$+6h_9 = 0$

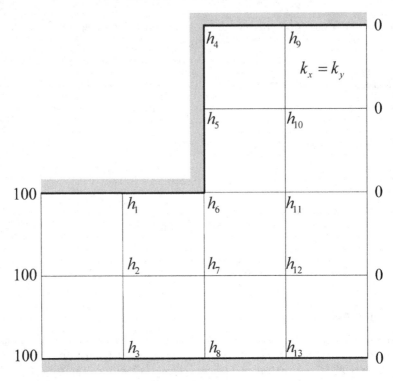

Figure 8.11: Finite difference grid from Example 8.6.

which, after collecting terms and simplification, can be written in standard matrix form as

$$
\begin{bmatrix}
2 & -1 & & & & -0.5 & & & & & & & \\
-1 & 4 & -1 & & & 0 & -1 & & & & & & \\
& -1 & 2 & & & 0 & 0 & -0.5 & & & & & \\
& & & 1 & -0.5 & 0 & 0 & 0 & -0.5 & & & & \\
& & & -0.5 & 5 & -0.5 & 0 & 0 & 0 & 1 & & & \\
-0.5 & 0 & 0 & 0 & -0.5 & 3 & -1 & 0 & 0 & 0 & -1 & & \\
& -1 & 0 & 0 & 0 & -1 & 4 & -1 & 0 & 0 & 0 & -1 & \\
& & -0.5 & 0 & 0 & 0 & -1 & 2 & 0 & 0 & 0 & 0 & -0.5 \\
& & & -0.5 & 0 & 0 & 0 & 0 & 2 & -1 & 0 & 0 & 0 \\
& & & & -1 & 0 & 0 & 0 & -1 & 4 & -1 & 0 & 0 \\
& & & & & -1 & 0 & 0 & 0 & -1 & 4 & -1 & 0 \\
& & & & & & -1 & 0 & 0 & 0 & -1 & 4 & -1 \\
& & & & & & & -0.5 & 0 & 0 & 0 & -1 & 2 \\
\end{bmatrix}
\begin{Bmatrix}
h_1 \\ h_2 \\ h_3 \\ h_4 \\ h_5 \\ h_6 \\ h_7 \\ h_8 \\ h_9 \\ h_{10} \\ h_{11} \\ h_{12} \\ h_{13}
\end{Bmatrix}
=
\begin{Bmatrix}
50 \\ 100 \\ 50 \\ 0 \\ 0 \\ 0 \\ 0 \\ 0 \\ 0 \\ 0 \\ 0 \\ 0 \\ 0
\end{Bmatrix}
$$

Only terms within the skyline have been included for clarity. All blank entries equal zero. Solution of these equations using, for example, Program 2.5 leads to

$$\begin{Bmatrix} h_1 \\ h_2 \\ h_3 \\ h_4 \\ h_5 \\ h_6 \\ h_7 \\ h_8 \\ h_9 \\ h_{10} \\ h_{11} \\ h_{12} \\ h_{13} \end{Bmatrix} = \begin{Bmatrix} 67.08 \\ 69.32 \\ 70.28 \\ 4.08 \\ 4.45 \\ 29.67 \\ 39.93 \\ 42.47 \\ 3.70 \\ 5.37 \\ 13.32 \\ 18.25 \\ 19.74 \end{Bmatrix}$$

8.5.2 Parabolic systems

For typical parabolic equations, for example the "conduction" or "consolidation" equation, we require boundary conditions together with initial conditions in time. The solution then marches along in time for as long as required. Unlike elliptic problems, the solution domain is "open" as shown in Figure 8.12 in the sense that the time variable can continue indefinitely.

A parabolic problem which often arises in civil engineering analysis is the consolidation equation in one dimension

$$c_v \frac{\partial^2 u}{\partial z^2} = \frac{\partial u}{\partial t} \tag{8.39}$$

where

$$c_v = \text{coefficient of consolidation}$$
$$z = \text{spatial coordinate}$$
$$t = \text{time}$$

In the context of heat flow problems, c_v would be replaced by the thermal diffusivity property, α.

It is sometimes convenient to nondimensionalize equation (8.39) so that a solution can be obtained which is more generally applicable.

In order to do this, let

$$Z = \frac{z}{D}, \quad U = \frac{u}{U_0}, \quad \text{and} \quad T = \frac{c_v t}{D^2} \tag{8.40}$$

where D is a reference length (e.g., a drainage path length); U_0 is a reference pressure (e.g., the initial pressure at $t = 0$); and T is a dimensionless time known as the "time factor".

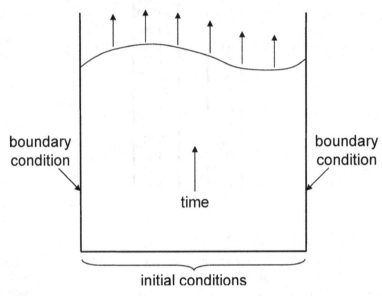

Figure 8.12: Open solution domain for parabolic systems.

The derivatives can be written as

$$\frac{\partial u}{\partial z} = \frac{\partial Z}{\partial z}\frac{\partial u}{\partial Z} = \frac{1}{D}\frac{\partial u}{\partial Z}$$

$$\frac{\partial^2 u}{\partial z^2} = \frac{\partial}{\partial z}\left(\frac{\partial u}{\partial z}\right) = \frac{\partial}{\partial Z}\left(\frac{\partial Z}{\partial z}\right)\left(\frac{\partial u}{\partial z}\right) = \frac{1}{D^2}\frac{\partial^2 u}{\partial Z^2}$$

(8.41)

also

$$\frac{\partial u}{\partial t} = \frac{\partial T}{\partial t}\frac{\partial u}{\partial T} = \frac{c_v}{D^2}\frac{\partial u}{\partial T}$$

(8.42)

and

$$u = UU_0$$

(8.43)

Substitution into equation (8.39) leads to the dimensionless form of the diffusion equation as follows

$$\frac{c_v}{D^2}\frac{\partial^2 (UU_0)}{\partial Z^2} = \frac{c_v}{D^2}\frac{\partial (UU_0)}{\partial T}$$

(8.44)

hence

$$\frac{\partial^2 U}{\partial Z^2} = \frac{\partial U}{\partial T}$$

(8.45)

8.5.2.1 Explicit finite differences

Equation (8.39) is readily expressed in finite difference form. Assuming a grid spacing of Δz in the z-direction and Δt in time, and using central differences for the second derivative in space we get

$$\frac{\partial^2 u}{\partial z^2} \approx \frac{1}{\Delta z^2}(u_{i-1,j} - 2u_{i,j} + u_{i+1,j}) \tag{8.46}$$

where subscripts i and j refer to the z and t variables respectively.

This "semi-discretization" of the space variable, applied over the whole space grid, leads to a set of *ordinary* differential equations in the time variable. Any of the methods described in Chapter 7 could be used to integrate these sets of equations but in this introductory treatment we shall apply finite differences to the time dimension as well. Thus returning to equation (8.39), using a simple forward difference scheme for the first derivative in time (equivalent to $\theta = 0$ in section 7.3.4.1) we get

$$\frac{\partial u}{\partial t} \approx \frac{1}{\Delta t}(u_{i,j+1} - u_{i,j}) \tag{8.47}$$

Equations (8.46) and (8.47) can be substituted into equation (8.39) and rearranged to give

$$u_{i,j+1} = u_{i,j} + \frac{c_v \Delta t}{\Delta z^2}(u_{i-1,j} - 2u_{i,j} + u_{i+1,j}) \tag{8.48}$$

The "molecule" concept described in the previous section is not so useful in parabolic problems such as this; however there is a simple pattern that can be visualized in the way new values are computed from equation (8.48). A new value of u at any given depth is a function only of u-values from the previous time step, immediately above, at the same level and immediately below the given depth.

This type of relationship is termed "explicit" because the value of u at the new time level is expressed solely in terms of values of u at the immediately preceding time level. However, as with all "explicit" approaches, numerical stability is conditional on a satisfactory combination of spatial and temporal step lengths being employed. Numerical instability occurs when a perturbation (or error) introduced at a certain stage in the stepping procedure grows uncontrollably at subsequent steps.

It can be shown that numerical stability is only guaranteed in this explicit method if

$$\frac{c_v \Delta t}{\Delta z^2} \leq \frac{1}{2} \tag{8.49}$$

Example 8.7

The insulated rod shown in Figure 8.13 is initially at 0° at all points along its length when a boundary condition of 100° is applied to the left end of the

rod and maintained at that value. Use an explicit finite difference approach
to compute the temperature variation as a function of position and time along
the rod.

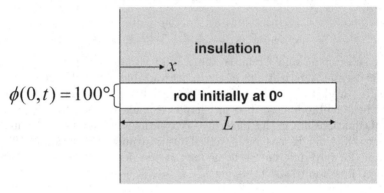

Figure 8.13: Heat diffusion problem from Example 8.7.

Solution 8.7

The governing equation for 1-d heat diffusion is given by

$$\alpha \frac{\partial^2 \phi}{\partial x^2} = \frac{\partial \phi}{\partial t}$$

where

$$\alpha = \text{thermal diffusivity}$$
$$\phi = \text{temperature}$$
$$x = \text{spatial coordinate}$$
$$t = \text{time}$$

with boundary conditions: $\phi(0,t) = 100° = \phi_0$ for $t > 0$

$$\frac{\partial \phi}{\partial x}(L,t) = 0 \text{ at insulated end}$$

and initial conditions: $\phi(x,0) = 0°$ for $0 \le x \le L$

By making the substitutions $X = \dfrac{x}{L}$, $\Phi = \dfrac{\phi}{\phi_0}$ and $T = \dfrac{\alpha t}{L^2}$, we can write
this equation in nondimensional form as

$$\frac{\partial^2 \Phi}{\partial X^2} = \frac{\partial \Phi}{\partial T}$$

with boundary conditions: $\Phi(0,T) = 1$ for $T > 0$

$$\frac{\partial \Phi}{\partial X}(1,T) = 0 \text{ at insulated end}$$

and initial conditions: $\Phi(X,0) = 0$ for $0 \le X \le 1$

Expressing the dimensionless equation in finite difference form, we get from equation (8.49)

$$\Phi_{i,j+1} = \Phi_{i,j} + \frac{\Delta T}{\Delta X^2}(\Phi_{i-1,j} - 2\Phi_{i,j} + \Phi_{i+1,j})$$

The stability requirement for this explicit formula is that $\frac{\Delta T}{\Delta X^2} \le \frac{1}{2}$. It may be noted that in the special case that $\frac{\Delta T}{\Delta X^2} = \frac{1}{2}$ the formula simplifies to

$$\Phi_{i,j+1} = \frac{1}{2}(\Phi_{i-1,j} + \Phi_{i+1,j})$$

In this example, let $\Delta X = 0.2$ and $\Delta T = 0.015$ in which case

$$\frac{\Delta T}{\Delta X^2} = 0.375$$

and the stability criterion is satisfied.

When modeling the differential equation with relatively large time steps, some compromise is recommended to model temperature changes at the left end of the rod, since initial conditions require that at $T = 0$, $\Phi = 0$ whereas an instant later at $T > 0$ boundary conditions at that location require $\Phi = 1$.

One option is to apply the full temperature change at $T = 0$; however in this example we have applied half at $T = 0.0$ and the remainder at $T = \Delta T$. The condition that $\frac{\partial \Phi}{\partial X} = 0$ at $X = 1$ is maintained using central differences, by including a hypothetical grid point at $X = 1.2$. The value of Φ at $X = 1.2$ is then set equal to the value of Φ at $X = 0.8$.

Consider, for example, the calculation required to find Φ at time $T = 0.015$ and depth $X = 0.2$. From the finite difference equation (8.48), the required value of Φ will depend on the values of Φ from the previous time step at depths of $X = 0, 0.2$ and 0.4, thus

$$\Phi = 0.000 + 0.0375(0.500 - 2(0.000) + 0.000) = 0.188$$

Another example later on would be the calculation of Φ at time $T = 0.120$ and depth $X = 0.6$, which would be given by

$$\Phi = 0.188 + 0.0375(0.384 - 2(0.188) + 0.079) = 0.221$$

A summary of the finite difference results is given in Table 8.2.

TABLE 8.2:   Tabulated values of $\Phi = f(X, T)$ from Example 8.7

X T	0.0	0.2	0.4	0.6	0.8	1.0	1.2
0.000	0.500	0.000	0.000	0.000	0.000	0.000	0.000
0.015	1.000	0.188	0.000	0.000	0.000	0.000	0.000
0.030	1.000	0.422	0.070	0.000	0.000	0.000	0.000
0.045	1.000	0.507	0.176	0.026	0.000	0.000	0.000
0.060	1.000	0.568	0.244	0.073	0.010	0.000	0.010
0.075	1.000	0.608	0.301	0.113	0.030	0.007	0.030
0.090	1.000	0.640	0.346	0.152	0.053	0.024	0.053
0.105	1.000	0.665	0.384	0.188	0.079	0.046	0.079
0.120	1.000	0.685	0.416	0.221	0.107	0.071	0.107
0.135	1.000	0.702	0.443	0.251	0.136	0.098	0.136
0.150	1.000	0.717	0.468	0.280	0.165	0.127	0.165

Example 8.8

A specimen of uniform saturated clay 240mm thick is placed in a conventional consolidation cell with drains at top and bottom. A sudden increment of vertical stress of 200 kN/m^2 is applied. If the coefficient of consolidation is $c_v = 10\text{m}^2/\text{yr}$ estimate the excess pore pressure distribution after 1 hour.

Solution 8.8

In this case we have elected to solve the consolidation equation in dimensional form in units of millimeters and hours, thus

$$c_v \frac{\partial^2 u}{\partial z^2} = \frac{\partial u}{\partial t}$$

where

$$c_v = \text{coefficient of consolidation} = 1142 \text{ mm}^2/\text{hr}$$
$$z = \text{spatial coordinate (mm)}$$
$$t = \text{time (hr)}$$

Since the consolidation test is symmetrical with double drainage we will solve half the problem assuming undrained conditions at the mid-plane ($z = 120 \text{ mm}$)

with boundary conditions: $u(0, t) = 0$ for $t > 0$

$$\frac{\partial u}{\partial z}(120, t) = 0 \text{ at the mid-plane}$$

and initial conditions: $u(x, 0) = 200 \text{ kN/m}^2$ for $0 \le z \le 120 \text{ mm}$

For the finite difference solution, let $\Delta z = 24\,\text{mm}$ and $\Delta t = 0.2\,\text{hr}$.

A check of the stability criterion gives $\dfrac{1142(0.2)}{24^2} = 0.397$ which is less than 0.5, indicating that the increment sizes are acceptable.

The finite difference equation from equation (8.48) becomes

$$u_{i,j+1} = u_{i,j} + 0.397(u_{i-1,j} - 2u_{i,j} + u_{i+1,j})$$

The condition that $\dfrac{\partial u}{\partial z} = 0$ at $z = 120\,\text{mm}$ is maintained by including a grid point at $z = 144\,\text{mm}$. From the central difference formula for a first derivative at $z = 120\,\text{mm}$, values at $z = 144\,\text{mm}$ are maintained at the same values as at $z = 96\,\text{mm}$ for all time steps. In this example, the drainage boundary condition at $z = 0$ is not introduced into the calculations until the end of the first time step at $t = 0.2\,\text{hr}$.

A summary of the finite difference results is given in Table 8.3, and both halves of the distribution of excess pore pressure with depth corresponding to $t = 1\,\text{hr}$ plotted in Figure 8.14.

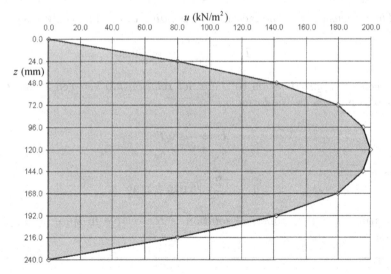

Figure 8.14: Excess pressure distribution after 1 hour from Example 8.8.

As a check, the Average Degree of Consolidation corresponding to the finite difference result after 1 hour is given as

$$U = \frac{\text{initial rectangular area} - \text{hatched area}}{\text{initial rectangular area}}$$

TABLE 8.3: Tabulated values of $u = f(z,t)\,(\text{kN/m}^2)$ from Example 8.8

z (mm)	0.0	24	48	72	96	120	144
t (hr)							
0.0	200.0	200.0	200.0	200.0	200.0	200.0	200.0
0.2	0.0	200.0	200.0	200.0	200.0	200.0	200.0
0.4	0.0	120.7	200.0	200.0	200.0	200.0	200.0
0.6	0.0	104.3	168.6	200.0	200.0	200.0	200.0
0.8	0.0	88.4	155.5	187.5	200.0	200.0	200.0
1.0	0.0	80.0	141.6	179.8	195.1	200.0	195.1

where

$$\text{initial rectangular area} = 240(200) = 48000$$
$$\text{hatched area} = 0.5(24)(0.0 + 2(80.0 + 141.6 + 179.8 + 195.1 +$$
$$200.0 + 195.1 + 179.8 + 141.6 + 80.0 + 0.0) = 33431$$

hence

$$U = \frac{48000 - 33431}{48000} = 0.30$$

The dimensionless time factor corresponding to $t = 1\,\text{hr}$ is given as

$$T = \frac{1142(1)}{120^2} = 0.079$$

An accurate approximation to the exact relationship between T and U (for $U \le 0.5$) is given by the formula

$$U = \frac{2}{\sqrt{\pi}}\sqrt{T}$$

thus

$$U = \frac{2}{\sqrt{\pi}}\sqrt{0.079} = 0.32$$

indicating that the numerical solution is reasonable in spite of the rather coarse grid.

Program 8.1: Explicit finite differences in 1D

```
PROGRAM nm81
!---Explicit Finite Differences in 1D---
 IMPLICIT NONE; INTEGER,PARAMETER::iwp=SELECTED_REAL_KIND(15,300)
 INTEGER::i,j,nres,nt,ntime,nz
 REAL(iwp)::area0,areat,beta,cv,dt,dz,layer,pt5=0.5_iwp,tmax,    &
   two=2.0_iwp,zero=0.0_iwp
```

```
REAL(iwp),ALLOCATABLE::a(:),b(:),c(:); CHARACTER(LEN=2)::bc
OPEN(10,FILE='nm95.dat'); OPEN(11,FILE='nm95.res')
WRITE(11,'(A)')"---Explicit Finite Differences in 1D---"
READ(10,*)layer,tmax,dz,dt,cv; beta=cv*dt/dz**2
IF(beta>pt5)THEN; WRITE(11,'("beta too big")'); STOP; ENDIF
nz=nint(layer/dz); nt=nint(tmax/dt)
ALLOCATE(a(nz+1),b(nz+1),c(nz+1)); READ(10,*)b,bc,nres,ntime
area0=pt5*dz*(b(1)+b(nz+1))+dz*SUM(b(2:nz))
WRITE(11,'(/,A,I3,A)')                                          &
  "    Time      Deg of Con  Pressure(grid pt",nres,")"
WRITE(11,'(3E12.4)')zero,zero,b(nres); a=zero
DO j=1,nt
  IF(bc=='uu'.OR.bc=='ud')a(1)=b(1)+two*beta*(b(2)-b(1))
  a(2:nz)=b(2:nz)+beta*(b(1:nz-1)-two*b(2:nz)+b(3:nz+1))
  IF(bc=='uu'.OR.bc=='du')                                      &
    a(nz+1)=b(nz+1)+two*beta*(b(nz)-b(nz+1)); b=a
  IF(j==ntime)c=b
  areat=pt5*dz*(b(1)+b(nz+1))+dz*SUM(b(2:nz))
  WRITE(11,'(3E12.4)')j*dt,(area0-areat)/area0,b(nres)
END DO
WRITE(11,'(/,A,E10.4,A)')                                       &
  "    Depth     Pressure(time=",ntime*dt,")"
WRITE(11,'(2E12.4)')(dz*(i-1),c(i),i=1,nz+1)
END PROGRAM nm81
```

To illustrate use of the program, consider a doubly drained 10m thick layer of clay subjected to an initial ($t = 0$) excess pore pressure distribution as follows

Depth (m)	0	2	4	6	8	10
u (kN/m^2)	60	54	41	29	19	15

If $c_v = 7.9\,\text{m}^2/\text{yr}$, estimate the average degree of consolidation of the layer after 1 year.

The input and output for this example are given in Data 8.1 and Results 8.1 respectively. The data indicate a layer thickness (layer) of 10m and a time duration for the calculation (tmax) of 1yr. Selected grid sizes in space and time are chosen to be dz=2 m and t=0.1 yr respectively. The coefficient of consolidation cv is read as $7.9\,\text{m}^2/\text{yr}$ followed by the initial pore pressure distribution read into vector b. The layer is doubly drained so the boundary condition variable bc is read as 'dd'. Other possible values that can be assigned to this variable are 'du', 'ud' and 'uu' where 'u' refers to undrained and 'd' refers to drained boundary conditions. The first letter refers to the boundary condition at the top of the layer and the second letter to the bottom. The output control asks for the time history to be printed for grid point number nres=5, which corresponds to a depth of 8m. The snapshot of the spatial distribution is requested at time step ntime=10 which is after 1 yr.

List 8.1:

Scalar integers:

i	simple counter
j	simple counter
nres	grid point number at which time history is to be printed
nt	number of calculation time step
ntime	time step at which spatial distribution is to be printed
nz	number of spatial grid points

Scalar reals:

area0	initial area of pressure profile
areat	hatched area at time t (see Figure 8.14)
beta	stability number
cv	coefficient of consolidation
dt	calculation time step
dz	spatial grid size
layer	total layer thickness
tmax	time to which calculation is to be taken
two	set to 2.0
zero	set to 0.0

Dynamic real arrays:

a	holds pressures at new time step
b	holds pressures at old time step
c	holds pressures after ntime time steps for printing

Scalar character:

bc	boundary condition variable

Depth and time duration	layer	tmax
	10.0	1.0
Spatial and temporal grid sizes	dz	dt
	2.0	0.1
Coefficient of Consolidation	cv	
	7.9	
Initial condition	b(i),i=1,nz+1	
	60.0 54.0 41.0 29.0 19.0 15.0	
Boundary conditions	bc	
	'dd'	

```
Output control                    nres    ntime
                                   5       10
```

Data 8.1: Explicit Finite Differences in 1d

Results 8.1 indicate that the Average Degree of Consolidation after 1 year is 0.63 or 63%. The maximum excess pore pressure remaining in the ground after 1 year is about $21\,\mathrm{kN/m^2}$ at a depth of 4m.

```
---Explicit Finite Differences in 1D---

   Time        Deg of Con   Pressure(grid pt  5)
0.0000E+00   0.0000E+00   0.1900E+02
0.1000E+00   0.2056E+00   0.2018E+02
0.2000E+00   0.2852E+00   0.1802E+02
0.3000E+00   0.3487E+00   0.1681E+02
0.4000E+00   0.4024E+00   0.1605E+02
0.5000E+00   0.4497E+00   0.1543E+02
0.6000E+00   0.4923E+00   0.1481E+02
0.7000E+00   0.5311E+00   0.1415E+02
0.8000E+00   0.5667E+00   0.1345E+02
0.9000E+00   0.5995E+00   0.1271E+02
0.1000E+01   0.6298E+00   0.1196E+02

   Depth       Pressure(time=0.1000E+01)
0.0000E+00   0.0000E+00
0.2000E+01   0.1359E+02
0.4000E+01   0.2114E+02
0.6000E+01   0.2013E+02
0.8000E+01   0.1196E+02
0.1000E+02   0.0000E+00
```

Results 8.1: Explicit Finite Differences in 1d

8.5.3 Hyperbolic systems

Hyperbolic systems involve propagation phenomena, for example as described by the wave equation. Consider the displacement v of a vibrating string which is given by

$$c^2 \frac{\partial^2 v}{\partial x^2} = \frac{\partial^2 v}{\partial t^2} \tag{8.50}$$

where

$$c^2 = \frac{T}{\rho}$$

T = tension in the string

ρ = mass density

x = spatial coordinate

t = time

Clearly these one-dimensional examples can be extended to two- or three-dimensions, in which case the Laplacian operator can be used, i.e.,

$$c^2 \Delta^2 v = \frac{\partial^2 v}{\partial t^2} \tag{8.51}$$

As with parabolic systems, it is sometimes convenient to nondimensionalize the one-dimensional wave equation. With reference to equation (8.50), by making the substitutions, $V = \frac{v}{v_0}$, $X = \frac{x}{L}$ and $T = \frac{ct}{L}$, where v_0 is an initial displacement and L is a reference length, we can write

$$\frac{\partial^2 v}{\partial x^2} = \left(\frac{\partial X}{\partial x}\right)^2 \frac{\partial^2 v}{\partial X^2} = \frac{1}{L^2}\frac{\partial^2 v}{\partial X^2}$$

$$\frac{\partial^2 v}{\partial t^2} = \left(\frac{\partial T}{\partial t}\right)^2 \frac{\partial^2 v}{\partial T^2} = \frac{c^2}{L^2}\frac{\partial^2 v}{\partial T^2} \tag{8.52}$$

hence

$$\frac{\partial^2 V}{\partial X^2} = \frac{\partial^2 V}{\partial T^2} \tag{8.53}$$

The finite difference form of this equation can be written as

$$\frac{1}{\Delta X^2}(V_{i-1,j} - 2V_{i,j} + V_{i+1,j}) = \frac{1}{\Delta T^2}(V_{i,j-1} - 2V_{i,j} + V_{i,j+1}) \tag{8.54}$$

where the subscripts i and j refer to changes in X and T respectively.

Rearrangement of equation (8.54) gives

$$V_{i,j+1} = \frac{\Delta T^2}{\Delta X^2}(V_{i-1,j} - 2V_{i,j} + V_{i+1,j}) - V_{i,j-1} + 2V_{i,j} \tag{8.55}$$

If we allow $\Delta X = \Delta T$, then equation (8.55) simplifies considerably to become

$$V_{i,j+1} = V_{i-1,j} - V_{i,j-1} + V_{i+1,j} \tag{8.56}$$

which fortuitously turns out to be the exact solution. Other values of the ratio $\frac{\Delta T}{\Delta X}$ could be used, but stability is only guaranteed if the ratio is less than one.

Example 8.9

Solve the wave equation

$$\frac{\partial^2 V}{\partial X^2} = \frac{\partial^2 V}{\partial T^2}$$

in the range $0 \leq X \leq 1$, $T \geq 0$ subject to the following initial and boundary conditions

$$\text{At } X = 0 \text{ and } X = 1, \quad U = 2\sin\left(\frac{\pi T}{5}\right) \text{ for } T > 0$$

$$\text{When } T = 0, \quad U = \frac{\partial U}{\partial T} = 0 \text{ for } 0 < X < 1$$

Solution 8.9

Choose $\Delta X = \Delta T = 0.1$ and take account of symmetry. The finite difference equation (8.56) requires information from two preceding steps in order to proceed. In order to compute the result corresponding to $T = 0.1$, the assumption is made that no initial change in the value of U has occurred for $X > 0$. This is equivalent to a simple Euler step based on the initial condition that $\frac{\partial U}{\partial T} = 0$. Only half the problem is considered due to the symmetry of the boundary conditions at $X = 0$ and $X = 1$ implying that $\frac{\partial U}{\partial X} = 0$ at $X = 0.5$. In order to enforce this derivative boundary condition, a solution has been included at $X = 0.6$ that is always held equal to that at $X = 0.4$. Application of equation (8.56) leads to the results shown in Table 8.4.

TABLE 8.4: Tabulated values of $U = f(Z,T)$ from Example 8.9

Z T	0.0	0.1	0.2	0.3	0.4	0.5	0.6
0.0	0.0	0.0	0.0	0.0	0.0	0.0	0.0
0.1	0.1256	0.0	0.0	0.0	0.0	0.0	0.0
0.2	0.2507	0.1256	0.0	0.0	0.0	0.0	0.0
0.3	0.3748	0.2507	0.1256	0.0	0.0	0.0	0.0
0.4	0.4974	0.3748	0.2507	0.1256	0.0	0.0	0.0
0.5	0.6180	0.4974	0.3748	0.2507	0.1256	0.0	0.1256
0.6	0.7362	0.6180	0.4974	0.3748	0.2507	0.2512	0.2507
0.7	0.8516	0.7362	0.6180	0.4974	0.5003	0.5013	0.5003

8.6   Finite element method

The Finite Element Method uses a quite different solution philosophy to finite differences, in that the governing differential equation is solved approximately over subregions of the solution domain. These subregions are called "finite elements" and are usually confined to simple shapes - triangles or quadrilaterals for plane areas and tetrahedra or hexahedra ("bricks") for volumes. The elements may have curved sides (faces) but are more usually limited to linear sides and plane faces.

Taking Figure 8.8 as an example, a finite element equivalent is shown in Figure 8.15.

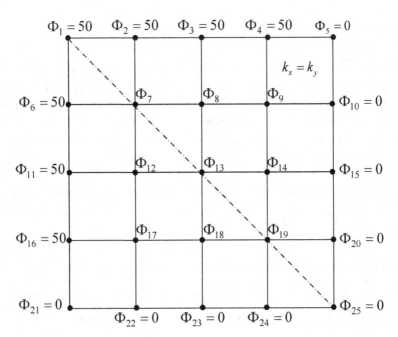

Figure 8.15: Finite element mesh from Example 8.3.

The area is divided into 16 equal-sized quadrilateral elements (all square in this case) and has 25 "nodes", equivalent to the finite difference "grid points". In this case it is simpler to treat all the nodal temperatures as "unknowns", even at the boundaries. A typical quadrilateral element is shown in Figure 8.16.

It has 4 nodes at the corners, and the variation of the unknown ϕ across the element is described in local coordinates by "shape functions" which we met

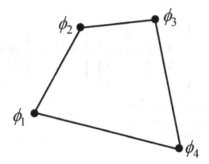

Figure 8.16: General 4-node quadrilateral element.

previously in Chapter 6, Section 6.6.2. Using these functions, the properties of the element are described by a matrix equation relating input $\{q\}$ at the nodes (fluxes etc. in this case) to the output $\{\phi\}$ (temperature) values in the general form

$$[\mathbf{k_c}]\{\phi\} = \{\mathbf{q}\} \qquad (8.57)$$

where $[\mathbf{k_c}]$ is the "conductivity matrix".

If the elements are rectangular as shown in Figure 8.17, the conductivity matrix is easily written down in closed form as

$$[\mathbf{k_c}] = \frac{k_x}{6} \frac{b}{a} \begin{bmatrix} 2 & 1 & -1 & -2 \\ 1 & 2 & -2 & -1 \\ -1 & -2 & 2 & 1 \\ -2 & -1 & 1 & 2 \end{bmatrix} + \frac{k_y}{6} \frac{a}{b} \begin{bmatrix} 2 & -2 & -1 & 1 \\ -2 & 2 & 1 & -1 \\ -1 & 1 & 2 & -2 \\ 1 & -1 & -2 & 2 \end{bmatrix} \qquad (8.58)$$

where k_x and k_y are thermal conductivities in the x- and y-directions.

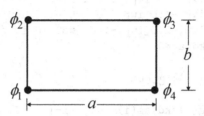

Figure 8.17: Rectangular 4-node element.

Further simplification for the current example with square elements ($a = b$) and isotropic properties ($k_x = k_y$) leads to the matrix equation

$$\frac{1}{6} \begin{bmatrix} 4 & -1 & -2 & -1 \\ -1 & 4 & -1 & -2 \\ -2 & -1 & 4 & -1 \\ -1 & -2 & -1 & 4 \end{bmatrix} \begin{Bmatrix} \phi_1 \\ \phi_2 \\ \phi_3 \\ \phi_4 \end{Bmatrix} = \begin{Bmatrix} q_1 \\ q_2 \\ q_3 \\ q_4 \end{Bmatrix} \qquad (8.59)$$

To solve the complete boundary value problem, the 16 $[\mathbf{k}_c]$ matrices must be assembled (involving summation at the nodes) to yield a global set of equations in matrix form

$$[\mathbf{K}_c]\{\mathbf{\Phi}\} = \{\mathbf{Q}\} \qquad (8.60)$$

relating the 25 nodal temperatures $\{\mathbf{\Phi}\}$ to the 25 nodal *net* fluxes $\{\mathbf{Q}\}$. The net flux vector $\{\mathbf{Q}\}$ is quite sparse, containing nonzero terms only at the boundary nodes where temperatures are fixed.

Although this example has no insulated (no-flow) boundaries, these are satisfied automatically in the finite element treatment should they be required. The condition $\phi =$ constant is enforced by the "penalty" approach we met in Chapter 2, Section 2.8.

Program 8.2: Simple FE analysis of Example 8.3

```
PROGRAM nm82
!---Simple FE analysis of Example 8.3---
 USE nm_lib; USE precision; IMPLICIT NONE
 INTEGER::g_num(4,16),i,iel,kdiag(25),num(4)
 REAL(iwp)::kc(4,4),loads(25); REAL(iwp),ALLOCATABLE::kv(:)
 OPEN(11,FILE='nm95.res')
 WRITE(11,'(A/)')"---Simple FE analysis of Example 8.3---"
 kc(1,:)=(/4.0_iwp,-1.0_iwp,-2.0_iwp,-1.0_iwp/)
 kc(2,:)=(/-1.0_iwp,4.0_iwp,-1.0_iwp,-2.0_iwp/)
 kc(3,:)=(/-2.0_iwp,-1.0_iwp,4.0_iwp,-1.0_iwp/)
 kc(4,:)=(/-1.0_iwp,-2.0_iwp,-1.0_iwp,4.0_iwp/)
 kc=kc/6.0_iwp; kdiag=0
 DO iel=1,25
   num=geometry(iel); g_num(:,iel)=num; CALL fkdiag(kdiag,num)
 END DO
 DO i=2,25; kdiag(i)=kdiag(i)+kdiag(i-1); END DO
 ALLOCATE(kv(kdiag(25))); kv=0.0_iwp; loads=0.0_iwp
 DO iel=1,25
   num=g_num(:,iel); CALL fsparv(kv,kc,num,kdiag)
 END DO
 kv(kdiag(:5))=kv(kdiag(:5))+1.E20_iwp
```

```
loads(:4)=kv(kdiag(:4))*50.0_iwp
kv(kdiag(6:21:5))=kv(kdiag(6:21:5))+1.E20_iwp
loads(6:16:5)=kv(kdiag(6:16:5))*50.0_iwp
kv(kdiag(10:25:5))=kv(kdiag(10:25:5))+1.E20_iwp
kv(kdiag(22:24))=kv(kdiag(22:24))+1.E20_iwp
CALL sparin(kv,kdiag); CALL spabac(kv,loads,kdiag)
WRITE(11,'(A)') " Node   Temperature"
DO i=1,25; WRITE(11,'(I5,F12.2)')i,loads(i); END DO
CONTAINS

FUNCTION geometry(iel)
!--Element Node Numbers for Example 8.3, 4-Node only 4x4 mesh--
IMPLICIT NONE
INTEGER,INTENT(IN)::iel; REAL(iwp)::geometry(4)
INTEGER::ip,iq; iq=(iel-1)/4+1; ip=iel-(iq-1)*4
geometry =(/iq*5+ip,(iq-1)*5+ip,(iq-1)*5+ip+1,iq*5+ip+1/)
END FUNCTION geometry

END PROGRAM nm82
```

There is no user input since the program deals only with the specific case of Example 8.3. The program begins with a specification of the typical $[\mathbf{k}_c]$ matrix (called kc). The linear equations leading to the solution are solved by the method described in Chapter 2, Program 2.5 so it is necessary to create the integer vector (kdiag) which defines the structure of the coefficient matrix (kv). This is done automatically using the library SUBROUTINE fkdiag. The specific geometry is attached as FUNCTION geometry. The assembly process is automated using SUBROUTINE fsparv.

It remains to enforce the boundary condition of $\phi = 50.0$ and $\phi = 0.0$ as indicated in Figure 8.15 along the sides. The resulting equations are solved using SUBROUTINEs sparin and spabac.

Output is listed as Results 8.2 and can be seen to be similar to those listed in Solution 8.3 from a finite difference analysis. The differences would become even smaller as the grids are refined. It may be noted that in the finite difference solution of this problem described in Example 8.3, the temperatures at the corners of the problems were unspecified.

```
---Simple FE analysis of Example 8.3---
```

Node	Temperature	Node	Temperature
1	50.00	14	12.41
2	50.00	15	0.00
3	50.00	16	50.00
4	50.00	17	21.47
5	0.00	18	12.41

6	50.00	19	6.05
7	43.15	20	0.00
8	35.80	21	0.00
9	21.47	22	0.00
10	0.00	23	0.00
11	50.00	24	0.00
12	35.80	25	0.00
13	23.57		

Results 8.2: Simple FE analysis of Example 8.3

List 8.2:

Scalar integers:
i   simple counter
iel   simple counter
kdiag diagonal term locations
num   element node numbers

Integer arrays:
g_num   global element node numbers matrix

Real arrays:
kc   element conductivity matrix
loads   global rhs and solution vector

Dynamic real arrays:
kv   global conductivity matrix

8.7   Exercises

1. Given the equation

$$\sqrt{x}\frac{\partial z}{\partial x} + z\frac{\partial z}{\partial y} = -z^2$$

use a numerical method to estimate $z(3.5, y)$ on the characteristic through $(3,0)$ given that $z = 1$ at all points on the x-axis.
Answer: 0.783

2. Given that $z(1,1) = 1$ satisfies the equation

$$z(xp - yzq) = y^2 - x^2$$

where

$$p = \frac{\partial z}{\partial x} \quad \text{and} \quad q = \frac{\partial z}{\partial y}$$

find the values of y and z when $x = 1.5$ on the characteristic passing through $(1,1)$.

Answer: $y = 0.702, \quad z = 0.562$

3. Use finite differences to estimate the steady state temperature distribution in the isotropic square plate shown in Figure 8.18.

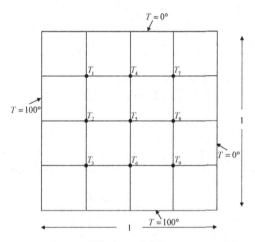

Figure 8.18

Answer: $T_1 = 85.71$, $T_2 = T_4 = 71.43$, $T_5 = 50.00$, $T_3 = T_7 = 50.00$, $T_9 = 14.29$, $T_6 = T_8 = 28.57$

4. Repeat question 3 assuming anisotropic conductivity properties in which $k_y = 3k_x$.

Answer: $T_1 = 42.44$, $T_2 = 36.76$, $T_3 = 28.15$, $T_4 = 33.93$, $T_5 = 25.00$, $T_6 = 16.07$, $T_7 = 21.85$, $T_8 = 13.24$, $T_9 = 7.56$

5. Figure 8.19 represents the cross-section of a long square bar of side length 8 cm subjected to pure torsion. The stress function ϕ is distributed across the section according to Poisson's equation

$$\frac{\partial^2 \phi}{\partial x^2} + \frac{\partial^2 \phi}{\partial y^2} + 2 = 0$$

Given that $\phi = 0$ on all the boundaries, use finite differences to estimate the value of ϕ at the internal grid points.

Answer: $\phi_1 = \phi_3 = \phi_7 = \phi_9 = 5.5$, $\phi_2 = \phi_4 = \phi_6 = \phi_8 = 7$, $\phi_5 = 9$

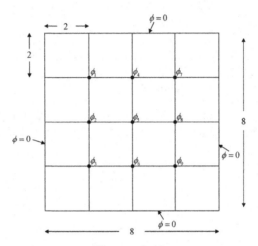

Figure 8.19

6. The variation of quantity T, with respect to Cartesian coordinates x and y, is defined by the fourth order differential equation

$$\frac{\partial^4 T}{\partial x^4} + \frac{\partial^4 T}{\partial y^4} = 0$$

Derive the finite difference form of this equation centered at E in Figure

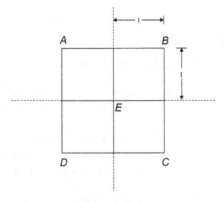

Figure 8.20

8.20 using a square grid of side length unity.

Given the boundary conditions

$$\text{AB: } T = 0, \quad \frac{\partial^2 T}{\partial y^2} = -10; \quad \text{CD: } T = 0, \quad \frac{\partial^2 T}{\partial y^2} = -20$$

$$\text{BC: } T = 0, \quad \frac{\partial T}{\partial x} = -50; \quad \text{DA: } T = 0, \quad \frac{\partial T}{\partial x} = 30$$

estimate the value of T at point E.

Answer: $T = 15.83$

7. The steady two-dimensional distribution in an isotropic heat conducting material is given by Laplace's equation

$$\frac{\partial^2 \phi}{\partial x^2} + \frac{\partial^2 \phi}{\partial y^2} = 0$$

where ϕ is temperature and x, y represent the Cartesian coordinate system. Using a finite difference form of this equation, find temperatures T_1, T_2, T_3 and T_4 for the metal plate shown in Figure 8.21. The thermal

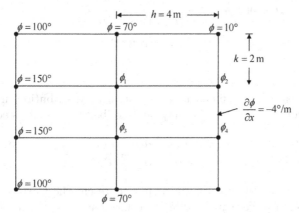

Figure 8.21

energy, Q, stored in the plate is calculated from

$$Q = \rho c \int T \, dV$$

where Q is in Joules (J), ρ is density ($=8000 \text{ kg/m}^3$), c is the specific heat capacity of the material ($=1000 \text{ J/kg/}^\circ \text{ C}$) and dV is an increment of volume (m^3). If the thickness of the plate is 50 mm, use the Repeated Trapezoid Rule to estimate Q.

Answer: $T_1 = T_3 = 77.33^\circ$, $T_2 = T_4 = 34.00^\circ$, $Q = 1.55GJ$

8. A square 2-d anisotropic domain is subjected to steady seepage with the boundary conditions indicated in Figure 8.22. Using the square finite difference grid indicated, set up four symmetric simultaneous equations in matrix form and solve for the unknown heads H_1 through H_4.

   Answer: $T_1 = 77.06$, $T_2 = 60.98$, $T_3 = 87.84$, $T_4 = 72.94$

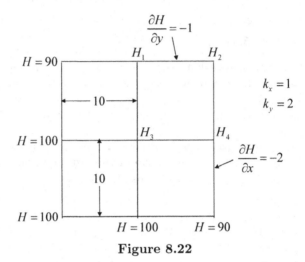

Figure 8.22

9. The steady three-dimensional temperature distribution in an isotropic heat conducting material may be described by Laplace's equation

$$\frac{\partial^2 \phi}{\partial x^2} + \frac{\partial^2 \phi}{\partial y^2} + \frac{\partial^2 \phi}{\partial z^2} = 0$$

   where ϕ is temperature and x, y, z represent the Cartesian coordinate system. Derive a finite difference formula to represent this equation, and use it with a grid size of 0.5 in all three directions to estimate the steady state temperature at the centroid of a unit cube given that

$$\phi(0, y, z) = 100 \qquad \frac{\partial \phi}{\partial x}(1, y, z) = -20 \qquad 0 \le y, z \le 1$$

$$\phi(x, 0, z) = 40 \qquad \phi(x, 1, z) = 70 \qquad 0 \le x, z \le 1$$

$$\phi(x, y, 0) = 40 \qquad \phi(x, y, 1) = 70 \qquad 0 \le x, y \le 1$$

   If the final boundary conditions were replaced by

$$\frac{\partial \phi}{\partial z}(x, y, 0) = \frac{\partial \phi}{\partial z}(x, y, 1) = 0 \qquad 0 \le x, y \le 1$$

how could the analysis be simplified?
Answer: $T = 60.29$. Change in bc leads to a 2-d analysis where $T = 66.43$

10. A horizontal clay stratum of thickness 5 m is subjected to a loading which produces a pressure distribution which varies from p kN/m^2 at the top to $0.5p$ kN/m^2 at the bottom.

   By subdividing the soil into five layers and using steps of one month, use a finite difference approximation to estimate the excess pore pressures distribution and hence the Average Degree of Consolidation after five months. Let $c_v = 3\,\text{m}^2/\text{yr}$ and assume double-drainage.
   Answer: 54%

11. The average coefficient of consolidation of a saturated clay layer 5m is $1.93\,\text{m}^2/\text{yr}$. The layer is drained at the top only and is subjected to an initial vertical stress distribution that decreases linearly from 100 kN/m^2 at the top to zero at the base. Select a reasonable finite difference scheme to estimate the average degree of consolidation after 3 months.
   Answer: Assuming drainage takes effect after one time step, $U = 0.28$

12. Show that the computational "molecules" for the square grids and conditions shown in Figures 8.23a and b applied to a steady seepage problem

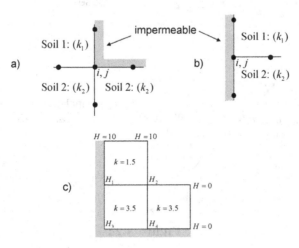

Figure 8.23

are given, respectively, by the algebraic equation

$$(2k_1+4k_2)\ H_{i,j}-(k_1+k_2)\ H_{i-1,j}-k_1\ H_{i,j+1}-k_2\ H_{i+1,j}-2k_2\ H_{i,j-1} = 0$$

and

$$(2k_1 + 2k_2)\, H_{i,j} - k_1\, H_{i,j+1} - (k_1 + k_2)\, H_{i+1,j} - k_2\, H_{i,j-1} = 0$$

Use these expressions and any others you may need to set up and solve the four symmetric simultaneous equations for the unknown total head values in Figure 8.23c.

Answer: $T_1 = 4.19$, $T_2 = 3.09$, $T_3 = 3.28$, $T_4 = 2.36$

13. Derive finite difference "molecules" for the two cases shown in Figure 8.24 assuming a square grid and isotropic properties.

    Hint: One way of dealing with a sloping impermeable boundary in a finite difference analysis is to replace it by a step function.
    Answer: Case 1: $3h_{i,j} - 2h_{i,j+1} - h_{i+1,j} = 0$, Case 2: $5h_{i,j} - 2h_{i,j+1} - 2h_{i+1,j} - h_{i,j-1} = 0$

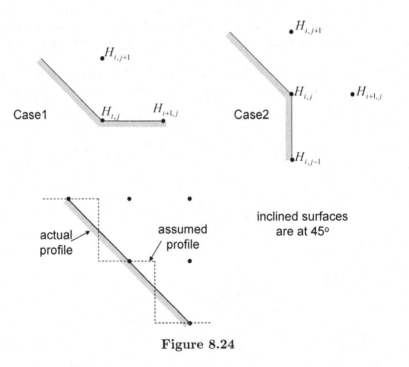

Figure 8.24

14. Use finite differences to solve for the total head at all the internal grid points of the isotropic steady flow problem shown in Figure 8.25.
    Answer: $T_1 = 49.58$, $T_2 = 48.65$, $T_3 = 41.65$, $T_4 = 37.76$, $T_5 = 37.14$, $T_6 = 25.29$, $T_7 = 24.07$, $T_8 = 23.72$

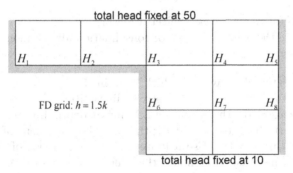

Figure 8.25

15. Use finite differences to solve for the total head at all the internal grid points of the isotropic steady flow problem shown in Figure 8.26. *Answer:* $T_1 = 18.67$, $T_2 = 18.40$, $T_3 = 17.50$, $T_4 = 17.18$, $T_5 = 11.18$, $T_6 = 11.16$

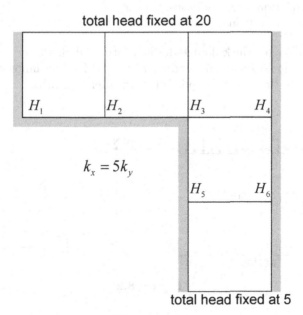

Figure 8.26

16. The average coefficient of consolidation of a saturated clay layer 5m is $3.21 \, \text{m}^2/\text{yr}$.

    The layer is drained at the top only, and is subjected to an initial vertical stress distribution that increases linearly from zero at the top to

100 kN/m^2 at the base. Select a reasonable finite difference scheme to
estimate the average degree of consolidation after 3 months.
Answer: $U = 0.06$

17. A saturated clay layer 8m thick is drained at the top only, and is subjected to an initial vertical stress distribution that increases linearly
from zero at depth 0m, to 150 kN/m^2 at depth 4m, and then decreases linearly back to zero at depth 8m. The coefficient of consolidation
is 0.193 m^2/month. Use a finite difference grid size of $\Delta z = $ 2m and
$\Delta t = $ 4 months, to estimate the average degree of consolidation after
one year.
Answer: $U = 0.14$

18. The average coefficient of consolidation of a saturated clay layer 5m
thick is 0.161 m^2/month. The layer is double-drained and subjected to
an initial vertical stress distribution that decreases linearly from 200
kN/m^2 at the top to 100 kN/m^2 at the base. Use a finite difference grid
with $\Delta z = $ 1m and $\Delta t = $ 1 month, to estimate the average degree of
consolidation after 4 months.
Answer: $U = 0.365$

19. A four meter thick double-drained saturated clay layer is subjected to
the ramp loading indicated in Figure 8.27. Use finite differences to
estimate the mid-plane pore pressure after 6 months.

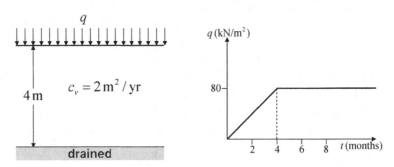

Figure 8.27

Answer: 66.5 kN/m^2 with $\Delta t = $ 1 month, $\Delta z = $ 1m
71.1 kN/m^2 with $\Delta t = $ 2 month, $\Delta z = $ 1m

20. A layer of clay 8m thick ($c_v = $ 2.4m^2/yr), drained at the top only, is
subjected to a triangular initial pore pressure distribution that varies
from zero at the ground surface to 100 kN/m^2 at the bottom. Use a
finite difference grid with $\Delta z = $ 2m and $\Delta t = $ 4 month to estimate the

average degree of consolidation of the layer after 2 years.
Answer: $U = 0.15$

21. Classify each of the following equations as elliptic, hyperbolic or parabolic.

a) $\dfrac{\partial^2 u}{\partial x^2} - \dfrac{\partial^2 u}{\partial y^2} = 0$,   b) $\dfrac{\partial u}{\partial x} + \dfrac{\partial^2 u}{\partial x \partial y} = 8$,

c) $\dfrac{\partial^2 u}{\partial x^2} - 2\dfrac{\partial^2 u}{\partial x \partial y} + 2\dfrac{\partial^2 u}{\partial y^2} = x + 3y$,

d) $\dfrac{\partial^2 u}{\partial x^2} + 3\dfrac{\partial^2 u}{\partial x \partial y} + 4\dfrac{\partial^2 u}{\partial x^2} + 5\dfrac{\partial u}{\partial x} - 2\dfrac{\partial u}{\partial y} + 4u = 2x - 6y$,

e) $\dfrac{\partial^2 u}{\partial x^2} - 7\dfrac{\partial^2 u}{\partial x \partial y} + \dfrac{\partial^2 u}{\partial y^2} = 0$,   f) $\dfrac{\partial^2 u}{\partial x^2} + \dfrac{\partial^2 u}{\partial x \partial y} - 6\dfrac{\partial^2 u}{\partial y^2} = 0$,

g) $\dfrac{\partial^2 u}{\partial x^2} + 6\dfrac{\partial^2 u}{\partial x \partial y} + 9\dfrac{\partial^2 u}{\partial y^2} = 0$

Answer: a) H, b) H, c) E, d) E, e) H, f) H, g) P

22. A function $u(x,t)$ is to satisfy the differential equation

$$\frac{\partial^2 u}{\partial x^2} = \frac{\partial^2 u}{\partial t^2}$$

in the solution domain $0 \leq x \leq 10$ and $0 \leq t \leq \infty$.
The following boundary/initial conditions apply

$$u(0,t) = \frac{\partial u}{\partial x}(10,t) = 0 \quad \text{for all } t$$

$$u(x,0) = x(20-x) \quad \text{for } 0 \leq x \leq 10$$

$$\frac{\partial u}{\partial t}(x,0) = 0 \quad \text{for } 0 \leq x \leq 10$$

Let $\Delta x = \Delta t = 2$ and hence estimate $u(x,20)$ for $x = 2, 4, 6$ and 8.
Answer: $u(2,20) = -36$, $u(4,20) = -64$, $u(6,20) = -84$, $u(8,20) = -96$

Appendix A

Descriptions of Library Subprograms

This Appendix describes the library SUBROUTINEs and FUNCTIONs attached to the majority of programs described in this book through the command USE nm_lib. The subprograms are listed in alphabetic order and include brief descriptions of their arguments. Arguments in **bold** are those returned by the subprograms with the attribute INTENT(OUT).

Source code for all these subprograms can be downloaded from the web site:

http://www.mines.edu/~vgriffit/NM

Name	Arguments	Description
check	x1,x0,tol	LOGICAL FUNCTION returns .TRUE. if scalar variables x1 and x0 have an absolute relative difference less than tol. Returns .FALSE. otherwise.
checkit	loads,oldlds, tol	LOGICAL FUNCTION returns .TRUE. if all components of vectors loads and oldlds have an absolute relative difference less than tol. Returns .FALSE. otherwise.
chobac	kb,**loads**	Performs forward and back-substitution on matrix kb factorized by cholin. **loads** holds rhs on entry and solution on exit.
cholin	kb	Performs Choleski factorization on a symmetric banded matrix **kb** stored as a lower rectangle by rows. **kb** overwritten by factorized coefficients on exit.
determinant	jac	FUNCTION returns the determinant of a 2×2 or 3×3 square matrix jac.
eliminate	a,b	FUNCTION returns the solution vector of linear equations where a is the coefficient matrix and b the right hand side.
fkdiag	**kdiag**,g	Returns maximum bandwidth **kdiag** for each row of a skyline storage system from g.
fsparv	**kv**,km, g,kdiag	Returns lower triangular global matrix **kv** stored as a vector in skyline form, from symmetric element matrix km and steering vector g. kdiag holds the locations of the diagonal terms.

Name	Arguments	Description
fun_der	fun,der, points,i	Returns the derivatives **der** of the shape functions **fun** with respect to local coordinates at the Gauss-Legendre point held in points(i,:).
gauss_laguerre	samp,wt	Returns the sampling points **samp** and weighting coefficients **wt** for Gauss-Laguerre numerical integration.
gauss_legendre	samp,wt	Returns the sampling points **samp** and weighting coefficients **wt** for Gauss-Legendre numerical integration.
inverse	matrix	FUNCTION returns the inverse of array matrix.
ldlfor	a,b	Performs forward substitution on factors in $\mathbf{LDL^T}$ factorization. **a** holds upper triangular factors stored as a lower triangle on which factorization is to be performed. **b** holds the rhs values overwritten by factored values.
ldlt	a,d	Factorizes a symmetric matrix using $\mathbf{LDL^T}$. **a** holds symmetric matrix coefficients on entry overwritten by $\mathbf{L}$ and $\mathbf{L^T}$ on exit. **d** is a vector holding the diagonal elements of $\mathbf{D}$.
lufac	a, lower,upper	Factorizes a matrix using $\mathbf{LU}$. **a** holds matrix coefficients. **lower** and **upper** hold $\mathbf{U}$ and $\mathbf{L}$, the lower and upper triangle factors respectively. **kdiag** holds the locations of the diagonal terms.
lupfac	a,row	Factorizes a matrix using $\mathbf{LU}$ with pivoting. **a** holds matrix coefficients on entry and $\mathbf{LU}$ factors after pivoting on exit. **row** holds reordered row numbers on exit.
lupsol	a,b, sol,row	Performs forward and back-substitution on factors in $\mathbf{LU}$ method with pivoting. **a** holds $\mathbf{LU}$ factors. **b** holds the rhs vector. **sol** holds the solution vector and **row** holds reordered row numbers.
newton_cotes	samp,wt	Returns the sampling points **samp** and weighting coefficients **wt** for Newton-Cotes numerical integration.
norm	x	FUNCTION returns the "L2 norm" of vector **x**.
spabac	kv,loads, kdiag	Returns solution **loads** which overwrites rhs by forward and back-substitution on (Choleski) factorised vector **kv** stored as a skyline. **kdiag** holds the locations of the diagonal terms.
sparin	kv,kdiag	Returns the (Choleski) factorised vector **kv** stored as a skyline. **kdiag** holds the locations of the diagonal terms.
subbac	a,b	Performs back-substitution on factors following forward substitution from $\mathbf{LDL^T}$ or $\mathbf{LU}$ methods. **a** holds upper triangular factors for back-substitution. **b** holds the rhs values following subfor on entry and solution vector on exit.
subfor	a,b	Performs forward-substitution on factors from $\mathbf{LU}$ meth **a** holds lower triangular factors for forward-substitution. **b** holds the rhs vector on entry and factored vector on e

Appendix B

Fortran 95 Listings of Library Subprograms

```fortran
LOGICAL FUNCTION check(x1,x0,tol)
!---Checks Convergence of Two Scalars---
USE precision; IMPLICIT NONE
!---.TRUE. if converged; no update of x0---
REAL(iwp)::x0,x1,tol
check=.NOT.ABS(x1-x0)/ABS(x0)>tol
RETURN
END FUNCTION check
```

```fortran
LOGICAL FUNCTION checkit(loads,oldlds,tol)
!---Checks Convergence of Two Vectors---
USE precision; IMPLICIT NONE
!---.TRUE. if converged; no update of oldlds---
REAL(iwp),INTENT(IN)::loads(:),oldlds(:),tol
REAL(iwp)::big; INTEGER::i,neq; LOGICAL::converged
neq=UBOUND(loads,1); big=.0_iwp; converged=.TRUE.
DO i=1,neq; IF(ABS(loads(i))>big)big=ABS(loads(i)); END DO
DO i=1,neq; IF(ABS(loads(i)-oldlds(i))/big>tol)converged=.FALSE.
END DO
checkit=converged
RETURN
END FUNCTION checkit
```

```fortran
SUBROUTINE chobac(kb,loads)
!---Choleski Forward and Back-substitution---
USE precision; IMPLICIT NONE
REAL(iwp),INTENT(IN)::kb(:,:)
REAL(iwp),INTENT(in out)::loads(:)
INTEGER::iw,n,i,j,k,l,m; REAL(iwp)::x
n=SIZE(kb,1); iw=SIZE(kb,2)-1
loads(1)=loads(1)/kb(1,iw+1)
DO i=2,n
  x=.0_iwp;k=1
```

```
    IF(i<=iw+1)k=iw-i+2
    DO j=k,iw; x=x+kb(i,j)*loads(i+j-iw-1); END DO
    loads(i)=(loads(i)-x)/kb(i,iw+1)
  END DO
  loads(n)=loads(n)/kb(n,iw+1)
  DO i=n-1,1,-1
    x=0.0_iwp; l=i+iw
    IF(i>n-iw)l=n; m=i+1
    DO j=m,l; x=x+kb(j,iw+i-j+1)*loads(j); END DO
    loads(i)=(loads(i)-x)/kb(i,iw+1)
  END DO
RETURN
END SUBROUTINE chobac
```

```
SUBROUTINE cholin(kb)
!---Choleski Factorization on a Lower Triangle Stored as a Band--
USE precision; IMPLICIT NONE
REAL(iwp),INTENT(in out)::kb(:,:); INTEGER::i,j,k,l,ia,ib,n,iw
REAL(iwp)::x
n=UBOUND(kb,1); iw=UBOUND(kb,2)-1
DO i=1,n
  x=.0_iwp
  DO j=1,iw; x=x+kb(i,j)**2; END DO
  kb(i,iw+1)=SQRT(kb(i,iw+1)-x)
  DO k=1,iw
    x=.0_iwp
    IF(i+k<=n)THEN
      IF(k/=iw)THEN
        DO l=iw-k,1,-1
          x=x+kb(i+k,l)*kb(i,l+k)
        END DO
      END IF
      ia=i+k; ib=iw-k+1
      kb(ia,ib)=(kb(ia,ib)-x)/kb(i,iw+1)
    END IF
  END DO
END DO
RETURN
END SUBROUTINE cholin
```

```
REAL FUNCTION determinant(jac)
!---Returns Determinant of a 1x1, 2x2 or 3x3 Jacobian Matrix---
USE precision; IMPLICIT NONE
REAL(iwp),INTENT(IN)::jac(:,:); REAL(iwp)::det; INTEGER::it
it=UBOUND(jac,1)
```

```
SELECT CASE(it)
CASE(1)
  det=1.0_iwp
CASE(2)
  det=jac(1,1)*jac(2,2)-jac(1,2)*jac(2,1)
CASE(3)
  det=jac(1,1)*(jac(2,2)*jac(3,3)-jac(3,2)*jac(2,3))
  det=det-jac(1,2)*(jac(2,1)*jac(3,3)-jac(3,1)*jac(2,3))
  det=det+jac(1,3)*(jac(2,1)*jac(3,2)-jac(3,1)*jac(2,2))
CASE DEFAULT
  WRITE(*,*)"Wrong dimension for Jacobian matrix"
END SELECT
determinant=det
RETURN
END FUNCTION determinant
```

```
FUNCTION eliminate(a,b)
!---Gaussian elimination with partial pivoting on n*n
! matrix a and rhs b---
USE precision; IMPLICIT NONE
REAL(iwp),INTENT(IN OUT)::a(:,:),b(:)
REAL(iwp)::eliminate(UBOUND(b,1))
REAL(iwp)::big,hold,fac
INTEGER::i,j,l,n,ihold; n=UBOUND(a,1)
!----------------------- Pivoting stage ---------------------
DO i=1,n-1
  big=ABS(a(i,i)); ihold=i
  DO j=i+1,n
    IF(ABS(a(j,i))>big)THEN
      big=ABS(a(j,i)); ihold=j
    END IF
  END DO
  IF(ihold/=i)THEN
    DO j=i,n
      hold=a(i,j); a(i,j)=a(ihold,j); a(ihold,j)=hold
    END DO
    hold=b(i); b(i)=b(ihold); b(ihold)=hold
  END IF
!----------------------- Elimination stage -------------------
  DO j=i+1,n
    fac=a(j,i)/a(i,i)
    DO l=i,n
      a(j,l)=a(j,l)-a(i,l)*fac
    END DO
    b(j)=b(j)-b(i)*fac
```

```
    END DO
  END DO
!------------------------ Backsubstitution -------------------
  DO i=n,1,-1
    hold=.0_iwp
    DO l=i+1,n; hold=hold+a(i,l)*b(l); END DO
    b(i)=(b(i)-hold)/a(i,i)
  END DO
  eliminate = b
RETURN
END FUNCTION eliminate
```

```
SUBROUTINE fkdiag(kdiag,g)
!---Computes the Skyline Profile---
 IMPLICIT NONE
 INTEGER,INTENT(IN)::g(:); INTEGER,INTENT(OUT)::kdiag(:)
 INTEGER::idof,i,iwp1,j,im,k; idof=SIZE(g)
 DO i=1,idof
   iwp1=1
   IF(g(i)/=0)THEN
     DO j=1,idof
       IF(g(j)/=0)THEN
          im=g(i)-g(j)+1; IF(im>iwp1)iwp1=im
       END IF
     END DO
     k=g(i); IF(iwp1>kdiag(k))kdiag(k)=iwp1
   END IF
 END DO
RETURN
END SUBROUTINE fkdiag
```

```
SUBROUTINE fsparv(kv,km,g,kdiag)
!---Assembles Element Matrices into a
! Symmetric Skyline Global Matrix---
 USE precision; IMPLICIT NONE
 INTEGER,INTENT(IN)::g(:),kdiag(:)
 REAL(iwp),INTENT(IN)::km(:,:); REAL(iwp),INTENT(OUT)::kv(:)
 INTEGER::i,idof,k,j,iw,ival; idof=UBOUND(g,1)
 DO i=1,idof
   k=g(i)
   IF(k/=0)THEN
     DO j=1,idof
       IF(g(j)/=0)THEN
          iw=k-g(j)
```

```
      IF(iw>=0)THEN
         ival=kdiag(k)-iw; kv(ival)=kv(ival)+km(i,j)
      END IF
    END IF
  END DO
 END IF
END DO
RETURN
END SUBROUTINE fsparv
```

```
SUBROUTINE fun_der(fun,der,points,i)
!---Computes Derivatives of Shape Functions w.r.t.
! Local Coordinates---
USE precision; IMPLICIT NONE
INTEGER,INTENT(IN)::i; REAL(iwp),INTENT(IN)::points(:,:)
REAL(iwp),INTENT(OUT)::der(:,:),fun(:); REAL(iwp)::eta,xi,zeta,&
  xi0,eta0,zeta0,etam,etap,xim,xip,c1,c2,c3
REAL(iwp)::t1,t2,t3,t4,t5,t6,t7,t8,t9,x2p1,x2m1,e2p1,e2m1,    &
  zetam,zetap
REAL(iwp),PARAMETER::zero=0.0_iwp,pt125=0.125_iwp,            &
  pt25=0.25_iwp,pt5=0.5_iwp,pt75=0.75_iwp,one=1.0_iwp,        &
  two=2.0_iwp,d3=3.0_iwp,d4=4.0_iwp,d5=5.0_iwp,d6=6.0_iwp,    &
  d8=8.0_iwp,d9=9.0_iwp,d10=10.0_iwp,d11=11.0_iwp,d12=12.0_iwp,&
  d16=16.0_iwp,d18=18.0_iwp,d27=27.0_iwp,d32=32.0_iwp,        &
  d36=36.0_iwp,d54=54.0_iwp,d64=64.0_iwp,d128=128.0_iwp
INTEGER::xii(20),etai(20),zetai(20),l,ndim,nod
ndim=UBOUND(der,1)
SELECT CASE(ndim)
CASE(1)        ! one dimension
  xi= points(i,1)
  der(1,1)=-pt5; der(1,2)= pt5
  fun=(/pt5*(one-xi),pt5*(one+xi)/)
CASE(2)        ! two dimensions
  xi= points(i,1); eta=points(i,2)
  etam=pt25*(one-eta); etap=pt25*(one+eta)
  xim= pt25*(one-xi); xip= pt25*(one+xi)
  der(1,1)=-etam; der(1,2)=-etap; der(1,3)=etap; der(1,4)=etam
  der(2,1)=-xim; der(2,2)=xim; der(2,3)=xip; der(2,4)=-xip
  fun=(/d4*xim*etam,d4*xim*etap,d4*xip*etap,d4*xip*etam/)
CASE(3)        ! three dimensions
  xi =points(i,1); eta =points(i,2); zeta=points(i,3)
  etam=one-eta; xim=one-xi; zetam=one-zeta
  etap=eta+one; xip=xi+one; zetap=zeta+one
  der(1,1)=-pt125*etam*zetam; der(1,2)=-pt125*etam*zetap
  der(1,3)= pt125*etam*zetap; der(1,4)= pt125*etam*zetam
```

```
      der(1,5)=-pt125*etap*zetam; der(1,6)=-pt125*etap*zetap
      der(1,7)= pt125*etap*zetap; der(1,8)= pt125*etap*zetam
      der(2,1)=-pt125*xim*zetam; der(2,2)=-pt125*xim*zetap
      der(2,3)=-pt125*xip*zetap; der(2,4)=-pt125*xip*zetam
      der(2,5)= pt125*xim*zetam; der(2,6)= pt125*xim*zetap
      der(2,7)= pt125*xip*zetap; der(2,8)= pt125*xip*zetam
      der(3,1)=-pt125*xim*etam; der(3,2)= pt125*xim*etam
      der(3,3)= pt125*xip*etam; der(3,4)=-pt125*xip*etam
      der(3,5)=-pt125*xim*etap; der(3,6)= pt125*xim*etap
      der(3,7)= pt125*xip*etap; der(3,8)=-pt125*xip*etap
      fun=(/pt125*xim*etam*zetam,pt125*xim*etam*zetap,        &
             pt125*xip*etam*zetap,pt125*xip*etam*zetam,        &
             pt125*xim*etap*zetam,pt125*xim*etap*zetap,        &
             pt125*xip*etap*zetap,pt125*xip*etap*zetam/)
    CASE DEFAULT
      WRITE(*,*)"Wrong number of dimensions"
    END SELECT
  RETURN
  END SUBROUTINE fun_der
```

```
SUBROUTINE gauss_laguerre(samp,wt)
!---Provides Weights and Sampling Points for Gauss-Laguerre---
 USE precision; IMPLICIT NONE
 REAL(iwp),INTENT(OUT)::samp(:,:),wt(:); INTEGER::nsp
 nsp=UBOUND(samp,1)
 SELECT CASE(nsp)
   CASE(1)
     samp(1,1)= 1.0_iwp
     wt(1)= 1.0_iwp
   CASE(2)
     samp(1,1)= 0.58578643762690495119_iwp
     samp(2,1)= 3.41421356237309504880_iwp
     wt(1)=     0.85355339059327376220_iwp
     wt(2)=     0.14644660940672623779_iwp
   CASE(3)
     samp(1,1)= 0.41577455678347908331_iwp
     samp(2,1)= 2.29428036027904171982_iwp
     samp(3,1)= 6.28994508293747919686_iwp
     wt(1)=     0.71109300992917301544_iwp
     wt(2)=     0.27851773356924084880_iwp
     wt(3)=     0.01038925650158613574_iwp
   CASE(4)
     samp(1,1)= 0.32254768961939231180_iwp
     samp(2,1)= 1.74576110115834657568_iwp
     samp(3,1)= 4.53662029692112798327_iwp
```

```
      samp(4,1)= 9.39507091230113312923_iwp
      wt(1)=     0.60315410434163360163_iwp
      wt(2)=     0.35741869243779968664_iwp
      wt(3)=     0.03888790851500538427_iwp
      wt(4)=     0.00053929470556132745_iwp
   CASE(5)
      samp(1,1)= 0.26356031971814091020_iwp
      samp(2,1)= 1.41340305910651679221_iwp
      samp(3,1)= 3.59642577104072208122_iwp
      samp(4,1)= 7.08581000585883755692_iwp
      samp(5,1)= 12.64080084427578265994_iwp
      wt(1)=     0.52175561058280865247_iwp
      wt(2)=     0.39866681108317592745_iwp
      wt(3)=     0.07594244968170759538_iwp
      wt(4)=     0.00361175867992204845_iwp
      wt(5)=     0.00002336997238577622_iwp
   CASE DEFAULT
      WRITE(*,*)"Wrong number of integrating points"
 END SELECT
RETURN
END SUBROUTINE gauss_laguerre
```

```
SUBROUTINE gauss_legendre(samp,wt)
!---Provides Weights and Sampling Points for Gauss-Legendre---
USE precision; IMPLICIT NONE
REAL(iwp),INTENT(OUT)::samp(:,:),wt(:); INTEGER::nsp,ndim
nsp=UBOUND(samp,1); ndim=UBOUND(samp,2)
SELECT CASE(ndim)
CASE(1)
SELECT CASE(nsp)
   CASE(1)
      samp(1,1)=  0.0_iwp
      wt(1)   =   2.0_iwp
   CASE(2)
      samp(1,1)= -0.57735026918962576449_iwp
      samp(2,1)=  0.57735026918962576449_iwp
      wt(1)=      1.0_iwp
      wt(2)=      1.0_iwp
   CASE(3)
      samp(1,1)= -0.77459666924148337704_iwp
      samp(2,1)=  0.0_iwp
      samp(3,1)=  0.77459666924148337704_iwp
      wt(1)=      0.55555555555555555556_iwp
      wt(2)=      0.88888888888888888889_iwp
      wt(3)=      0.55555555555555555556_iwp
```

```
   CASE(4)
     samp(1,1)= -0.86113631159405257524_iwp
     samp(2,1)= -0.33998104358485626481_iwp
     samp(3,1)=  0.33998104358485626481_iwp
     samp(4,1)=  0.86113631159405257524_iwp
     wt(1)=      0.34785484513745385737_iwp
     wt(2)=      0.65214515486254614271_iwp
     wt(3)=      0.65214515486254614271_iwp
     wt(4)=      0.34785484513745385737_iwp
   CASE(5)
     samp(1,1)= -0.90617984593866399282_iwp
     samp(2,1)= -0.53846931010568309105_iwp
     samp(3,1)=  0.0_iwp
     samp(4,1)=  0.53846931010568309105_iwp
     samp(5,1)=  0.90617984593866399282_iwp
     wt(1)=      0.23692688505618908751_iwp
     wt(2)=      0.47862867049936646804_iwp
     wt(3)=      0.56888888888888888889_iwp
     wt(4)=      0.47862867049936646804_iwp
     wt(5)=      0.23692688505618908751_iwp
   CASE DEFAULT
     WRITE(*,*)"Wrong number of integrating points"
   END SELECT
 CASE(2)
   SELECT CASE(nsp)
   CASE(1)
     samp(1,1)=  0.0_iwp
     samp(1,2)=  0.0_iwp
     wt(1)=      4.0_iwp
   CASE(4)
     samp(1,1)= -0.57735026918962576449_iwp
     samp(2,1)=  0.57735026918962576449_iwp
     samp(3,1)= -0.57735026918962576449_iwp
     samp(4,1)=  0.57735026918962576449_iwp
     samp(1,2)= -0.57735026918962576449_iwp
     samp(2,2)= -0.57735026918962576449_iwp
     samp(3,2)=  0.57735026918962576449_iwp
     samp(4,2)=  0.57735026918962576449_iwp
     wt(1)=      1.0_iwp
     wt(2)=      1.0_iwp
     wt(3)=      1.0_iwp
     wt(4)=      1.0_iwp
   CASE(9)
     samp(1,1)= -0.77459666924148337704_iwp
     samp(2,1)=  0.0_iwp
```

```
      samp(3,1)=  0.77459666924148337704_iwp
      samp(4,1)= -0.77459666924148337704_iwp
      samp(5,1)=  0.0_iwp
      samp(6,1)=  0.77459666924148337704_iwp
      samp(7,1)= -0.77459666924148337704_iwp
      samp(8,1)=  0.0_iwp
      samp(9,1)=  0.77459666924148337704_iwp
      samp(1,2)= -0.77459666924148337704_iwp
      samp(2,2)= -0.77459666924148337704_iwp
      samp(3,2)= -0.77459666924148337704_iwp
      samp(4,2)=  0.0_iwp
      samp(5,2)=  0.0_iwp
      samp(6,2)=  0.0_iwp
      samp(7,2)=  0.77459666924148337704_iwp
      samp(8,2)=  0.77459666924148337704_iwp
      samp(9,2)=  0.77459666924148337704_iwp
      wt(1) =     0.30864197530864197531_iwp
      wt(2) =     0.49382716049382716049_iwp
      wt(3) =     0.30864197530864197531_iwp
      wt(4) =     0.49382716049382716049_iwp
      wt(5) =     0.79012345679012345679_iwp
      wt(6) =     0.49382716049382716049_iwp
      wt(7) =     0.30864197530864197531_iwp
      wt(8) =     0.49382716049382716049_iwp
      wt(9) =     0.30864197530864197531_iwp
    CASE(16)
      samp(1,1) =-0.86113631159405257524_iwp
      samp(2,1) =-0.33998104358485626481_iwp
      samp(3,1) = 0.33998104358485626481_iwp
      samp(4,1) = 0.86113631159405257524_iwp
      samp(5,1) =-0.86113631159405257524_iwp
      samp(6,1) =-0.33998104358485626481_iwp
      samp(7,1) = 0.33998104358485626481_iwp
      samp(8,1) = 0.86113631159405257524_iwp
      samp(9,1) =-0.86113631159405257524_iwp
      samp(10,1)=-0.33998104358485626481_iwp
      samp(11,1)= 0.33998104358485626481_iwp
      samp(12,1)= 0.86113631159405257524_iwp
      samp(13,1)=-0.86113631159405257524_iwp
      samp(14,1)=-0.33998104358485626481_iwp
      samp(15,1)= 0.33998104358485626481_iwp
      samp(16,1)= 0.86113631159405257524_iwp
      samp(1,2) =-0.86113631159405257524_iwp
      samp(2,2) =-0.86113631159405257524_iwp
      samp(3,2) =-0.86113631159405257524_iwp
```

```
    samp(4,2) =-0.86113631159405257524_iwp
    samp(5,2) =-0.33998104358485626481_iwp
    samp(6,2) =-0.33998104358485626481_iwp
    samp(7,2) =-0.33998104358485626481_iwp
    samp(8,2) =-0.33998104358485626481_iwp
    samp(9,2) = 0.33998104358485626481_iwp
    samp(10,2)= 0.33998104358485626481_iwp
    samp(11,2)= 0.33998104358485626481_iwp
    samp(12,2)= 0.33998104358485626481_iwp
    samp(13,2)= 0.86113631159405257524_iwp
    samp(14,2)= 0.86113631159405257524_iwp
    samp(15,2)= 0.86113631159405257524_iwp
    samp(16,2)= 0.86113631159405257524_iwp
    wt(1)=       0.12100299328560200551_iwp
    wt(2)=       0.22685185185185185185_iwp
    wt(3)=       0.22685185185185185185_iwp
    wt(4)=       0.12100299328560200551_iwp
    wt(5)=       0.22685185185185185185_iwp
    wt(6)=       0.42529330301069429082_iwp
    wt(7)=       0.42529330301069429082_iwp
    wt(8)=       0.22685185185185185185_iwp
    wt(9)=       0.22685185185185185185_iwp
    wt(10)=      0.42529330301069429082_iwp
    wt(11)=      0.42529330301069429082_iwp
    wt(12)=      0.22685185185185185185_iwp
    wt(13)=      0.12100299328560200551_iwp
    wt(14)=      0.22685185185185185185_iwp
    wt(15)=      0.22685185185185185185_iwp
    wt(16)=      0.12100299328560200551_iwp
  CASE(25)
    samp(1,1) =-0.90617984593866399282_iwp
    samp(2,1) =-0.53846931010568309105_iwp
    samp(3,1) = 0.0_iwp
    samp(4,1) = 0.53846931010568309105_iwp
    samp(5,1) = 0.90617984593866399282_iwp
    samp(6,1) =-0.90617984593866399282_iwp
    samp(7,1) =-0.53846931010568309105_iwp
    samp(8,1) = 0.0_iwp
    samp(9,1) = 0.53846931010568309105_iwp
    samp(10,1)= 0.90617984593866399282_iwp
    samp(11,1)=-0.90617984593866399282_iwp
    samp(12,1)=-0.53846931010568309105_iwp
    samp(13,1)= 0.0_iwp
    samp(14,1)= 0.53846931010568309105_iwp
    samp(15,1)= 0.90617984593866399282_iwp
```

```
samp(16,1)=-0.90617984593866399282_iwp
samp(17,1)=-0.53846931010568309105_iwp
samp(18,1)= 0.0_iwp
samp(19,1)= 0.53846931010568309105_iwp
samp(20,1)= 0.90617984593866399282_iwp
samp(21,1)=-0.90617984593866399282_iwp
samp(22,1)=-0.53846931010568309105_iwp
samp(23,1)= 0.0_iwp
samp(24,1)= 0.53846931010568309105_iwp
samp(25,1)= 0.90617984593866399282_iwp
samp(1,2) =-0.90617984593866399282_iwp
samp(2,2) =-0.90617984593866399282_iwp
samp(3,2) =-0.90617984593866399282_iwp
samp(4,2) =-0.90617984593866399282_iwp
samp(5,2) =-0.90617984593866399282_iwp
samp(6,2) =-0.53846931010568309105_iwp
samp(7,2) =-0.53846931010568309105_iwp
samp(8,2) =-0.53846931010568309105_iwp
samp(9,2) =-0.53846931010568309105_iwp
samp(10,2)=-0.53846931010568309105_iwp
samp(11,2)= 0.0_iwp
samp(12,2)= 0.0_iwp
samp(13,2)= 0.0_iwp
samp(14,2)= 0.0_iwp
samp(15,2)= 0.0_iwp
samp(16,2)= 0.53846931010568309105_iwp
samp(17,2)= 0.53846931010568309105_iwp
samp(18,2)= 0.53846931010568309105_iwp
samp(19,2)= 0.53846931010568309105_iwp
samp(20,2)= 0.53846931010568309105_iwp
samp(21,2)= 0.90617984593866399282_iwp
samp(22,2)= 0.90617984593866399282_iwp
samp(23,2)= 0.90617984593866399282_iwp
samp(24,2)= 0.90617984593866399282_iwp
samp(25,2)= 0.90617984593866399282_iwp
wt(1) =     0.05613434886242863595_iwp
wt(2) =     0.1134_iwp
wt(3) =     0.13478507238752090312_iwp
wt(4) =     0.1134_iwp
wt(5) =     0.05613434886242863595_iwp
wt(6) =     0.1134_iwp
wt(7) =     0.22908540422399111713_iwp
wt(8) =     0.27228653255075070182_iwp
wt(9) =     0.22908540422399111713_iwp
wt(10)=     0.1134_iwp
```

```
   wt(11)=     0.13478507238752090305_iwp
   wt(12)=     0.27228653255075070171_iwp
   wt(13)=     0.32363456790123456757_iwp
   wt(14)=     0.27228653255075070171_iwp
   wt(15)=     0.13478507238752090305_iwp
   wt(16)=     0.1134_iwp
   wt(17)=     0.22908540422399111713_iwp
   wt(18)=     0.27228653255075070182_iwp
   wt(19)=     0.22908540422399111713_iwp
   wt(20)=     0.1134_iwp
   wt(21)=     0.05613434886242863595_iwp
   wt(22)=     0.1134_iwp
   wt(23)=     0.13478507238752090312_iwp
   wt(24)=     0.1134_iwp
   wt(25)=     0.05613434886242863595_iwp
  CASE DEFAULT
  WRITE(*,*)"Wrong number of integrating points"
  END SELECT
CASE(3)
  SELECT CASE(nsp)
  CASE(1)
    samp(1,1)=  0.0_iwp
    samp(1,2)=  0.0_iwp
    samp(1,3)=  0.0_iwp
    wt(1)=      8.0_iwp
  CASE(8)
    samp(1,1)=-0.57735026918962576449_iwp
    samp(2,1)= 0.57735026918962576449_iwp
    samp(3,1)=-0.57735026918962576449_iwp
    samp(4,1)= 0.57735026918962576449_iwp
    samp(5,1)=-0.57735026918962576449_iwp
    samp(6,1)= 0.57735026918962576449_iwp
    samp(7,1)=-0.57735026918962576449_iwp
    samp(8,1)= 0.57735026918962576449_iwp
    samp(1,2)=-0.57735026918962576449_iwp
    samp(2,2)=-0.57735026918962576449_iwp
    samp(3,2)=-0.57735026918962576449_iwp
    samp(4,2)=-0.57735026918962576449_iwp
    samp(5,2)= 0.57735026918962576449_iwp
    samp(6,2)= 0.57735026918962576449_iwp
    samp(7,2)= 0.57735026918962576449_iwp
    samp(8,2)= 0.57735026918962576449_iwp
    samp(1,3)=-0.57735026918962576449_iwp
    samp(2,3)=-0.57735026918962576449_iwp
    samp(3,3)= 0.57735026918962576449_iwp
```

```
        samp(4,3)= 0.57735026918962576449_iwp
        samp(5,3)=-0.57735026918962576449_iwp
        samp(6,3)=-0.57735026918962576449_iwp
        samp(7,3)= 0.57735026918962576449_iwp
        samp(8,3)= 0.57735026918962576449_iwp
        wt(1)=      1.0_iwp
        wt(2)=      1.0_iwp
        wt(3)=      1.0_iwp
        wt(4)=      1.0_iwp
        wt(5)=      1.0_iwp
        wt(6)=      1.0_iwp
        wt(7)=      1.0_iwp
        wt(8)=      1.0_iwp
CASE(27)
        samp(1,1)= -0.77459666924148337704_iwp
        samp(2,1)=  0.0_iwp
        samp(3,1)=  0.77459666924148337704_iwp
        samp(4,1)= -0.77459666924148337704_iwp
        samp(5,1)=  0.0_iwp
        samp(6,1)=  0.77459666924148337704_iwp
        samp(7,1)= -0.77459666924148337704_iwp
        samp(8,1)=  0.0_iwp
        samp(9,1)=  0.77459666924148337704_iwp
        samp(10,1)=-0.77459666924148337704_iwp
        samp(11,1)= 0.0_iwp
        samp(12,1)= 0.77459666924148337704_iwp
        samp(13,1)=-0.77459666924148337704_iwp
        samp(14,1)= 0.0_iwp
        samp(15,1)= 0.77459666924148337704_iwp
        samp(16,1)=-0.77459666924148337704_iwp
        samp(17,1)= 0.0_iwp
        samp(18,1)= 0.77459666924148337704_iwp
        samp(19,1)=-0.77459666924148337704_iwp
        samp(20,1)= 0.0_iwp
        samp(21,1)= 0.77459666924148337704_iwp
        samp(22,1)=-0.77459666924148337704_iwp
        samp(23,1)= 0.0_iwp
        samp(24,1)= 0.77459666924148337704_iwp
        samp(25,1)=-0.77459666924148337704_iwp
        samp(26,1)= 0.0_iwp
        samp(27,1)= 0.77459666924148337704_iwp
        samp(1,2)= -0.77459666924148337704_iwp
        samp(2,2)= -0.77459666924148337704_iwp
        samp(3,2)= -0.77459666924148337704_iwp
        samp(4,2)= -0.77459666924148337704_iwp
```

```
      samp(5,2)= -0.77459666924148337704_iwp
      samp(6,2)= -0.77459666924148337704_iwp
      samp(7,2)= -0.77459666924148337704_iwp
      samp(8,2)= -0.77459666924148337704_iwp
      samp(9,2)= -0.77459666924148337704_iwp
      samp(10,2)= 0.0_iwp
      samp(11,2)= 0.0_iwp
      samp(12,2)= 0.0_iwp
      samp(13,2)= 0.0_iwp
      samp(14,2)= 0.0_iwp
      samp(15,2)= 0.0_iwp
      samp(16,2)= 0.0_iwp
      samp(17,2)= 0.0_iwp
      samp(18,2)= 0.0_iwp
      samp(19,2)= 0.77459666924148337704_iwp
      samp(20,2)= 0.77459666924148337704_iwp
      samp(21,2)= 0.77459666924148337704_iwp
      samp(22,2)= 0.77459666924148337704_iwp
      samp(23,2)= 0.77459666924148337704_iwp
      samp(24,2)= 0.77459666924148337704_iwp
      samp(25,2)= 0.77459666924148337704_iwp
      samp(26,2)= 0.77459666924148337704_iwp
      samp(27,2)= 0.77459666924148337704_iwp
      samp(1,3)= -0.77459666924148337704_iwp
      samp(2,3)= -0.77459666924148337704_iwp
      samp(3,3)= -0.77459666924148337704_iwp
      samp(4,3)=  0.0_iwp
      samp(5,3)=  0.0_iwp
      samp(6,3)=  0.0_iwp
      samp(7,3)=  0.77459666924148337704_iwp
      samp(8,3)=  0.77459666924148337704_iwp
      samp(9,3)=  0.77459666924148337704_iwp
      samp(10,3)=-0.77459666924148337704_iwp
      samp(11,3)=-0.77459666924148337704_iwp
      samp(12,3)=-0.77459666924148337704_iwp
      samp(13,3)= 0.0_iwp
      samp(14,3)= 0.0_iwp
      samp(15,3)= 0.0_iwp
      samp(16,3)= 0.77459666924148337704_iwp
      samp(17,3)= 0.77459666924148337704_iwp
      samp(18,3)= 0.77459666924148337704_iwp
      samp(19,3)=-0.77459666924148337704_iwp
      samp(20,3)=-0.77459666924148337704_iwp
      samp(21,3)=-0.77459666924148337704_iwp
      samp(22,3)= 0.0_iwp
```

```
      samp(23,3)= 0.0_iwp
      samp(24,3)= 0.0_iwp
      samp(25,3)= 0.774596666924148337704_iwp
      samp(26,3)= 0.774596666924148337704_iwp
      samp(27,3)= 0.774596666924148337704_iwp
      wt(1)     = 0.171467764060356652955_iwp
      wt(2)     = 0.274348422496570644472_iwp
      wt(3)     = 0.171467764060356652955_iwp
      wt(4)     = 0.274348422496570644472_iwp
      wt(5)     = 0.438957475994513031555_iwp
      wt(6)     = 0.274348422496570644472_iwp
      wt(7)     = 0.171467764060356652955_iwp
      wt(8)     = 0.274348422496570644472_iwp
      wt(9)     = 0.171467764060356652955_iwp
      wt(10)    = 0.274348422496570644472_iwp
      wt(11)    = 0.438957475994513031555_iwp
      wt(12)    = 0.274348422496570644472_iwp
      wt(13)    = 0.438957475994513031555_iwp
      wt(14)    = 0.702331961591220850483_iwp
      wt(15)    = 0.438957475994513031555_iwp
      wt(16)    = 0.274348422496570644472_iwp
      wt(17)    = 0.438957475994513031555_iwp
      wt(18)    = 0.274348422496570644472_iwp
      wt(19)    = 0.171467764060356652955_iwp
      wt(20)    = 0.274348422496570644472_iwp
      wt(21)    = 0.171467764060356652955_iwp
      wt(22)    = 0.274348422496570644472_iwp
      wt(23)    = 0.438957475994513031555_iwp
      wt(24)    = 0.274348422496570644472_iwp
      wt(25)    = 0.171467764060356652955_iwp
      wt(26)    = 0.274348422496570644472_iwp
      wt(27)    = 0.171467764060356652955_iwp
    CASE DEFAULT
      WRITE(*,*)"Wrong number of integrating points"
    END SELECT
  CASE DEFAULT
    WRITE(*,*)"Not a valid dimension"
  END SELECT
RETURN
END SUBROUTINE gauss_legendre
```

```
FUNCTION inverse(matrix)
!---Returns Inverse of Small Matrix by Gauss-Jordan---
 USE precision; IMPLICIT NONE
 REAL(iwp),INTENT(IN)::matrix(:,:)
```

```
  REAL(iwp)::inverse(UBOUND(matrix,1),UBOUND(matrix,2))
  REAL(iwp)::temp(UBOUND(matrix,1),UBOUND(matrix,2))
  INTEGER::i,k,n; REAL(iwp)::con; n=UBOUND(matrix,1)
  temp=matrix
  DO k=1,n
    con=temp(k,k); temp(k,k)=1.0_iwp
    temp(k,:)=temp(k,:)/con
    DO i=1,n
      IF(i/=k) THEN
        con=temp(i,k);temp(i,k)=0.0_iwp
        temp(i,:)=temp(i,:)-temp(k,:)*con
      END IF
    END DO
  END DO
  inverse=temp
END FUNCTION inverse
```

```
SUBROUTINE ldlfor(a,b)
!---Forward Substitution on Upper Triangle Stored as a Lower
! Triangle---
 USE precision; IMPLICIT NONE
 REAL(iwp),INTENT(IN)::a(:,:); REAL(iwp),INTENT(IN OUT)::b(:)
 INTEGER::i,j,n; REAL(iwp)::total; n= UBOUND(a,1)
 DO i=1,n
   total=b(i)
   IF(i>1)THEN
     DO j=1,i-1; total=total-a(j,i)*b(j); END DO
   END IF
   b(i)=total/a(i,i)
 END DO
RETURN
END SUBROUTINE ldlfor
```

```
SUBROUTINE ldlt(a,d)
!---LDLT Factorization of a Square Matrix---
 USE precision; IMPLICIT NONE
 REAL(iwp),INTENT(IN OUT)::a(:,:); real(iwp),INTENT(OUT)::d(:)
 INTEGER::i,j,k,n; REAL(iwp)::x,small=1.E-10_iwp; n=UBOUND(a,1)
 DO k=1,n-1
   d(1)=a(1,1)
   IF(ABS(a(k,k))>small)THEN
     DO i=k+1,n
       x=a(i,k)/a(k,k)
       DO j=k+1,n; a(i,j)=a(i,j)-a(k,j)*x; END DO
       d(i)=a(i,i)
```

```
      END DO
    ELSE; WRITE(11,*)"Zero pivot found in row ",k
    END IF
  END DO
RETURN
END SUBROUTINE ldlt
```

```
SUBROUTINE lufac(a,lower,upper)
!-----LU Factorisation of a Square Matrix-----
 USE precision; IMPLICIT NONE
 REAL(iwp),INTENT(IN)::a(:,:)
 REAL(iwp),INTENT(OUT)::lower(:,:),upper(:,:)
 INTEGER::i,j,k,l,n; REAL(iwp)::total,zero=.0_iwp; n=UBOUND(a,1)
 upper=zero; lower=zero; upper(1,:)=a(1,:)
 DO i=1,n; lower(i,i)=1.0_iwp; end do
 DO k=1,n-1
   IF(ABS(upper(k,k))>1.e-10_iwp)THEN
     DO i=k+1,n
!---Lower Triangular Components---
       DO j=1,i-1
         total=zero
         DO l=1,j-1
           total= total-lower(i,l)*upper(l,j)
         END DO
         lower(i,j)=(a(i,j)+total)/upper(j,j)
       END DO
!---Upper Triangular Components---
       DO j=1,n
         total=zero
         DO l=1,i-1
           total=total-lower(i,l)*upper(l,j)
         END DO
         upper(i,j)=a(i,j)+total
       END DO
     END DO
   ELSE
     WRITE(11,*)"Zero pivot found in row", k; EXIT
   END IF
 END DO
RETURN
END SUBROUTINE lufac
```

```
SUBROUTINE lupfac(a,row)
!---LU Factorization of a Square Matrix with Pivoting---
 USE precision; IMPLICIT NONE
```

```
REAL(iwp),INTENT(IN OUT)::a(:,:); INTEGER,INTENT(OUT)::row(:)
INTEGER::i,j,k,ip,ie,ih,irow,n
REAL(iwp)::pval,pivot,small=1.E-10_iwp
n=UBOUND(a,1); DO i=1,n; row(i)=i; END DO
DO i=1,n-1
  ip=i; pval=a(row(ip),ip)
  DO j=i+1,n
    IF(ABS(a(row(j),i))>ABS(pval))THEN
      ip=j; pval=a(row(j),i)
    END IF
  END DO
  IF(ABS(pval)<small)THEN
    WRITE(11,*)"Singular equations detected"; STOP
  END IF
  ih=row(ip); row(ip)=row(i); row(i)=ih
  DO j=i+1,n
    ie= row(j); pivot=a(ie,i)/pval; a(ie,i)=pivot; irow=row(i)
    DO k=i+1,n
      a(ie,k)=a(ie,k)-a(irow,k)*pivot
    END DO
  END DO
END DO
IF(ABS(a(row(n),n))< small)THEN
  WRITE(11,*)"Singular equations detected"; STOP
END IF
RETURN
END SUBROUTINE lupfac
```

```
SUBROUTINE lupsol(a,b,sol,row)
!---Forward and Back-substitution with Pivots---
USE precision; IMPLICIT NONE
REAL(iwp),INTENT(IN)::a(:,:); REAL(iwp),INTENT(IN)::b(:)
REAL(iwp),INTENT(OUT)::sol(:); INTEGER,INTENT(IN)::row(:)
INTEGER::i,j,n,irow
REAL(iwp)::total,temp(UBOUND(a,1)); n=UBOUND(a,1)
temp=b
DO i=1,n
  irow=row(i); total=b(irow)
  IF(i>1)THEN
    DO j=1,i-1
      total=total-a(irow,j)*temp(row(j))
    END DO
    temp(irow)=total
  END IF
END DO
```

```
  DO i=n,1,-1
    irow=row(i); total=temp(irow)
    IF(i<n)THEN
      DO j=i+1,n
        total=total-a(irow,j)*temp(row(j))
      END DO
    END IF
    temp(irow)=total/a(irow,i)
  END DO
  sol=temp(row(:))
RETURN
END SUBROUTINE lupsol
```

```
SUBROUTINE newton_cotes(samp,wt)
!---Provides Weights and Sampling Points for Newton-Cotes---
USE precision; IMPLICIT NONE
REAL(iwp),INTENT(OUT)::samp(:,:),wt(:); INTEGER::nsp
nsp=UBOUND(samp,1)
SELECT CASE(nsp)
  CASE(1)
    samp(1,1)=-1.0_iwp
    wt(1)=     2.0_iwp
  CASE(2)
    samp(1,1)=-1.0_iwp
    samp(2,1)= 1.0_iwp
    wt(1)=     1.0_iwp
    wt(2)=     1.0_iwp
  CASE(3)
    samp(1,1)=-1.0_iwp
    samp(2,1)= 0.0_iwp
    samp(3,1)= 1.0_iwp
    wt(1)=     0.33333333333333333333_iwp
    wt(2)=     1.33333333333333333333_iwp
    wt(3)=     0.33333333333333333333_iwp
  CASE(4)
    samp(1,1)=-1.0_iwp
    samp(2,1)=-0.33333333333333333333_iwp
    samp(3,1)= 0.33333333333333333333_iwp
    samp(4,1)= 1.0_iwp
    wt(1)=     0.25_iwp
    wt(2)=     0.75_iwp
    wt(3)=     0.75_iwp
    wt(4)=     0.25_iwp
  CASE(5)
    samp(1,1)=-1.0_iwp
```

```
      samp(2,1)=-0.5_iwp
      samp(3,1)= 0.0_iwp
      samp(4,1)= 0.5_iwp
      samp(5,1)= 1.0_iwp
      wt(1)=      0.15555555555555555556_iwp
      wt(2)=      0.71111111111111111111_iwp
      wt(3)=      0.26666666666666666667_iwp
      wt(4)=      0.71111111111111111111_iwp
      wt(5)=      0.15555555555555555556_iwp
   CASE DEFAULT
     WRITE(*,*)"Wrong number of integrating points"
   END SELECT
RETURN
END SUBROUTINE newton_cotes
```

```
REAL FUNCTION norm(x)
!---Returns L2 Norm of Vector x---
 USE precision; IMPLICIT NONE
 REAL(iwp),INTENT(IN)::x(:)
 norm=SQRT(SUM(x**2))
END FUNCTION norm
```

```
SUBROUTINE spabac(a,b,kdiag)
!---Choleski Forward and Back-substitution on a Skyline Matrix---
 USE precision; IMPLICIT NONE
 REAL(iwp),INTENT(IN)::a(:); REAL(iwp),INTENT(in out)::b(:)
 INTEGER,INTENT(IN)::kdiag(:); INTEGER::n,i,ki,l,m,j,it,k
 REAL(iwp)::x
 n=UBOUND(kdiag,1); b(1)=b(1)/a(1)
 DO i=2,n
   ki=kdiag(i)-i; l=kdiag(i-1)-ki+1; x=b(i)
   IF(l/=i)THEN
     m=i-1
     DO j=l,m; x=x-a(ki+j)*b(j); END DO
   END IF
   b(i)=x/a(ki+i)
 END DO
 DO it=2,n
   i=n+2-it; ki=kdiag(i)-i; x=b(i)/a(ki+i)
   b(i)=x; l=kdiag(i-1)-ki+1
   IF(l/=i)THEN
     m=i-1
     DO k=l,m; b(k)=b(k)-x*a(ki+k); END DO
   END IF
```

```
END DO
 b(1)=b(1)/a(1)
RETURN
END SUBROUTINE spabac
```

```
SUBROUTINE sparin(a,kdiag)
!---Choleski Factorization of Symmetric Skyline Matrix---
 USE precision; IMPLICIT NONE
 REAL(iwp),INTENT(IN OUT)::a(:); INTEGER,INTENT(IN)::kdiag(:)
 INTEGER::n,i,ki,l,kj,j,ll,m,k; REAL(iwp)::x
 n=UBOUND(kdiag,1); a(1)=SQRT(a(1))
 DO i=2,n
   ki=kdiag(i)-i; l=kdiag(i-1)-ki+1
   DO j=l,i
     x=a(ki+j); kj=kdiag(j)-j
     IF(j/=1)THEN
       ll=kdiag(j-1)-kj+1; ll=MAX(l,ll)
       IF(ll/=j)THEN
         m=j-1
         DO k=ll,m; x=x-a(ki+k)*a(kj+k); END DO
       END IF
     END IF
     a(ki+j)=x/a(kj+j)
   END DO
   a(ki+i)=SQRT(x)
 END DO
RETURN
END SUBROUTINE sparin
```

```
SUBROUTINE subbac(a,b)
!---Back-substitution on an Upper Triangle---
 USE precision; IMPLICIT NONE
 REAL(iwp),INTENT(IN)::a(:,:); REAL(iwp),INTENT(IN OUT)::b(:)
 INTEGER::i,j,n; REAL(iwp)::total; n=UBOUND(a,1)
 DO i=n,1,-1
   total=b(i)
   IF(i<n)THEN
     DO j=i+1,n; total=total-a(i,j)*b(j); END DO
   END IF
   b(i)=total/a(i,i)
 END DO
RETURN
END SUBROUTINE subbac
```

```
SUBROUTINE subfor(a,b)
!---Forward-substitution on a Lower Triangle---
 USE precision; IMPLICIT NONE
 REAL(iwp),INTENT(IN)::a(:,:); REAL(iwp),INTENT(IN OUT)::b(:)
 INTEGER::i,j,n; REAL(iwp)::total; n=UBOUND(a,1)
 DO i=1,n
   total=b(i)
   IF(i>1)THEN
     DO j=1,i-1; total=total-a(i,j)*b(j); END DO
   END IF
   b(i)=total/a(i,i)
 END DO
RETURN
END SUBROUTINE subfor
```

Appendix C

References and Additional Reading

1. Abramowitz, M. and Stegun, I.A., *Handbook of mathematical functions*. U.S. Government Printing Office, Washington, D.C. (1964)

2. Ames, W.F, *Mathematics for mechanical engineers*. CRC Press, Boca Raton (2000)

3. Ayyub, B.M. and McCuen, R.H., *Numerical methods for engineers*. Prentice-Hall, Upper Saddle River, NJ (1996)

4. Baker, T.H. and Phillips, C. (eds), *The numerical solution of nonlinear problems*. Clarendon Press, Oxford (1981)

5. Bathe, K.J. and Wilson, E.L., *Numerical methods in finite element analysis*. Prentice-Hall, Englewood Cliffs, NJ (1976)

6. Burden, R.L. and Faires, J.D., *Numerical analysis, 8th ed.* Thomson Brooks/Cole, Belmont, CA (2005)

7. Byrne, G. and Hall, C. (eds), *Numerical solution of systems of nonlinear algebraic equations*. Academic Press, New York (1973)

8. Chaitin-Chatelin, F. and Ahués, M., *Eigenvalues of matrices*. Wiley, New York (1993)

9. Chapra, S.C. and Canale, R.P., *Numerical methods for engineers*. 5th ed., McGraw-Hill, Boston (2006)

10. Cheney, E.W. and Kincaid, D., *Numerical mathematics and computing, 5th ed.* Thomson-Brooks/Cole, Belmont, CA (2004)

11. Collatz, L., *The Numerical treatment of differential equations, 3rd ed.* Springer-Verlag, New York (1966)

12. Conte, S.D. and De Boor, C., *Elementary numerical analysis, 3rd ed.* McGraw-Hill, London (1980)

13. Crandall, S.H., *Engineering analysis*. McGraw-Hill, New York (1956)

14. Dahlquist, G. and Björck, A., *Numerical methods*. Prentice-Hall, Englewood Cliffs, NJ (1974)

15. Davis, P.J. and Rabinowitz, P., *Methods of numerical integration, 2nd ed.* Academic Press, New York (1984)

16. Dijkstra, E. W., *A discipline of programming.* Prentice-Hall, Englewood Cliffs, NJ (1976)

17. Dongarra, J., Bunch, J., Moler, G. and Stewart, G., LINPACK User's Guide. SIAM (1979)

18. Dongarra, J.J. and Walker, D., *Software libraries for linear algebra computations on high performance computers. Siam Rev*, 37(2), pp.151-180 (1995)

19. EISPACK, Fortran subroutines for computing the eigenvalues and eigenvectors (1973)
    `www.netlib.org/eispack`

20. Evans, D.J. (ed), *Sparsity and its applications.* Cambridge University Press, Cambridge (1985)

21. Fairweather, G., *Finite element Galerkin methods for differential equations.* M. Dekker, New York (1978)

22. Forsythe, G.E. and Wasow, W.R., *Finite difference methods for partial differential equations.* Wiley, New York (1960)

23. Fox, L., *An introduction to numerical linear algebra.* Clarendon Press, Oxford (1964)

24. Frazer, R.A., Duncan, W.J. and Collar, A.R., *Elementary matrices and some applications to dynamics and differential equations.* The University Press, Cambridge (1938)

25. Froberg, C.E., *Introduction to numerical linear algebra, 2nd ed.* Addison-Wesley, Reading, MA (1970)

26. Froberg, C.E., *Numerical mathematics.* Addison-Wesley, Redwood City, CA (1985)

27. Garbow, B., Boyle, J., Dongarra, J. and Molar, C., *Matrix Eigensystem Routines - EISPACK Guide Extension.* Lecture Notes on Computer Science, 51, Springer-Verlag, New York (1977)

28. Gear, C.W., *Numerical initial value problems in ordinary differential equations.* Prentice-Hall, Englewood Cliffs, NJ (1971)

29. Gill, P.E., Murray, W. and Wright, M.H., *Practical optimization.* Academic Press, San Diego (1986)

30. Givens, W., *Numerical computation of the characteristic values of a real symmetric matrix.* Oak Ridge National Laboratory Report ORNL-1574 (1954)

31. Gladwell, I. and Wait, R. (eds), *A survey of numerical methods for partial differential equations.* Oxford University Press, Oxford (1979)

32. Gladwell, I. and Sayers, D.K. (eds), *Computational techniques for ordinary differential equations.* Academic Press, New York (1980)

33. Golub, G.H. and Van Loan, C.F., *Matrix computations, 3rd ed.* Johns Hopkins University Press, Baltimore (1996)

34. Gourlay, A.R. and Watson, G.A., *Computation methods for matrix eigenproblems.* Wiley, London (1973)

35. Greenbaum, A., *Iterative methods for solving linear systems.* SIAM, Philadelphia (1997)

36. Griffiths, D.V., Generalized numerical integration of moments. *Int. J. Numer. Methods Eng.*, vol.32, no.1, pp.129-147 (1991)

37. Higham, N.J., *Accuracy and stability of numerical algorithms, 2nd ed.* SIAM, Philadelphia (2002)

38. Householder, A.S. *The numerical treatment of a single nonlinear equation.* McGraw-Hill, New York (1970)

39. Householder, A.S., *The theory of matrices in numerical analysis.* Dover Publications, New York (1975)

40. HSL, Aspen HSL (2004)
    www.aspentech.com/hsl/

41. IMSL, Absoft (2004)
    www.absoft.com/Products/Libraries/imsl.html#description

42. Jennings, A., *Matrix computation for engineers and scientists, 2nd ed.* Wiley, New York (1992)

43. Johnson L.W. and Riess, R.D., *Numerical analysis, 2nd ed.* Addison-Wesley, Reading, MA (1982)

44. Kelley, C.T., *Iterative methods for linear and nonlinear equations.* SIAM, Philadelphia (1995)

45. Lanczos, C., An iteration method for the solution of the eigenvalue problems of linear differential and integral operators. *J. Res. Nat. Bur. Stand.*, 45, 255-282 (1950)

46. LAPACK, Library of Fortran 77 routines for solving common problems in numerical linear algebra (1990)
    `www.cisl.ucar.edu/softlib/LAPACK.html`

47. Lapidus, L. and Pinder, G.F., *Numerical solution of partial differential equations in science and engineering*. Wiley, New York (1982)

48. Lapidus, L. and Seinfeld, J.H., *Numerical solution of ordinary differential equations*. Academic Press, New York (1971)

49. LINPACK, Fortran 77 routines for solving common problems in numerical linear algebra (1979)
    `www.cisl.ucar.edu/softlib/LINPACK.html`

50. Mitchell, A.R., *Computational methods in partial differential equations*. Wiley, New York (1969)

51. NAG, Numerical Algorithms Group (2005)
    `www.nag.co.uk`

52. Ortega, J.M. and Rheinboldt, W.C., *Iterative solution of nonlinear equations in several variables*. SIAM, Philadelphia (1970)

53. Parlett, B.N., *The symmetric eigenvalue problem*. Prentice-Hall, Englewood Cliffs, NJ (1980)

54. Peters, G. and Wilkinson, J., Practical problems arising in the solution of polynomial equations. *J. Inst. Math. and its Appl.* 8, 16-35 (1971)

55. Ralston, A. and Rabinowitz, P., *A first course in numerical analysis, 2nd ed.* McGraw-Hill, New York (1965)

56. Rheinboldt, W.C., *Methods for solving systems of nonlinear equations, 2nd ed.* SIAM, Philadelphia (1998)

57. Rutishauser, H., Solution of eigenvalues problems with the LR transformation. *Nat. Bur. Standards Appl. Math. Ser.*, 49, 47-81 (1985)

58. Schumaker, L.L., *Spline functions: Basic theory*. Wiley, New York (1981)

59. Shampine, L.F., Allen, R.C. and Pruess, S., *Fundamentals of numerical computing*. Wiley, New York (1997)

60. Shoup, T.E., *A practical guide to computer methods for engineers*. Prentice-Hall, Englewood Cliffs, NJ (1979)

61. Smith, G.D., *Numerical solution of partial differential equations, 3rd ed.* Oxford University Press, Oxford (1985)

62. Smith, I.M., *Programming in Fortran 90*. Wiley, Chichester (1995)

63. Smith, I.M. and Griffiths, D.V., *Programming the finite element method*, *4th ed.* Wiley, Chichester (2004)

64. Smith, I.M. and Kidger, D.J., Properties of the 20-node brick, *Int. J. Numer. Anal. Methods Geomech.*, vol.15, no.12, pp.871-891 (1991)

65. Smith, I.M. and Margetts, L., The convergence variability of parallel iterative solvers, *Engineering Computations*, vol.23, no.2, pp.154-165 (2006)

66. Stewart, G.W., *Introduction to matrix computations*. Academic Press, New York (1973)

67. Stroud, A.H. and Secrest, D., *Gaussian quadrature formulas*. Prentice-Hall, Englewood Cliffs, NJ (1966)

68. Traub, J.F., *Iterative methods for the solution of equations*. Prentice-Hall, Englewood Cliffs, NJ (1964)

69. Wilkinson, J.H., *Rounding errors in algebraic processes*. Prentice-Hall, Englewood Cliffs, NJ (1963)

70. Wilkinson, J.H., *The algebraic eigenvalue problem*. Clarendon Press, Oxford (1965)

71. Wilkinson, J.H., Reinsch, C. and Bauer, F.L. *Linear algebra*. Springer-Verlag, New York (1971)

72. Zienkiewicz, O.C., Taylor, R.L. and Zhu, J.Z., *The finite element method*, *6th ed.* Elsevier Butterworth-Heinemann, Oxford (2005)

Index

Adaptive Gauss-Legendre rules
    Program 6.3, 280
Adaptive integration rules, 278
Aitken's method, 112
ALLOCATABLE, 5
    ALLOCATE, 6
Average degree of consolidation, 423

Back-substitution, 16, 20
Banded matrices, 33, 410
Bandwidth, 33
    half-bandwidth, 33, 35
Best fit, 226
BiCGSTAB method, 69
    Program 2.13, 70
    Program 2.15 (left), 77
    Program 2.16 (right), 79
Bisection method, 97
    Program 3.2, 98
Boundary conditions
    derivative, 367
    irregular, 405
Boundary value problems, 319, 362
Buckling load, 154

Characteristic polynomial, 131
Characteristic polynomial method,
    180
    Program 4.10, 184
Characteristics (in PDEs), 395
check, 445, 447
checkit, 445, 447
chobac, 445, 447
Cholesky's method, 155
    LL^T factorization, 31
    Program 2.4, 34
    Program 2.5 (skyline storage),
    36

Program 2.7 (prescribed solu-
    tions), 43
cholin, 445, 448
Coefficient of determination, 229
Computational molecules
    biharmonic, 403
    finite difference formulas, 401
    Laplacian, 403
Conditional stability, 419
Conductivity matrix, 431
Conjugate gradients method, 64
    Program 2.12, 65
Convergence, 49, 68
    iteration limit, 51
    tolerance, 51
Correlation coefficient, 229
Cubic spline interpolation, 207
    Program 5.3, 211
Curve fitting, 193, 226

Determinant, 30
determinant, 445, 448
Diagonal matrix, 159
Diagonally dominant, 39, 46
Difference methods, 198
    backward difference table, 200
    forward difference table, 200
Direct methods, 15
    comparison with indirect, 81
Dominant error term, 219
Downloading programs, 14

Eigenvalue equations, 131
    characteristic value, 131
    characteristic vector, 131
    deflation, 148
    eigenvalue, 131
    eigenvector, 131

general symmetrical matrices, 187

    properties, 134

    solution methods, 136

    standard form, 154

    transformation methods, 136

    vector iteration, 136

Eigenvectors

    Euclidean norm, 133

    L2 norm, 133

    normalization, 132

    orthogonality, 132, 133, 148

`eliminate`, 445, 449

Elliptic systems, 404

Errors

    absolute, 9, 279

    cancellation, 11

    propagation, 360

    relative, 9, 226, 279

    roundoff, 9

    truncation, 10

Euler method, 321

Extrapolation methods, 103

Factorization, 143

    LDL^T, 26

    LU, 20

False position method, 100

    Program 3.3, 102

Fill-in, 36, 52

Finite difference method, 362

    explicit, 419

    formulas, 362

    grid, 403

    partial differential equations, 401

    Program 8.1, 424

Finite element method, 52, 430

    nodes, 430

    Program 8.2, 432

    shape functions, 430

Fortran 95, 1

    `CASE`, 7

    conditional statements, 7

    `DO` loops, 7

    example program, 4

    format specifications, 6

    free format, 5

    `FUNCTION`, 8

    `IMPLICIT NONE`, 4

    input, 5

    intrinsic functions, 7

    `iwp`, 5, 8

    output, 5

    precision routines, 11

    `RECURSIVE FUNCTION`, 284

    `SELECTED_REAL_KIND`, 9

    `SUBROUTINE`, 8

    `TYPE`, 5

    `USE`, 4

    user-supplied functions, 8

Forward-substitution, 16

`fsparv`, 445, 450

`fun_der`, 446, 451

Gauss-Chebyshev rules, 288

`gauss_laguerre`, 446, 452

Gauss-Laguerre rules, 285

    Program 6.4, 287

    sampling points and weights, 286

`gauss_legendre`, 446, 452

Gauss-Legendre rules, 265

    accuracy, 277

    one-point(Midpoint), 265

    Program 6.2, 275

    sampling points and weights, 269

    three-point, 270

    two-point, 267

Gauss-Seidel method, 52

    Program 2.9, 54

Gaussian elimination, 15

    pivots, 20

    Program 2.1, 17

Generalized eigenvalue equation, 150

    conversion to standard form, 154

    Program 4.4 (solution), 151

    Program 4.5 (conversion), 155

Geometric matrix, 151
Given's method, 167
Gradient methods, 61
Graphical output, 12

Hardware, 2
Hilbert matrix, 41
Householder's method, 167
    Program 4.7, 169
Hybrid rules, 290
Hyperbolic systems, 427
    wave equation, 428

Ill-conditioning, 41, 42
Initial value problems, 319
    initial conditions, 417
    summary, 361
Interpolation by forward difference
    Program 5.2, 204
Interpolation methods, 97, 193
    differences, 197
    Lagrangian, 194
    polynomials, 193, 215
inverse, 446, 461
Iterative substitution, 91
    Program 3.1 (single equation),
      92
    Program 3.6 (systems), 114
    systems, 113

Jacobi diagonalization, 160
    Program 4.6, 163
Jacobi iteration, 47
    Program 2.8, 49
Jacobian matrix, 118, 121, 302
    the Jacobian, 118, 301

kdiag, 445, 450

Lagrangian interpolation
    Program 5.1, 196
Lagrangian polynomials, 194
Lanczos's method, 171
    Program 4.8, 173
LDL^T factorization, 26
    Program 2.3, 28

ldlfor, 446, 462
ldlt, 446, 462
Leading diagonal, 33
Least squares method, 226
    Program 5.4, 232
Libraries, 2
    EISPACK, 3
    HSL, 3
    IMSL, 3
    LINPACK, 3
    NAG, 2
Limits of integration
    changing, 271
Linear equations, 15
    θ-methods, 343
    first order, 343
    indirect methods, 46
    iterative methods, 46
    Program 7.2 (using θ-method),
      345
    second order, 344
Linear regression, 226
    multiple, 226
Linearization of data, 229
Lobatto rules, 290
Lower triangular matrix, 16, 20
LR transformation, 176
    Program 4.9, 177
LU factorization, 8, 176
    Program 2.2, 23
    Program 2.6 (pivoting), 40
lufac, 446, 463
lupfac, 446, 463
lupsol, 446, 464

Mass matrix, 151
Mid-point method, 325
Modified Euler method, 323
Modified Newton-Raphson, 107, 121
    Program 3.5 (single equation),
      109
    Program 3.8 (systems), 123
Multi-step methods, 320
Multiple integrals, 292
    Gauss-Legendre rules, 296

Newton-Cotes rules, 294
    Program 6.5, 302
    quadrilateral region, 299
    rectangular region, 293
Multiple roots, 94

newton_cotes, 446, 465
Newton-Cotes rules, 247
    accuracy, 253
    higher order, 252
    Program 6.1, 262
    rectangle rule, 247
    repeated, 255
    sampling points and weights,
        255
    Simpson's rule, 250
    summary, 254
    trapezoid rule, 248
Newton-Raphson, 104
    Program 3.4 (single equation),
        106
    Program 3.7 (systems), 119
    systems, 116
nm_lib library, 39
Nonlinear equations, 89
    systems, 112
Nontrivial solution, 131
norm, 446, 466
Numerical differentiation, 214, 216
Numerical differentiation formulas
    backward, 222
    central, 221
    forward, 220
Numerical integration, 245
Numerical stability, 360

One-step methods, 320, 321
    accuracy, 328
    Program 7.1, 337
Ordinary differential equations, 317
    degree, 318
    linear, 318
    order, 317
    reduction to first order, 330
Orthogonal matrix, 159

Parabolic systems, 417
    dimensionless form, 417
Parallel processing, 52, 82
Partial differential equations, 393
    biharmonic operator, 394
    common types, 394
    degree, 394
    elliptic, 400
    first order, 394
    hyperbolic, 400
    Laplace operator, 394
    linear, 393
    parabolic, 400
    second order, 399
PCG method, 73
    Program 2.14, 74
Penalty method, 43, 166, 432
Pivoting, 38
    partial, 39
Polynomial substitution, 250, 290
Positive definite, 30, 31
    semi-definite, 30
Power method, 136
Precision, 8
    32-bit word, 42
    64-bit word, 42
    double, 42
    module, 42
    precision, 1, 42
Preconditioning, 72
    diagonal, 73
    left, 73
    right, 73
Predictor-corrector methods, 320,
        349
    accuracy, 354
    Adams-Bashforth-Moulton, 352
    corrector, 350
    estimation of errors, 355
    Milne-Simpson, 351
    predictor, 290, 350
    Program 7.3, 356
Prescribed solutions, 42
Principal minors, 181

Quadratic form, 30
Quadrature, 245

Rectangle rule, 247
   error, 253
   repeated, 256
References and additional reading,
   469
Residual, 61
Rotation angle, 161
Rotation matrix, 159
Rules, 245
Runge-Kutta methods, 327
   fourth order, 327
Running programs, 1
   run2, 2

Sampling points, 246
   outside integration range, 290
Self-starting methods, 320
Shape functions, 301
Shifted inverse iteration, 143
   Program 4.3, 145
Shifted vector iteration, 140
   Program 4.2, 140
Shooting methods, 368
   linear, 374
   nonlinear, 375
   Program 7.4, 371
Simpson's rule, 250
   error, 254
   repeated, 260
Singular matrix, 30
Software
   Excel, 12
   Free Software Foundation, 2
   g95, 2
   Mathematica, 1
   MATLAB, 1, 8
spabac, 446, 466
sparin, 446, 467
Stationary methods, 61
Steepest descent method, 61
   Program 2.11, 62
Stiff equations, 359

Stiffness matrix, 151
Storage of matrices, 35
   skyline, 36
Sturm sequence, 181
subbac, 446, 467
subfor, 446, 467
Subprogram library, 445
Subprogram listings, 447
Successive overrelaxation, 57
   Program 2.10, 58
Symmetrical coefficients, 25

Taylor series, 104, 219, 329
Time stepping
   explicit, 404
   implicit, 404
Transformation methods, 158
Transpose, 26
Trapezoid rule, 248
   error, 253
   repeated, 258
Tridiagonal matrix, 167, 180, 209
   determinant, 180
   determinantal equations, 180

Unsymmetrical coefficients, 68
Upper triangular matrix, 16, 20,
   176

Vector iteration, 136
   Program 4.1, 137
Very sparse systems, 52

Weighted residual methods, 376
   collocation, 380
   Galerkin, 382
   least squares, 381
   subdomain, 380
   summary, 383
Weighting coefficients, 246
   fixed, 289

Zero determinant, 131